Applied Computational Fluid Dynamics and Turbulence Modeling

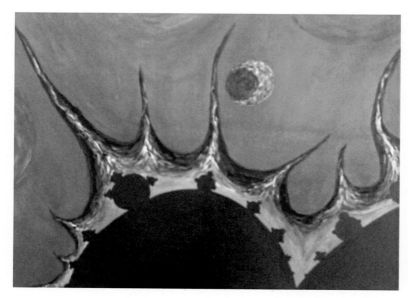

The author used acrylic dyes to paint a multicolored fractal surrounded by worlds (shown above), as a reflection of the fractal nature of the universe and its ethereal flows and motions. Then, the Any Capture software was used to "colorize" the painting with diverse shades of blue and white, as these colors are reminiscent of raging, turbulent water. Five copies of the image were made, each being rotated by 90° and scaled by the golden ratio (1.618), for a subtle embodiment of chaotic fluid motion. The five images also represent the author's total number of family members, thus generating a fractal family. Fractals and chaos are deeply embedded in turbulent eddy structure, for it is not possible to have turbulence without chaos and the fractal shapes that it generates. Though fractals represent a repeating pattern that gets smaller and smaller indefinitely, however, eddies cannot decrease in size forever, and are thus limited in lifespan. Nevertheless, the circle of life flows indefinitely through the fractal family, and much more so through Him who is above all the universe's fractals and flows; the source of all the eternal ebbs and flows that comprise the totality of everything.

Sal Rodriguez

Applied Computational Fluid Dynamics and Turbulence Modeling

Practical Tools, Tips and Techniques

 Springer

Sal Rodriguez
Sandia National Laboratories
Albuquerque, NM, USA

ISBN 978-3-030-28693-4   ISBN 978-3-030-28691-0   (eBook)
https://doi.org/10.1007/978-3-030-28691-0

This Springer imprint is published by the registered company Springer Nature Switzerland AG
The registered company address is: Gewerbestrasse 11, 6330 Cham, Switzerland

The author would like to dedicate the book to the late and great David C. Wilcox.

The book is also dedicated to my precious family and friends, as well as to the idealists, the dreamers, and the naysayers.
Salvador "Sal" B. Rodriguez

Preface

The book is intended for undergraduate senior students, graduate students, and professionals interested in modeling fluid dynamics with commercial or national laboratory software, such as Fluent, STAR-CCM, COMSOL, Fuego, and OpenFOAM, or perhaps using their own computational tool. The book's intent is to provide a hands-on approach that focuses on applied theory and applications. The book provides hundreds of tips, techniques, and tricks of the computational fluid dynamics (CFD) and turbulence trade that have been accumulated by the author, as well as published in many recent reports, articles, and books across the literature.

Key tips and guidelines in this book focus on "how to," with step-by-step instructions and examples so the reader will be able to proficiently

- Calculate eddy length, time, and velocity scales for the integral, Taylor, and Kolmogorov eddies
- Compute the Reynolds number for many useful engineering systems
- Avoid dozens of CFD simulation pitfalls
- Develop "bullet proof" meshes
- Apply strategies associated with time steps, node distance, geometry domain, and computational stability
- Improve data visualization images
- Select turbulence models
- Compute the y distance for $y^+ = 1$, 7, and 30
- Use guidelines associated with laminar and turbulent flow, fully developed flow, natural circulation, boundary conditions, and initial conditions
- Apply dozens upon dozens of RANS, LES, and DNS modeling tips

In addition, the book includes the entire coding for MATLAB scripts that calculate key laminar and turbulence parameters. This includes the LIKE algorithm for estimating the integral eddy length scale, turbulence intensity, turbulent kinetic energy, and turbulent dissipation. In turn, these parameters are used to estimate many other useful turbulence variables, such as the integral, Taylor, and Kolmogorov eddy velocity, length, and time scales, as well as the distance y from the wall for any value of y^+. Another MATLAB script calculates the peak velocity for laminar and turbulent natural circulation flows. The script also calculates the convective heat transfer coefficient and key dimensionless numbers such as Prandtl, Grashof, Nusselt, and Raleigh. Both horizontal and vertical geometries are calculated, although the emphasis is on vertical flows. The laminar regime is valid for $0.001 < Pr < 1,000$ and has been validated for a horizontal pipe with air and water and a vertical plate with air. Finally, the book includes a MATLAB function that is useful for calculating the physical properties for liquid lead, lead bismuth eutectic, bismuth, and sodium.

The book assumes the reader has at least taken ordinary and partial differential equations. Though not absolutely necessary, a basic fluid dynamics course would be helpful. The material is useful for senior projects in mechanical, civil, chemical, and nuclear engineering, as a first year of graduate level CFD and turbulence modeling, and for applied commercial and research applications. The major goal of the book is to provide as many useful and practical guidelines for applications that involve fluid flow, such as engineering (mechanical, civil, chemical, and nuclear), aerospace, naval, energy systems, military, micro- and nanodevices, medical, and the environmental sciences.

There are most certainly many great books written for CFD (e.g., J. Anderson's *Computational Fluid Dynamics* and *Computational Fluid Dynamics for Engineers* by B. Andersson et al.), as well as turbulence modeling (e.g., John Hinze's *Turbulence: An Introduction to Its Mechanism and Theory* and David Wilcox's *Turbulence Modeling for CFD*, Third Edition). So, why add another book to the already crammed field? The author's goal and motivation are to fill an important gap, a special niche, that is sorely lacking in the field: to provide a comprehensive set of practical, hands-on guidelines that can guide beginners, intermediate users, and, on occasion, perhaps even experts. The book is therefore a comprehensive summary of "how to" and therefore has detailed descriptions of what works and what does not in the field of CFD and turbulence. The book is based on the author's 34 years of engineering work at three national laboratories and tested through various courses at the University of New Mexico.

Finally, the book contains some recent, practical theories recently published by the author, as well as unpublished material. This includes methods to estimate the peak laminar and turbulent natural circulation velocities, the "LIKE" algorithm, the estimation of stable turbulent-flow time steps based on the turbulence viscosity, and practical engineering applications for engineered surfaces (e.g., dimples and swirl). There is a recent, strong interest in the use of dimples to increase heat transfer and reduce fluid drag in compact heat exchangers, impellers, fins, power plant equipment, vehicles, aerospace, and, of course, golf balls.

Albuquerque, NM, USA   Sal Rodriguez

Acknowledgments

The author is indebted to many friends and colleagues who encouraged and supported this work, as well as to his students for their helpful questions and comments. The reviewers deserve a million thanks for providing many useful suggestions that greatly improved the book, as well as for catching many pesky typos, despite their ubiquitous nature; first and foremost, the following are thanked: Dr. Nima Fathi, Dr. Patrick McDaniel, Dr. Mark Kimber, and Victor Figueroa. The author is also indebted to the professional Springer staff, especially Denise Penrose, David Packer, and Chandhini Kuppusamy.

Contents

Author Biography

Sal Rodriguez has a BS in Nuclear Engineering, an MS in Mechanical Engineering and in Applied Mathematics, and a PhD in Philosophy and Apologetics and in Nuclear Engineering. He has 34 years' experience in Nuclear Engineering and Energy Systems. His primary interests include turbulent flow, computational fluid dynamics (CFD), swirl, and advanced nuclear concepts such as small modular reactors and micro reactors. His recent research advances include the development of the "right-sized" dimpling algorithm for reduced flow drag and increased heat transfer, the development of the first dynamic code for modeling hydrogen production in integral nuclear reactor systems, aerodynamic drag reduction for race cars, an isotropic turbulence decay model, and the development of a vortex-unification theory. He has taught CFD, turbulence modeling, and advanced turbulence modeling at the University of New Mexico and Sandia National Laboratories. He also conducts turbulence seminars and workshops at Sandia National Laboratories, where he is a Principal Member of the Technical Staff. His combined CFD and turbulence modeling experience at three national laboratories (Idaho National Laboratory, Los Alamos National Laboratory, and Sandia National Laboratories) is reflected in this book. Other research interests include the advanced manufacturing of drag-reducing surfaces with enhanced heat transfer, as well as the development and laser-engineered manufacturing of refractory high-entropy alloys (RHEAs) for nuclear reactor and aerodynamic applications.

Chapter 1
Introduction

Choose a job you love, and you will never have to work a day in your life.

—Confucius

Abstract The motivation for learning and mastering CFD and turbulence modeling is based on their ever-increasing potential to solve problems, especially those that cannot be resolved experimentally due to physical limitations, economic cost, safety, time constraints, or environmental regulatory procedures. Current and future employment trends, computational capacity growth, and experimentation costs favor CFD. The integration of CFD, theory, advanced manufacturing, and experiments will lead towards even more economical and faster development of prototypes and systems for more streamlined concept-to-market development. As Moore's law eventually ceases to apply, quantum and nano-computing devices, as well as quantum algorithms, will continue the computational growth trend for many decades to come, if not centuries. The near future holds the potential for simulations that are thousands of times faster than is currently feasible. This fascinating subject is elaborated further in Sect. 5.2.

1.1 Motivation for CFD and Turbulence Modeling

Computational fluid dynamics (CFD) is not just "eye candy" designed to amaze and provide great careers. When done correctly, CFD can be a powerful tool for the prediction of system behavior and the design of more efficient and innovative systems; CFD can be used to validate system performance metrics, increase operational safety, yield higher profit margins, and provide many other desirable attributes.

In terms of job potential, CFD is geared for significant growth that is fueled by a powerhouse of more efficient and powerful algorithms, Moore's law, and advances in parallel computing and visualization (College Grad 2017). But, depending on which expert is consulted, Moore's law has already reached an asymptote (Alfonsi 2011) or will reach it sometime after 2017. Indeed, as of 2004, the computing

© Springer Nature Switzerland AG 2019
S. Rodriguez, *Applied Computational Fluid Dynamics and Turbulence Modeling*,
https://doi.org/10.1007/978-3-030-28691-0_1

industry is relying on multicore and manycore processors to increase computing performance (Alfonsi 2011). In any case, as Moore's law eventually ceases to apply, quantum and nano-computing devices, as well as quantum algorithms, will continue the growth trend for many decades to come, if not centuries (Rudinger 2017; Singer 2019). Though still in their infancy, quantum algorithms can already solve linear systems of equations (Singer 2019), which, of course, are essential for CFD solvers. And to further sweeten the prospect toward CFD's bright future, it is expected that quantum algorithms will result in an exponential decrease of the time required to solve systems of linear equations (Singer 2019).

Fortunately, CFD offers solutions in many areas that have substantial growth potential, including:

- Flow dynamics and heat transfer (pipes, pumps, fans, turbines, heat exchangers, boilers, combustors)
- Aerodynamics (automobiles, aerospace, missiles, ships, torpedoes, submarines)
- Micro- and nanotechnology (lab on a chip, cooling and heating, heat pipes)
- Engine design (efficiency, power, heat transfer, combustion)
- Power plants (chemical, solar, nuclear, wind, bio, coal; waterless power production, smart grids, micro grids)
- Heating, ventilation, and air conditioning systems (from homes to large buildings)
- Internal combustion
- Casting, soldering, liquid plastics and metal flow, 3D printing
- Weather prediction
- Rheology (deformation and flow in rivers, lakes, oceans)
- Cooling applications for dense circuitry, micro and quantum computers
- Aerosol dispersal
- Biological and chemical attack modeling
- Medical applications (blood flow, medical devices)
- Petrochemical, industrial applications
- Military
- And much, much more

Certainly, there are many great books already published in CFD (Anderson 1995; Andersson et al. 2012; Hinze 1987), computational methods (Ferziger and Peric 2002), as well as turbulence modeling (Wilcox 2006), with many other great books too numerous to cite here, but that are cited throughout the relevant chapters in this book. Therefore, to avoid repetition, the goal and motivation for this book are to fill an important gap, a specialized niche that is greatly lacking in the field: to provide a comprehensive set of practical, hands-on guidelines and examples designed to assist beginners, intermediate users, and, on occasion, perhaps even experts. This book is therefore a comprehensive summary of what works and what does not work in the field of CFD and turbulence, based primarily on CFD and turbulence modeling experience at three national laboratories and academia, as well as recent advances in fluid modeling found in the literature.

1.2 Computational Advantages of CFD over the Experimental Approach

CFD provides system designers and analysts a tool that is not only cheaper than experiments but also generates more data, including data that is not currently measurable with current instrumentation. For example, consider a pipe undergoing flow, with one million computational nodes. If CFD is coupled with heat transfer, then each computational node is just like a thermocouple, a pressure transducer, a flow meter, and so forth. When was the last time that a million sensors were applied onto a single experiment? Thus, CFD allows analysts and designers to probe deeper and wider into the details of system behavior than experimentation ever could (Clark et al. 1979). If done correctly, this enables the development of more efficient energy systems that are cost-competitive and environmentally friendlier.

Consider CFD as a tool that allows analysts to dynamically in time "x-ray" the system of interest, enabling them to view exactly how the fluid is behaving in real time, find any design flaws and inefficiencies, and thereby provide insights that can improve the design. And as computational power grows, more refined analysis can be conducted, especially via large eddy simulation (LES) and direct numerical simulation (DNS). For example, Fig. 1.1 shows turbulent flow in a pipe with a sharp contraction and expansion, using the dynamic Smagorinsky LES turbulence model and five million hexahedral elements. The flow is from the left to the right, and the simulation shows large eddy structures that provide insights regarding pressure losses in pipe components. As another example, Fig. 1.2 shows a DNS simulation of a water jet with 63 million hexahedral elements and flowing against gravity. The jet's external and internal eddy structure is shown on the left-hand side and right-hand side, respectively. As computational power increases, it is not unusual to have fluid combustion calculations involving DNS with billions of elements (Cheng 2008).

Another practical advantage of CFD over experiments is its ability to modify nature selectively, for the purpose of investigating physical behavior that is otherwise impossible to gauge. The possibilities are practically limitless. For instance, with a single stroke on the input file, an analyst can "delete" gravity, or double its

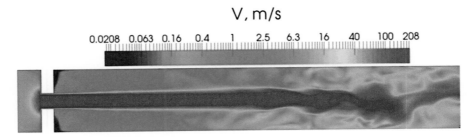

Fig. 1.1 LES turbulence modeling of a pipe with a sharp contraction and expansion

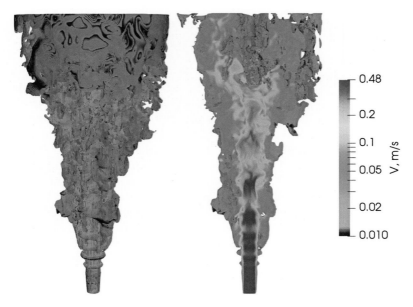

Fig. 1.2 DNS of jet flowing against gravity. Left, full jet; right: cross section

strength, to investigate the impact of gravity on a buoyant flow. As another example, Andre Bakker performed CFD of a tyrannosaurus rex and found out that the dinosaur had less fluid drag than modern vehicles (Bakker 2002)! Anderson cites an engineering situation where wind tunnel results were inconclusive for an airfoil (Anderson 1995). The experimental data was not clear as to whether the flow was laminar or turbulent. However, CFD simulations corroborated that the flow was indeed turbulent by using a judicious choice of turbulence models. As final example, consider a safety analysis on a scaled nuclear reactor to investigate its behavior under severe conditions. For this hypothetical situation, the transient is initiated with a large, double-ended guillotine break that may occur as a result of a strong earthquake. As the analysis proceeds, the reactor core can be completely destroyed in the virtual world, without releasing radiation, without causing damage to the environment, and at a mere fraction of an experiment's cost, assuming anyone would even pay for such scaled experiment!

Thus, CFD simulations and its plethora of sensitivity studies and modeling techniques can provide an upper edge to analysts. This includes system analysis that is experimentally impossible due to physical limitations, economic cost, safety, time constraints, or regulatory procedures.

1.2.1   Simulation Economics

With faster computation year after year and more efficient numerical and data visualization algorithms, it is not surprising that CFD becomes less expensive with the passage of time, while experiments become more expensive because of regulations. Anderson noted that there is a downward trend associated with computation cost, with approximately 1/10th reduction every eight years (Anderson 1995). On the other hand, the cost for experiments increases year after year due to inflationary pressures on experimentalist's salaries, expensive equipment, and costly regulations. Certainly, many environmental, safety, and health (ES&H) concerns are justified, while others challenge common sense. Nevertheless, whether justified or not, ES&H regulations typically increase with time.

1.2.2   Future Capabilities and Opportunities

Near-term goals for CFD analysts and researchers include the generation of secure, environmentally benign energy at a competitive cost, as well as the production of waterless power. For example, the typical evaporative water loss though cooling towers in a nuclear plant is estimated at 80,000,000 gallons per day. With diminishing natural energy resources and limited freshwater supplies, waterless power production will provide a solution that replaces cooling towers with passive, waterless heat transfer mechanisms. CFD will always be at the forefront of advanced nuclear reactor designs, more efficient power sources with decreased environmental footprint, improved aerodynamics, medical devices, nano- and microtechnology, and many other industrial processes. This is particularly so as designs become more complex, and the impact of Multiphysics is incorporated into designs.

In the area of transportation, CFD continues to be at the forefront of aerodynamic designs that reduce fluid drag, thereby increasing distance traveled per unit energy consumed. These improvements include race cars (Fig. 1.3) (SimScale 2017; Rodriguez et al. 2017), commercial vehicles, and the Department of Energy (DOE) SuperTruck Initiative (DOE STI 2017) (Fig. 1.4). Started in 2009, the DOE SuperTruck Initiative aims to develop and demonstrate a 50% improvement in freight efficiency. As of 2017, current SuperTruck efforts and commercialized technology include bumpers, as well as roof, gap, and chassis fairings. Figure 1.4 shows a SuperTruck built by Cummins/Peterbilt. Surface dimpling (akin to golf ball dimpling) for additional drag reduction is being considered to reduce parasitic drag loss and thus increase fuel efficiency. These low-drag surfaces are also suitable for commercial and military aircraft, missiles, rockets, submarines, and watercraft. Recent wind tunnel data verifies the many beneficial properties of engineered surfaces in transonic (Kontis and Lada 2005; Kontis et al. 2008), supersonic (Sekaran and Naik 2011), and hypersonic (Babinsky and Edwards 1997; Abney et al. 2013) surfaces and airfoils.

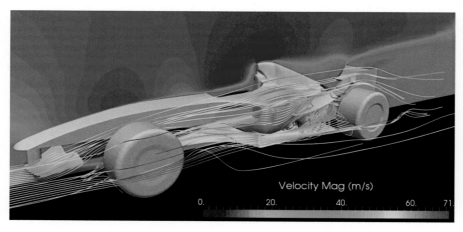

Fig. 1.3 Simulation of Formula 1 race car. (Courtesy of SimScale 2017; A. Arafat, modeler)

Fig. 1.4 Aerodynamic Cummins/Peterbilt SuperTruck. (Source: US Department of Energy)

Other novel CFD applications of recent interest which are in the area of micro- and nanofluidics include advanced heat pipe designs, micro heat pipes, propulsion systems, micro cooling systems, quantum computers, micro- and nanobots, sensors, and medical devices such as "lab on a chip." During the latter part of the twentieth century and the early part of the twenty-first century, the confluence of CFD and advanced manufacturing (AM) demonstrated significant design cost savings that are without precedent in the field of engineering and manufacturing. For example, the development of prototypes that used to require weeks to months can now be achieved within a day or so (AT Kearney 2015; Rodriguez and Chen 2017).

In the future, the integration of CFD, theory, AM (i.e., 3D printing), and experiments will lead toward even more economical and faster development of prototypes and systems that are closer to being market-ready. The technological quad represented by these tools, and their interaction, is shown conceptually in Fig. 1.5.

Fig. 1.5 The technology quad: theory, CFD, AM, and experiments

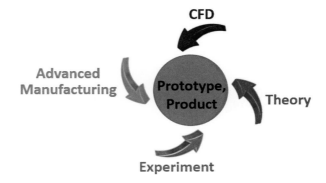

It is conceived that theory will guide an initial design concept, CFD will be used to investigate its virtual behavior, AM will be used to manufacture prototypes, and experiments will validate their design performance (or lack thereof!). Then, the process will be iterated, until the design reaches the desired design performance metrics (e.g., lower flow drag, higher thermal efficiency, and so forth). Other worthy design-guiding criteria include an optimal rate of return on investment, safety metrics, minimized environmental impact, etc. In summary, it is envisioned that theory will guide the CFD simulations, which will drive the AM design. Then, AM will be used to expedite cost-effective prototypes for experimentation, functional testing, and marketing. This, in turn, will serve to enhance the theory, and so forth, in a continuous-improvement cycle with much synergism and profitability.

1.3   Jobs and Future Employment Trends

Certainly, CFD is a multidisciplinary field with wide usage in the fields of mechanical, nuclear, chemical, electrical, and civil engineering. However, CFD also has substantial applications in areas such as physics, mathematics, chemistry, biology, medical, computation, nanotechnology, and AM. It is therefore not surprising that CFD involves many different subjects, as shown in Fig. 1.6. The more subject matters the CFD analyst masters, the more rewarding their careers will be, literally and figuratively.

Future job trends are positive, especially when experience is combined with the most advanced modeling tools and other technologies, such as AM, computer-aided design, and optimization. Just in the area of mechanical engineering, there were 277,500 jobs in 2014, with a projected growth to 292,100 by 2024 (College Grad 2017). Median annual salary ranged in 2014 from $76,190 (machinery manufacturing) to $94,640 (research and development). For comparison, a worker that earns the minimum wage of $7.25/h (national standard as of 2017), and that works a typical

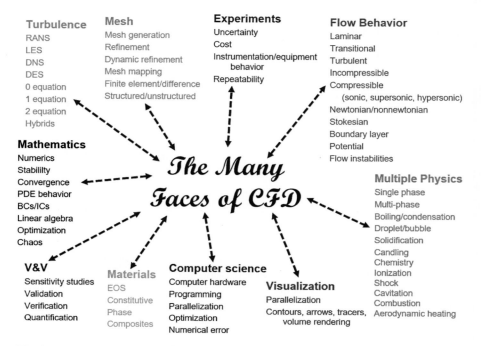

Fig. 1.6 CFD's multidisciplinary footprint

work year of 2000 h, would make $14,500; this is 5.25 to 6.5 times *less* than the salary of a *median* mechanical engineer.

College Grad also offered several insights for future job employment trends (College Grad 2017):

> Prospects for mechanical engineers overall are expected to be good. They will be best for those with training in the latest software tools, particularly for computational design and simulation. Such tools allow engineers and designers to take a project from the conceptual phase directly to a finished product, eliminating the need for prototypes. Engineers who have experience or training in three-dimensional printing also will have better job prospects.

For the above reasons and more, those who take the time to master CFD and turbulence modeling will reap many rewards. *In summary, the possibilities offered by CFD are limitless, if done correctly. If only… In what follows, it will be endeavored to provide such guidance.*

References

Abney, A., et al. (2013). Hypersonic boundary-layer transition experiments in the Boeing/AFOSR Mach-6 Quiet Tunnel, 51st AIAA aerospace sciences meeting.

Alfonsi, G. (2011). On direct numerical simulation of turbulent flows. *ASME, Applied Mechanics Reviews, 64,* 020802.

Anderson, J. (1995). *Computational fluid dynamics: The basics with applications*. New York: McGraw-Hill International Editions.

Andersson, B., et al. (2012). *Computational fluid dynamics for engineers*. Cambridge, UK: Cambridge University Press.

AT Kearney. (2015). 3D printing: A manufacturing revolution. International Research Journal of Engineering and Technology (pp. 1–16). Atlanta, Georgia.

Babinsky, H., & Edwards, J. A. (1997). Large scale roughness influence on turbulent hypersonic boundary layers approaching compression corners. *Journal of Spacecraft and Rockets, 34*(1), 70–75.

Bakker, A. (2002). Applied computational fluid dynamics, Lecture 1, Introduction to CFD, www.bakker.org

Cheng, R. (2008). *Low-swirl burner for heating & power generation*. Berkeley: Lawrence Berkeley National Laboratory.

Clark, R. A., Ferziger, J. H., & Reynolds, W. C. (1979). Evaluation of subgrid-scale models using an accurately simulated turbulent flow. *Journal of Fluid Mechanics, 91*(Part 1), 1–16.

College Grad. https://collegegrad.com/careers/all.shtml. Accessed on 16 May 2017.

DOE STI. SuperTruck initiative. https://energy.gov/eere/articles/supertruck-making-leaps-fuel-efficiency, https://energy.gov/sites/prod/files/2016/06/f33/EERE_SuperTruck_FS_R121%20FINAL.pdf. Accessed on 16 May 2017.

Ferziger, J., & Peric, M. (2002). *Computational methods for fluid dynamics* (3rd ed.). Berlin/New York: Springer.

Hinze, J. (1987). *Turbulence* (McGraw-Hill classic textbook reissue series). New York: McGraw-Hill.

Kontis, K., & Lada, C. (2005). Flow control effectiveness at high speed flows, Proceedings of the fifth European symposium on aerothermodynamics for space vehicles.

Kontis, K., Lada, C., & Zare-Behtash, H. (2008). Effect of dimples on glancing shock wave turbulent boundary layer interactions. *Shock Waves, 17*, 323.

Rodriguez, S., & Chen, A. (2017). *Advanced fire sprinkler design and validation*. Albuquerque, New Mexico: Sandia National Laboratories.

Rodriguez, S., Grieb, N., & Chen, A. (2017). Dimpling aerodynamics for a Ford Mustang GT, Sandia National Laboratories, SAND2017-6522.

Rudinger, K. (2017). Quantum computing is and is not amazing, Sandia National Laboratories, SAND2017-7067C.

Sekaran, M., & Naik, B. A. (2011). Dimples – An effective way to improve in performance, The first international conference on interdisciplinary research and development, Thailand, May 31 through June 1, 2011.

SimScale. The SimScale home page. www.simscale.com. Accessed on 16 May 2017.

Singer, N. (2019). *Quantum computing steps further ahead with new projects*. Albuquerque, New Mexico: Sandia National Laboratories, Sandia LabNews.

Wilcox, D. C. (2006 k-ω model). *Turbulence modeling for CFD*, 3rd Ed., La Cañada: DCW Industries, printed on 2006 and 2010.

Chapter 2
Overview of Fluid Dynamics and Turbulence

I am an old man now, and when I die and go to heaven there are two matters on which I hope for enlightenment. One is quantum electrodynamics, and the other is the turbulent motion of fluids. And about the former I am rather optimistic.

—Sir Horace Lamb, circa 1930

Abstract The first part of the chapter provides a review for mass, momentum, and energy conservation, as well as turbulence theory and modeling. Then, the historical development and importance of the Reynolds number (*Re*) is firmly established, leading to guidelines for calculating *Re* for many useful engineering geometries. Fully developed laminar and turbulent flow is described, as well as insights regarding the turbulent kinematic viscosity. Finally, isotropic turbulence and Taylor eddy theory are introduced, to lay a foundation regarding their merit and practical applications in CFD modeling; this development culminates in Sect. 3.7, where numerous practical drag reduction and heat transfer applications are described in more detail.

This chapter is intended as a review for mass, momentum, and energy conservation as well as turbulence theory and modeling. It is therefore ideal for those desiring a quick review or perhaps who have never had a fluid dynamics course. Though the chapter contains many insights, readers with a background in fluid dynamics and turbulence may skip the chapter and proceed into the delights of computational fluid dynamics (CFD) and turbulence modeling in the chapters that follow.

The interested reader is referred to Anderson (1995) for a well-documented, detailed comparison of integral vs. differential control volumes as well as the conservative (Eulerian; fixed-space tracking) vs. nonconservative form (Lagrangian; follows the mass, i.e., moves with the flow). For this book, the Eulerian differential approach is considered, mostly because of its straightforward relationship with regard to coding; however, as so eloquently shown by Anderson, transformations from one form onto another are fairly straightforward, despite the seemingly different formulations.

© Springer Nature Switzerland AG 2019
S. Rodriguez, *Applied Computational Fluid Dynamics and Turbulence Modeling*,
https://doi.org/10.1007/978-3-030-28691-0_2

The interested reader is also strongly recommended to review the excellent description, derivation, and utility of the mass, momentum, and energy conservation equations provided by the unrivaled classic, *Transport Phenomena* (Bird et al. 1960, 2007). An excellent and unparalleled review of turbulence modeling is found in the classics *Turbulence* (Hinze 1987) and *Turbulence Modeling for CFD* (Wilcox 2006).

Conservation of physical quantities is of crucial importance in CFD. In its most basic form, this involves either the conservation of mass or momentum as the governing equation but more often than not both. If the system involves energy changes, then the conservation of energy comes into play as well. For example, during natural circulation, a heated wall causes the fluid mass to undergo motion as a result of changes in density, which in turn is a function of temperature. Additional conservation mechanisms must be included if more complex physics are involved. As an example, the conservation of species must be considered as well during combustion. Structural analysis can be added when the fluid dynamics and structural displacements are coupled, e.g., turbulence motion in a flow can cause a nuclear reactor fuel bundle to oscillate (Rodriguez and Turner 2012); see Fig. 2.1. When additional physics are added beyond the conventional mass and momentum CFD governing equations, the simulations are typically called "multiphysics."

Note: unless otherwise specified, the units used in this text are Système International (SI), with length based on meters (m), mass in kilograms (kg), time in seconds (s), and temperature in Kelvin (K).

Conservation of a given quantity basically means that it is neither destroyed nor created, but is instead transformed from one form onto another. For example, a falling rock through air involves the change of potential energy onto kinetic energy, as well as energy that opposes the air resistance. And once the rock strikes the ground, some kinetic energy is transferred to deform the ground and the rock, while some of the energy transfers onto sound energy. Thus, mass, momentum, and energy are in continuous change as engineering processes advance. Quite briefly, mass is neither created nor destroyed but is instead transformed (no particle physics

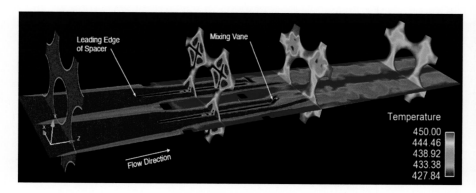

Fig. 2.1 Mass, momentum, energy, and structural Multiphysics analysis of a nuclear fuel bundle

processes are considered here). The same conservation principle applies to momentum and energy, meaning these quantities are neither created nor destroyed but are instead transformed. These ideas are expounded in the sections that follow.

2.1   Introduction to Conservation of Mass

To say that mass is conserved means that a fundamental principle of nature is satisfied, namely, that the system's mass quantity remains unchanged. In its simplest form, a liquid undergoes motion, so it "flows." As the liquid flows, it continues to have the same amount of mass despite the degree of rotation, stretching, or translation. For example, in a closed system, 1 kg of water can flow as gently or as vigorously as desired, and for as long as desired, but in the end, there will still be 1 kg of mass, and its composition is still H_2O. Even if the water boiled while flowing, there would still be 1 kg total of H_2O liquid and vapor.

Certainly, lead can be transformed into gold with a powerful particle accelerator, as high-speed nuclei shear off three protons from the lead atoms. However, in the context of CFD, atomic species are preserved, though chemical species can react, such as during combustion. In this context, mass conservation means that the quantity of mass is preserved at the atomic level. Consider the following simple chemical reaction for the production of water, $2H_2(g) + O_2(g) \rightarrow 2H_2O(l)$. This says that two hydrogen molecules combine with one oxygen molecule to form two water molecules. Stated differently in terms of atomic species conservation, four hydrogen atoms combine with two oxygen atoms to form water that has a total of four hydrogen and two oxygen atoms. Thus, the gaseous hydrogen and oxygen molecules are transformed into liquid water, but the atomic species is preserved. This is a conservation of mass for a reactive fluid.

Another way to view conservation of mass is to consider 1 g of wood. Burn it for a few seconds, and then weigh its charred remains. Assume the ashes weigh 0.1 g. Where did the remaining 0.9 g go? The missing wood mass has formed into combustion by-products. For this simplified example, there are now 0.9 g of combustion by-products in the form of soot, CO_2, CO, water vapor, and other chemicals. *However, despite the chemical reactions, no mass was lost or gained, and no atomic species were transformed—mass was conserved.*

Consider a system in a Cartesian space, as shown in Fig. 2.2. Let the system have multiple mass inlets and outlets. If more mass is coming into the system than is entering, then the system will accumulate mass and vice versa. Said more succinctly, the system's mass accumulation with respect to (WRT) time is the net inflow minus the net outflow. When viewed from the so-called "conservative" (Eulerian) invariant form, this is expressed as

Fig. 2.2 3D Cartesian
system showing the u, v, and
w velocity components

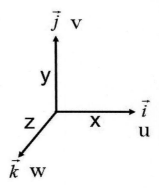

$$\frac{d\rho}{dt} + \vec{\nabla} \cdot \left(\rho\vec{V}\right) = 0 \qquad\qquad (2.1A)$$

or

$$\frac{d\rho}{dt} = -\vec{\nabla} \cdot \left(\rho\vec{V}\right) = -\rho\left(\vec{\nabla} \cdot \vec{V}\right) - \left(\vec{V} \cdot \vec{\nabla}\right)\rho. \qquad (2.1B)$$

In this context, $\vec{\nabla} \cdot \left(\rho\vec{V}\right)$ can be thought of as the overall rate of mass addition per unit volume due to convection associated with velocity and density gradients (Bird et al. 2007).

The above equation indicates that the density change WRT time is balanced by the fluid velocity divergence and density gradients, respectively. Here, without loss of generality (WLOG), and due to its simplicity, the velocity is expressed in Cartesian coordinates as a function of the x, y, and z coordinates as well as time, t. Thus, the velocity can be expressed as

$$\vec{V} = \vec{V}(x, y, z, t) = u\,\vec{i} + v\,\vec{j} + w\,\vec{k} \qquad\qquad (2.2)$$

where

u = velocity distribution in the x-direction = $u(x,y,z,t)$
v = velocity distribution in the y-direction = $v(x,y,z,t)$
w = velocity distribution in the z-direction = $w(x,y,z,t)$
ρ = density = $\rho(x,y,z,t)$

Another way to view the velocity divergence is to consider it as the time rate of change for the moving volume of fluid, per unit volume, with overall units in inverse time. The velocity divergence therefore represents a fractional change per unit time—it describes how much a fraction of mass changes WRT time.

Equation 2.1B can be expanded explicitly into its divergence and gradient operators, respectively.

$$\frac{\partial \rho}{\partial t} = -\rho \left(\frac{\partial u}{\partial x} + \frac{\partial v}{\partial y} + \frac{\partial w}{\partial z} \right) - \left(u \frac{\partial \rho}{\partial x} + v \frac{\partial \rho}{\partial y} + w \frac{\partial \rho}{\partial z} \right). \qquad (2.1C)$$

The terms are interpreted as follows:

$\frac{\partial \rho}{\partial t}$ = mass change per unit volume WRT time; mass accumulation per unit volume.

$\rho \left(\frac{\partial u}{\partial x} + \frac{\partial v}{\partial y} + \frac{\partial w}{\partial z} \right)$ = mass transfer per unit volume due to velocity gradients.

The velocity gradient, $\frac{\partial u}{\partial x}$, represents the change of the velocity component in the x-direction (i.e., u) WRT x and so forth.

Therefore, the product of the density and the velocity gradient, $\rho \frac{\partial u}{\partial x}$, is the transfer of mass per unit volume associated with the x-direction as a result of motion arising from the u velocity component, etc. Finally,

$\left(u \frac{\partial \rho}{\partial x} + v \frac{\partial \rho}{\partial y} + w \frac{\partial \rho}{\partial z} \right)$ = mass convection per unit volume.

Example 2.1 Derive the conservation of mass equation for an *incompressible* fluid that has reached steady state (SS).

Solution While the system was evolving WRT time, there were density changes. But, as the density approaches SS, it becomes constant as a function of time, so its derivative is zero; thus, $\frac{\partial \rho}{\partial t} = 0$. In fact, for any SS system, the time derivative is zero. In other words, "SS" explicitly means that the variable does not change WRT time.

Because the fluid is incompressible, the divergence of the velocity is zero (this will be discussed in detail later in this chapter). Therefore,

$$\vec{\nabla} \cdot \vec{V} = \frac{\partial u}{\partial x} + \frac{\partial v}{\partial y} + \frac{\partial w}{\partial z} \equiv 0.$$

Consequently, the first term on the right-hand side (RHS) of Eq. 2.1C is zero,

$$-\rho \left(\frac{\partial u}{\partial x} + \frac{\partial v}{\partial y} + \frac{\partial w}{\partial z} \right) = 0.$$

This leaves behind a single term in parenthesis, which happens to be zero because the other terms in the mass conservation equation are zero as well,

$$-\left(u \frac{\partial \rho}{\partial x} + v \frac{\partial \rho}{\partial y} + w \frac{\partial \rho}{\partial z} \right) = 0.$$

If the flow is only one dimensional (1D), or can reasonably be approximated as 1D, then the above simplifies even more. WLOG chose the x direction, so the y and z terms drop out:

$$u\frac{\partial \rho}{\partial x} + v\frac{\partial \rho}{\partial y} + w\frac{\partial \rho}{\partial z} = 0.$$

Dividing by u, $\frac{\partial \rho}{\partial x} = 0$, which implies that if the derivative were integrated, then ρ = constant. (Note that only one direction is being considered, so the partial derivative becomes an ordinary derivative.) Thus, if the flow is 1D, SS, and incompressible, then its density must be constant.

A simple yet very useful relationship for a SS system with fixed density and velocity is

$$\dot{m} = \rho V A \qquad (2.3)$$

where

$V = $ *average* fluid velocity
$A = $ flow area

If the SS system has multiple n inlets and/or p outlets, or perhaps various flow areas, velocities, and/or densities, then it is convenient to apply the following formulation:

$$\sum_{1}^{n} \dot{m}_{\text{in}} = \sum_{1}^{p} \dot{m}_{\text{out}} \qquad (2.4A)$$

where

$\dot{m}_{\text{in}} = $ total inlet mass flow rate
$\dot{m}_{\text{out}} = $ total outlet mass flow rate

That is, the above expression expands as follows by using Eqs. 2.3 and 2.4A:

$$\begin{aligned}
\rho_{1,\text{in}} V_{1,\text{in}} A_{1,\text{in}} + \rho_{2,\text{in}} V_{2,\text{in}} A_{2,\text{in}} + \cdots + \rho_{n,\text{in}} V_{n,\text{in}} A_{n,\text{in}} \\
= \rho_{1,\text{out}} V_{1,\text{out}} A_{1,\text{out}} + \rho_{2,\text{out}} V_{2,\text{out}} A_{2,\text{out}} + \cdots + \rho_{p,\text{out}} V_{p,\text{out}} A_{p,\text{out}}.
\end{aligned} \qquad (2.4B)$$

Example 2.2 Consider a system with two inlets and three outlets, each with a different density, velocity, and flow area. Express the SS mass flow rate explicitly in terms of ρ, V, and A.

Solution From Eq. 2.4A,

$$\dot{m}_{1,\text{in}} + \dot{m}_{2,\text{in}} = \dot{m}_{1,\text{out}} + \dot{m}_{2,\text{out}} + \dot{m}_{3,\text{out}}.$$

Setting up the individual flows,

$$\begin{aligned}
\rho_{1,\text{in}} V_{1,\text{in}} A_{1,\text{in}} + \rho_{2,\text{in}} V_{2,\text{in}} A_{2,\text{in}} = \rho_{1,\text{out}} V_{1,\text{out}} A_{1,\text{out}} + \rho_{2,\text{out}} V_{2,\text{out}} A_{2,\text{out}} \\
+ \rho_{3,\text{out}} V_{3,\text{out}} A_{3,\text{out}}.
\end{aligned}$$

If the system consisted solely of an inlet and outlet (e.g., pipe, duct), the SS conservation of mass reduces to

$$\rho_1 V_1 A_1 = \rho_2 V_2 A_2. \qquad (2.4C)$$

Example 2.3 A cylindrical pipe with a contraction reduces in diameter from 0.2 to 0.1 m. The fluid is water at 300 K and 1 atmosphere and is flowing at 1 kg/s. The system is SS, isothermal, and incompressible. What is the exit velocity?

Solution Because this problem involves mass flow rate and velocity, Eqs. 2.3 and 2.4A will be needed as well as the water density. Note: there are many reliable resources for physical properties, including REFPROP (Lemmon et al. 2010) and CoolProp (CoolProp 2017). Based on the above pressure and temperature, the water density is 996.6 kg/m^3. Label the pipe inlet with subscript "1,in" and the outlet with subscript "1,out." The exit area for a cylinder is

$$A_{1,out} = \pi \frac{D^2}{4} = \pi \frac{(0.1 \text{ m})^2}{4} = 7.85\text{E}{-}3 \text{ m}^2$$

where D is the diameter. For convenience, the more succinct exponent notation "En" will be used fairly often instead of "×10^n." Thus, 7.85E$-$3 is the same as 7.85×10^{-3}.

Based on conservation of mass, the mass flow rate going into the contraction is

$$\dot{m}_{1,in} = \rho_{1,in} V_{1,in} A_{1,in}$$

and the flow coming out is

$$\dot{m}_{1,out} = \rho_{1,out} V_{1,out} A_{1,out}.$$

Because the exit mass flow rate was not specified, conservation of mass under SS requires that it has the same value as the inlet mass flow rate, which was given. Therefore,

$$\dot{m}_{1,out} = \rho_{1,out} V_{1,out} A_{1,out} = \dot{m}_{1,in}.$$

Solving for the exit velocity,

$$V_{1,out} = \frac{\dot{m}_{1,in}}{\rho_{1,out} A_{1,out}} = \frac{1 \text{ kg/s}}{(996.6 \text{ kg/m}^3)(7.85\text{E}{-}3 \text{ m}^2)} = 0.128 \text{ m/s}.$$

Note: unless the analyst is working for the military or systems that absolutely must have high precision (e.g., the Hubble telescope lens), three or four significant digits is typically all that is needed to get good engineering results; even two digits are reasonable, especially when applying correlations whose accuracy is oftentimes less than two digits! A good example of this is the Dittus-Boelter equation, which has an accuracy of ~±25% (Holman 1990), and yet, its value is oftentimes quoted using three or more significant digits. A colloquial story warns engineers to *not* "measure with a micrometer, draw a line with a ruler, and cut with an ax."

Example 2.4 Consider the same geometry and fluid density as in Example 2.3. But now, the fluid exits the contraction at an average velocity of 5 m/s. What is the average inlet velocity and mass flow rate?

Solution Calculate $A_{1,\text{in}}$ because it will be needed later:

$$A_{1,\text{in}} = \pi \frac{D^2}{4} = \pi \frac{(0.2 \text{ m})^2}{4} = 3.14\text{E}{-2} \text{ m}^2.$$

At this point $\rho_{1,\text{in}}$ and $A_{1,\text{in}}$ are known, but the inlet mass flow rate and the average inlet velocity are not (the goal is to use $\dot{m}_{1,\text{in}} = \rho_{1,\text{in}} V_{1,\text{in}} A_{1,\text{in}}$). Thus, there are two unknowns and only one equation, so one more independent equation is needed. Applying Eq. 2.4C, it is clear that

$$\dot{m}_{1,\text{in}} = \dot{m}_{1,\text{out}},$$

which means that

$$\rho_{1,\text{in}} V_{1,\text{in}} A_{1,\text{in}} = \rho_{1,\text{out}} V_{1,\text{out}} A_{1,\text{out}}.$$

Solving for the inlet velocity, and noting that the density ratio drops out because $\rho_{1,\text{in}} = \rho_{1,\text{out}}$, then

$$V_{1,\text{in}} = \frac{\rho_{1,\text{out}} V_{1,\text{out}} A_{1,\text{out}}}{\rho_{1,\text{in}} A_{1,\text{in}}} = \frac{V_{1,\text{out}} A_{1,\text{out}}}{A_{1,\text{in}}} = \frac{(5 \text{ m/s})(7.85\text{E}{-3} \text{ m}^2)}{3.14\text{E}{-2} \text{ m}^2} = 1.25 \text{ m/s}.$$

Because of conservation of mass ($\dot{m}_{1,\text{in}} = \dot{m}_{1,\text{out}}$), the mass flow rate can now be calculated from either $\dot{m}_{1,\text{in}} = \rho_{1,\text{in}} V_{1,\text{in}} A_{1,\text{in}}$ or $\dot{m}_{1,\text{out}} = \rho_{1,\text{out}} V_{1,\text{out}} A_{1,\text{out}}$. This implies that

$$\dot{m}_1 = \rho_1 V_1 A_1 = (996.6 \text{ kg/m}^3)(1.25 \text{ m/s})(3.14\text{E}{-2} \text{ m}^2) = 39.1 \text{ kg/s}.$$

The inlet mass flow rate makes intuitive sense, because the same amount of mass flows throughout the system, thereby requiring that the fluid velocity be faster at the exit (5.0 vs. 1.25 m/s) because the water flows through a smaller flow area (7.85E−3 vs. 3.14E−2 m²). In other words, the flow is constricted. This same effect occurs in rivers, where the wider streams flow slower than the narrower parts of the river.

2.2 Introduction to Conservation of Momentum

The key idea behind momentum conservation is that it is neither destroyed nor created but that rather it is transformed from one type of momentum onto another. *This is really just Newton's second law of motion being applied to a fluid*:

$$\vec{F} = m\vec{a}, \tag{2.5}$$

where

F = force
m = mass
a = acceleration

That is, the fluid experiences various body and surface forces that inevitably generate motion. Body forces refers to those forces that act directly upon the entire mass enclosed by the volume (system) in question, such as gravity and electromagnetic forces. Surface forces refer to those forces that act on the surface of the volume. Surface forces include two categories: pressure as a result of the action from the surrounding fluid and the viscous shear/normal stresses caused by frictional forces. This is shown conceptually in Fig. 2.3.

If the body and surface forces are considered, then the invariant conservation of momentum is

$$\rho \frac{\partial \vec{V}}{\partial t} + \rho \vec{V} \cdot \vec{\nabla} \vec{V} = -\vec{\nabla} \cdot \vec{\vec{\tau}} - \vec{\nabla} P + \rho \vec{g} \tag{2.6}$$

where

P = pressure
g = gravity
τ = shear stress

Fig. 2.3 Terms associated with forces and momentum conservation

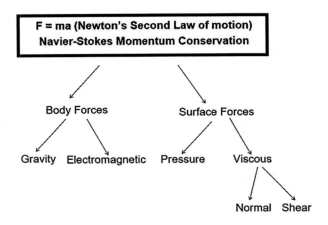

Certainly, conservation of momentum (AKA Navier-Stokes) in the above vector/ tensor formulation can look quite formidable to those who perhaps just recently engaged in fluid dynamics. But do not forget that it is simply a glorified version of "$F = ma$"; each term of the momentum equation represents the spatial or temporal way in which forces are transformed from one form to another:

$\rho \frac{\partial \vec{V}}{\partial t} =$ rate of momentum accumulation per unit volume. When there is no net change WRT time, this term is zero, and the system is said to be at SS.

$\rho \vec{V} \cdot \vec{\nabla} \vec{V} =$ rate of momentum change due to convection per unit volume. This is how velocity gradients push the liquid. This is the "accelerator" and source of instabilities that lead to turbulence if the term is sufficiently high. Note that if the flow is fully developed (FD), the velocity distribution does not change WRT to position; in this special case, the partial of the velocity component is zero for such direction.

$\vec{\nabla} \cdot \vec{\tau} =$ rate of momentum change per unit volume due to viscous forces. This term forms the "brakes" that slow down a fluid and dampen instabilities; viscous forces inhibit turbulence.

$\vec{\nabla} P =$ rate of momentum change per unit volume based on pressure gradients. This term can arise from pumps, heat sources, etc.

$\rho \vec{g} =$ momentum rate of change based on gravity, per unit volume. This term is important during buoyant flows.

As a helpful way of understanding the shear stresses, note that $\tau_{ij} =$ stress in the j direction exerted on the plane perpendicular to the i axis, where $\tau_{i,j\,=\,i} =$ normal stress (i. e., τ_{ii}) and $\tau_{i,j\,\neq\,i} =$ shear stress.

At this point, fluid behavior can be modeled once a Newtonian or non-Newtonian shear function is specified. For Newtonian fluids, the stress is linearly proportional to the velocity gradients,

$$\tau_{ij} = \text{fluid shear} = \begin{vmatrix} \tau_{xx} & \tau_{xy} & \tau_{xz} \\ \tau_{yx} & \tau_{yy} & \tau_{yz} \\ \tau_{zx} & \tau_{zy} & \tau_{zz} \end{vmatrix} = 2\mu s_{ij} - \frac{2}{3}\left(\vec{\nabla} \cdot \vec{V}\right)\delta_{ij} \qquad (2.7A)$$

where

$$s_{ij} = \text{laminar shear} = \frac{1}{2}\left(\frac{\partial u_i}{\partial x_j} + \frac{\partial u_j}{\partial x_i}\right). \qquad (2.7B)$$

Combining Eqs. 2.7A and 2.7B,

$$
\tau_{ij} = \begin{vmatrix} 2\mu\dfrac{\partial u}{\partial x} - \dfrac{2}{3}\vec{\nabla}\cdot\vec{V} & \mu\left(\dfrac{\partial u}{\partial y} + \dfrac{\partial v}{\partial x}\right) & \mu\left(\dfrac{\partial u}{\partial z} + \dfrac{\partial w}{\partial x}\right) \\[2ex] \mu\left(\dfrac{\partial v}{\partial x} + \dfrac{\partial u}{\partial y}\right) & 2\mu\dfrac{\partial v}{\partial y} - \dfrac{2}{3}\vec{\nabla}\cdot\vec{V} & \mu\left(\dfrac{\partial v}{\partial z} + \dfrac{\partial w}{\partial y}\right) \\[2ex] \mu\left(\dfrac{\partial w}{\partial x} + \dfrac{\partial u}{\partial z}\right) & \mu\left(\dfrac{\partial w}{\partial y} + \dfrac{\partial v}{\partial z}\right) & 2\mu\dfrac{\partial w}{\partial z} - \dfrac{2}{3}\vec{\nabla}\cdot\vec{V} \end{vmatrix} \quad \text{(compressible fluid).}
$$

$$(2.7C)$$

Note that the diagonal terms include the divergence of the velocity vector. So, Eq. 2.7C is suitable for compressible fluids; but if the divergence term is zero, the expression becomes suitable for incompressible fluids.

$$
\tau_{ij} = \begin{vmatrix} 2\mu\dfrac{\partial u}{\partial x} & \mu\left(\dfrac{\partial u}{\partial y} + \dfrac{\partial v}{\partial x}\right) & \mu\left(\dfrac{\partial u}{\partial z} + \dfrac{\partial w}{\partial x}\right) \\[2ex] \mu\left(\dfrac{\partial v}{\partial x} + \dfrac{\partial u}{\partial y}\right) & 2\mu\dfrac{\partial v}{\partial y} & \mu\left(\dfrac{\partial v}{\partial z} + \dfrac{\partial w}{\partial y}\right) \\[2ex] \mu\left(\dfrac{\partial w}{\partial x} + \dfrac{\partial u}{\partial z}\right) & \mu\left(\dfrac{\partial w}{\partial y} + \dfrac{\partial v}{\partial z}\right) & 2\mu\dfrac{\partial w}{\partial z} \end{vmatrix} \quad \text{(incompressible fluid).}
$$

$$(2.7D)$$

Note that most fluids are Newtonian, including all gases (Ar, He, H_2, N_2, CO, CO_2, air, etc.), homogeneous non-polymeric liquids, simple oils, and water (Bird et al. 1960). However, there are also many non-Newtonian fluids, including:

- Blood, synovial fluids, DNA
- Liquid polymers, oobleck (cornstarch + H_2O), slime (e.g., cloud slime), silly putty
- Shampoos, conditioners, toothpaste
- Many salt solutions
- Many foods, including melted chocolate, ketchup, mustard, custard, mayonnaise, cold caramel topping, syrup, molasses, whipped cream
- Silica nanoparticles in polyethylene alcohol
- Drilling mud, clay suspensions, sludge (e.g., Hanford waste tanks)
- Water-manure slurries with fibrous content greater than 10–20%
- Paints (e.g., latex)
- Paper pulp in aqueous solution
- Some lubricants

In general, sludge, slurries, polymers, pastes, and suspensions are non-Newtonian. It has been suggested that any fluid whose molecular weight is 5000 or less is Newtonian (Bird et al. 2007); unfortunately, they did not include a citation. If such molecular rule of thumb is used, then most polymers, DNA, and complex molecules are non-Newtonian as well. Therefore, caution must be exercised when dealing with fluids for which the analyst is unfamiliar. Whether a fluid is Newtonian or not has a significant impact on the velocity distribution; as always, caveat emptor!

Here, a little detour is in order, with the purpose of clarifying whether a fluid is compressible or not. After this brief detour, the discussion will continue by expanding and then applying the vector Navier-Stokes equation into its three Cartesian components, u, v, and w.

Quite often, much confusion arises regarding a fluid's "compressibility" or lack thereof. If the fluid is incompressible, then by definition,

$$\vec{\nabla} \cdot \vec{V} = \left(\frac{\partial u}{\partial x} + \frac{\partial v}{\partial y} + \frac{\partial w}{\partial z} \right) \equiv 0. \tag{2.8}$$

Note that the strict (and mathematically correct) definition of "incompressibility" is that the divergence of the velocity vector is zero. It is simply not correct to assume that because density remains constant, such fluid must therefore be incompressible. For example, water at room temperature can hardly be compressed such that its density changes. However, if that same water were moved at a sufficiently high velocity, then it would be compressible by definition! What happened here? The issue is that "compressibility" is purely the net result of the sum of the three velocity differentials found in the velocity divergence, Eq. 2.8. But, when the fluid velocity approaches its material-specific sound speed, the fluid approximates a "compressible" state. Thus, with a few exceptions, compressibility can be *approximated* as follows: a fluid is said to be compressible if its speed reaches or exceeds 30% of its sound speed. This is expressed as the ratio of the fluid speed and its sound speed and is defined by the Mach number (*Ma*).

$$\text{Mach number} = Ma \equiv \frac{u_{\text{fluid}}}{u_{\text{sound}}} \tag{2.9A}$$

$$Ma_{\text{crit}} \begin{cases} >0.3 \Rightarrow \text{compressible} \\ <0.3 \Rightarrow \text{incompressible} \end{cases}. \tag{2.9B}$$

Now consider a velocity vector field. If the gradient of the velocity is zero, must each of its terms be zero? The answer is no. It is the *sum of the individual differentials in the velocity divergence* that must be zero for a fluid to be considered incompressible; thus, each term can be nonzero, so long as their *sum* is zero. At this point, perhaps a few examples can help clarify these issues.

Example 2.5 Helium at 300 K and 1 atmosphere flows through a pipe. A positive displacement pump increases the gas's velocity until it becomes compressible. At what velocity will the fluid become compressible?

Solution From REFPROP, u_{sound} for helium is 1019.6 m/s. Solve for u_{fluid} from Eqs. 2.9A and 2.9B:

$$u_{\text{fluid}} = u_{\text{sound}} Ma_{\text{crit}} = (1019.6 \text{ m/s})(0.3) = 305.9 \text{ m/s}.$$

Example 2.6 Redo the previous example, except now replace the helium with water. At what velocity will the *room temperature water* become compressible?

Solution From REFPROP, u_{sound} for water is 1501.5 m/s. From Eqs. 2.9A and 2.9B,

$$u_{\text{fluid}} = u_{\text{sound}} Ma_{\text{crit}} = (1501.5 \text{ m/s})(0.3) = 450.5 \text{ m/s}.$$

Granted, 450.5 m/s is a high velocity obtained in the case of Example 2.6, but such velocity is certainly achievable in a water nozzle or jet. Most importantly, the water in this example was at room temperature! Therefore, with fluids having a wide gamut of sound velocities based on pressure and temperature, it is always a good idea to check *Ma*, as the equations for incompressible fluids behave differently than those of compressible fluids! A quick check is especially important when dealing with fluids that have a high velocity or fluids for which the analyst is unfamiliar. Consider these situations as red flags, and for which doubts can be quickly checked; this is part of a good CFD analyst's due diligence. It is not atypical to have someone ask in engineering design meetings, as well as thesis and dissertation defenses, "What is the Mach number?", to which the reply is quite often, "Uh, I don't know, but it should be small...why? Aren't we just modeling water?!?"

Example 2.7 Show that a system with the following velocity distribution is incompressible:

$$\vec{V} = 2x^2 y \vec{i} + 3xy^2 \vec{j} - 10xyz \vec{k}.$$

Solution The flow's compressibility, or lack thereof, can be investigated by obtaining the divergence of the velocity:

$$\vec{\nabla} \cdot \vec{V} = \frac{\partial u}{\partial x} + \frac{\partial v}{\partial y} + \frac{\partial w}{\partial z} = \frac{\partial (2x^2 y)}{\partial x} + \frac{\partial (3xy^2)}{\partial y} + \frac{\partial (-10xyz)}{\partial z} = 4xy + 6xy - 10xy$$

$$= 0.$$

Because the divergence is zero, the fluid is incompressible. Note that all three terms are nonzero! But the compressibility criterion is based on the sum of the divergence terms.

Example 2.8 Is a flow field with $\vec{V} = 4x \vec{i} + 3y \vec{j} - 6z \vec{k}$ compressible or incompressible?

Solution $\vec{\nabla} \cdot \vec{V} = \frac{\partial(4x)}{\partial x} + \frac{\partial(3y)}{\partial y} + \frac{\partial(-6z)}{\partial z} = 4 + 3 - 6 = 1$. Therefore, this flow is compressible.

Returning to the vector/tensor formulation for Navier-Stokes, it can be split onto three Cartesian components, x, y, and z, resulting in three momentum equations: one for the x-direction velocity u (Eq. 2.10A), another for the y-direction velocity v (Eq. 2.10B), and the third for the z-direction velocity, w (Eq. 2.10C):

$$\rho\frac{\partial u}{\partial t} + \rho\left(u\frac{\partial u}{\partial x} + v\frac{\partial u}{\partial y} + w\frac{\partial u}{\partial z}\right) = \left(\frac{\partial \tau_{xx}}{\partial x} + \frac{\partial \tau_{xy}}{\partial y} + \frac{\partial \tau_{xz}}{\partial z}\right) - \frac{\partial P}{\partial x} + \rho g_x, \quad (2.10A)$$

$$\rho\frac{\partial v}{\partial t} + \rho\left(u\frac{\partial v}{\partial x} + v\frac{\partial v}{\partial y} + w\frac{\partial v}{\partial z}\right) = \left(\frac{\partial \tau_{yx}}{\partial x} + \frac{\partial \tau_{yy}}{\partial y} + \frac{\partial \tau_{yz}}{\partial z}\right) - \frac{\partial P}{\partial y} + \rho g_y, \quad (2.10B)$$

and

$$\rho\frac{\partial w}{\partial t} + \rho\left(u\frac{\partial w}{\partial x} + v\frac{\partial w}{\partial y} + w\frac{\partial w}{\partial z}\right) = \left(\frac{\partial \tau_{zx}}{\partial x} + \frac{\partial \tau_{zy}}{\partial y} + \frac{\partial \tau_{zz}}{\partial z}\right) - \frac{\partial P}{\partial z}$$

$$+ \rho g_z. \quad (2.10C)$$

For convenience, Eqs. 2.10A, 2.10B, and 2.10C can be expressed with explicit representations for a Newtonian fluid as

$$\rho\frac{\partial u}{\partial t} + \rho\left(u\frac{\partial u}{\partial x} + v\frac{\partial u}{\partial y} + w\frac{\partial u}{\partial z}\right) = \mu\left(\frac{\partial^2 u}{\partial x^2} + \frac{\partial^2 u}{\partial y^2} + \frac{\partial^2 u}{\partial z^2}\right) - \frac{\partial P}{\partial x} + \rho g_x, \quad (2.11A)$$

$$\rho\frac{\partial v}{\partial t} + \rho\left(u\frac{\partial v}{\partial x} + v\frac{\partial v}{\partial y} + w\frac{\partial v}{\partial z}\right) = \mu\left(\frac{\partial^2 v}{\partial x^2} + \frac{\partial^2 v}{\partial y^2} + \frac{\partial^2 v}{\partial z^2}\right) - \frac{\partial P}{\partial y} + \rho g_y, \quad (2.11B)$$

and

$$\rho\frac{\partial w}{\partial t} + \rho\left(u\frac{\partial w}{\partial x} + v\frac{\partial w}{\partial y} + w\frac{\partial w}{\partial z}\right) = \mu\left(\frac{\partial^2 w}{\partial x^2} + \frac{\partial^2 w}{\partial y^2} + \frac{\partial^2 w}{\partial z^2}\right) - \frac{\partial P}{\partial z}$$

$$+ \rho g_z. \quad (2.11C)$$

where μ is the dynamic viscosity, while the kinematic viscosity is expressed as $\nu = \frac{\mu}{\rho}$.

Dividing by the density and substitution of the kinematic viscosity yields

$$\frac{\partial u}{\partial t} + \left(u\frac{\partial u}{\partial x} + v\frac{\partial u}{\partial y} + w\frac{\partial u}{\partial z}\right) = \nu\left(\frac{\partial^2 u}{\partial x^2} + \frac{\partial^2 u}{\partial y^2} + \frac{\partial^2 u}{\partial z^2}\right) - \frac{1}{\rho}\frac{\partial P}{\partial x} + g_x, \quad (2.12A)$$

$$\frac{\partial v}{\partial t} + \left(u \frac{\partial v}{\partial x} + v \frac{\partial v}{\partial y} + w \frac{\partial v}{\partial z} \right) = \nu \left(\frac{\partial^2 v}{\partial x^2} + \frac{\partial^2 v}{\partial y^2} + \frac{\partial^2 v}{\partial z^2} \right) - \frac{1}{\rho} \frac{\partial P}{\partial y} + g_y, \quad (2.12B)$$

and

$$\frac{\partial w}{\partial t} + \left(u \frac{\partial w}{\partial x} + v \frac{\partial w}{\partial y} + w \frac{\partial w}{\partial z} \right) = \nu \left(\frac{\partial^2 w}{\partial x^2} + \frac{\partial^2 w}{\partial y^2} + \frac{\partial^2 w}{\partial z^2} \right) - \frac{1}{\rho} \frac{\partial P}{\partial z} + g_z. \quad (2.12C)$$

Equations 2.12A, 2.12B, and 2.12C assume that the fluid system is Newtonian, Cartesian, *laminar*, and has a dynamic viscosity that is not a function of space. As will be shown later, the above equations can be used for direct numerical simulation (DNS) involving *turbulent* flow if sufficiently small temporal and spatial scales are considered—more DNS details and guidelines are presented in Sect. 5.2.

The three independent equations discussed above in Cartesian coordinates, along with the necessary equations to obtain closure (e.g., an equation of state for P and auxiliary material properties for ρ and μ), can be solved for the three unknown velocity magnitudes, u, v, and w. The velocity magnitudes can then be used to compute the entire velocity vector, $\vec{V} = u\vec{i} + v\vec{j} + w\vec{k}$.

Again, the Navier-Stokes equation is simply an elegant form of Newton's second law of motion, as applied to fluids! In other words, the net change WRT time for momentum is equal to the sum of all the body and surface forces. Notice that the momentum conservation terms are in SI units of kg/m²-s², e.g.,

$$\rho \frac{\partial u}{\partial t} \ [=] \ \left(\frac{\text{kg}}{\text{m}^3} \right) \frac{\left(\frac{\text{m}}{\text{s}} \right)}{\text{s}} = \frac{\text{kg}}{\text{m}^2 \text{s}^2}. \quad (2.13A)$$

Consider any of the momentum terms (they are homogenous in terms of units). WLOG, choose the transient term and multiply it by volume. Then,

$$\left(\rho \frac{\partial u}{\partial t} \right) \cdot \text{Vol} \ [=] \ \left(\frac{\text{kg}}{\text{m}^3} \frac{\frac{\text{m}}{\text{s}}}{\text{s}} \right) \text{m}^3 = \text{kg} \frac{\text{m}}{\text{s}^2}, \text{which is a force.} \quad (2.13B)$$

Momentum p is defined as

$$\vec{p} = m\vec{V} \ [=] \ \text{kg} \frac{\text{m}}{\text{s}}. \quad (2.14)$$

The change of momentum WRT time is

$$\frac{d\vec{p}}{dt} \ [=] \ \left(\text{kg} \frac{\text{m}}{\text{s}} \right) \frac{1}{\text{s}} = \text{kg} \frac{\text{m}}{\text{s}^2}. \quad (2.15)$$

The above expression is a force as well, because the SI unit for force is measured in Newtons, which is expressed in kg-m/s². Thus, if the terms are multiplied by volume, the Navier-Stokes terms are in units of force. Furthermore, the change of momentum WRT time is also a force. Thus, the Navier-Stokes equation can therefore be viewed as a glorified form of Newton's second law of motion.

Example 2.9 Derive the SS equation for 1D momentum in the z direction for an inviscid fluid (viscosity approaches zero or is zero) with no pressure force and negligible gravitational effects.

Solution Because the flow is SS, has no pressure force, and has negligible gravitation, the following terms are deleted,

$$\cancel{\frac{\partial w}{\partial t}} + \left(u\frac{\partial w}{\partial x} + v\frac{\partial w}{\partial y} + w\frac{\partial w}{\partial z} \right) = \nu \left(\frac{\partial^2 w}{\partial x^2} + \frac{\partial^2 w}{\partial y^2} + \frac{\partial^2 w}{\partial z^2} \right) - \frac{1}{\cancel{\rho}}\cancel{\frac{\partial P}{\partial z}} + \cancel{g_z}.$$

Because the flow is only 1D in the z direction, u and v are zero. Furthermore, there are no gradients in the x or y direction, so those derivatives are zero as well. Therefore,

$$\left(\cancel{u}\frac{\partial w}{\partial x} + \cancel{v}\frac{\partial w}{\partial y} + w\frac{\partial w}{\partial z} \right) = \nu \left(\cancel{\frac{\partial^2 w}{\partial x^2}} + \cancel{\frac{\partial^2 w}{\partial y^2}} + \frac{\partial^2 w}{\partial z^2} \right).$$

The expression therefore simplifies to

$$w\frac{\partial w}{\partial z} = \nu\frac{\partial^2 w}{\partial z^2}.$$

Because w is solely a function of z, the partial derivative is just an ordinary derivative,

$$\nu\frac{d^2 w}{dz^2} - w\frac{dw}{dz} = 0.$$

Example 2.10 Derive the transient partial differential equation (PDE) for 1D momentum in the y direction for a viscous fluid with no pressure force and negligible gravitational effect.

Solution Because the flow is only 1D in the y direction, u and w are zero, so those terms can be dropped. Furthermore, there are no gradients in the x or z directions, so any derivatives WRT those coordinates are also zero. Starting with Eq. 2.12B,

$$\frac{\partial v}{\partial t} + \left(u\frac{\partial v}{\partial x} + v\frac{\partial v}{\partial y} + w\frac{\partial v}{\partial z} \right) = \nu \left(\frac{\partial^2 v}{\partial x^2} + \frac{\partial^2 v}{\partial y^2} + \frac{\partial^2 v}{\partial z^2} \right) - \frac{1}{\rho}\frac{\partial P}{\partial y} + g_y,$$

and simplifying, then the sought-after differential is

$$\frac{\partial v}{\partial t} + v \frac{\partial v}{\partial y} = \nu \frac{\partial^2 v}{\partial y^2}.$$

Now, if the viscosity is negligible, then $v\frac{\partial v}{\partial y} \ll \nu \frac{\partial^2 v}{\partial y^2}$, which allows for further reduction to

$$\frac{\partial v}{\partial t} = \nu \frac{\partial^2 v}{\partial y^2},$$

where $v = v(y,t)$.

The Navier-Stokes equation can be simplified into many, very useful engineering and design equations. For example, for 1D SS flows under a constant pressure gradient with negligible gravitational and inertial terms (e.g., a Hagen-Poiseuille type of flow), the x-momentum equation simplifies to

$$\nu \frac{\partial^2 u}{\partial x^2} - \frac{1}{\rho} \frac{\partial P}{\partial x} = 0. \tag{2.16}$$

For *laminar* Hagen-Poiseuille internal flow in a smooth pipe, the conservation equation for momentum in cylindrical coordinates yields

$$\Delta P = 32 \frac{\mu L u}{D^2}. \tag{2.17A}$$

It is noteworthy that the above expression shows that the pressure drop is a *linear* function of the velocity for a laminar flow; by contrast, the pressure drop is a *nonlinear* function of the velocity for turbulent flows. *This once again attests to the linear nature of laminar flows and the nonlinear behavior of turbulence. So, for example, to say that a laminar flow transitioned onto turbulence is the same as saying that a linear system just became nonlinear and* vice versa.

At this point, it is convenient to define a dimensionless quantity f_{Darcy} as the Darcy friction factor,

$$f_{\text{Darcy}} \equiv \frac{64}{Re}, \tag{2.18A}$$

which is applicable in the laminar regime for smooth, internal flow. As a cautionary note, the literature also uses the Fanning friction factor. This can lead to problems, as the Darcy friction factor is four times the Fanning friction factor. So, in the laminar regime for smooth internal flow (e.g., pipe, duct),

$$f_{\text{Fanning}} \equiv \frac{16}{Re} = \frac{f_{\text{Darcy}}}{4}. \tag{2.18B}$$

Fig. 2.4 Pressure force driving a fluid through a cylindrical pipe

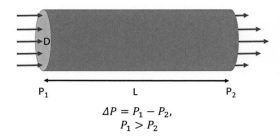

$$\Delta P = P_1 - P_2,$$
$$P_1 > P_2$$

Now, to see the utility of the laminar Hagen-Poiseuille formulation discussed above, recall the definition for Re and f_{Darcy}. Then expand Hagen-Poiseuille as follows, until a familiar expression for pressure drop is teased out,

$$\Delta P = 32\frac{\mu L u}{D^2} = \frac{64}{2}\frac{L}{D}\frac{\mu u}{D} = \frac{64}{2}\frac{L}{D}\frac{\mu u}{D}\frac{u\rho}{u\rho} = \frac{64}{Re}\frac{L}{D}\frac{\rho u^2}{2} = f_{\text{Darcy}}\frac{L}{D}\frac{\rho u^2}{2}. \quad (2.17\text{B})$$

Example 2.11 Pressure is imposed at the inlet of a long, isothermal, horizontal tube with a 0.002 m internal diameter and is 0.006 m long, as shown in Fig. 2.4. Water enters at atmospheric pressure and 350 K. There is a 1 Pa pressure drop across the tube. What is the expected mass flow rate and average velocity?

Solution Assume the flow is incompressible and laminar for now (neither the velocity nor the mass flow rate are known, so Re and Ma are unknown at this point). Because the mass flow rate is sought, it is reasonable to consider

$$\dot{m} = \rho u A \text{ (unknown} : \dot{m}, u).$$

A pressure drop is given, so Eq. 2.17A can be used *if* the flow is laminar,

$$\Delta P = 32\frac{\mu L u}{D^2} \text{ (unknown} : u).$$

At this point, there are two equations and two unknowns (mass flow rate and u); ρ and μ can be obtained based on the water pressure and temperature. Thus, $\mu = 368.6\text{E}{-}6$ Pa-s and $\rho = 973.9$ kg/m^3. The speed of sound for water can be readily obtained from a database (e.g., Engineering ToolBox) (ETB 2017). For this application, its value is 1555 m/s. The pipe has a circular cross section, so the flow area is $\pi D^2/4 = 2.83\text{E}{-}5$ m^2.

Solving for u from the pressure drop expression,

$$u = \frac{D^2\Delta P}{32\mu L} = \frac{(0.002 \text{ m})^2(1 \text{ Pa})}{32(368.6\text{E}{-}6\text{Pa} \cdot \text{s})(0.006 \text{ m})} = 0.057 \text{ m/s}$$

Now that u is known, the mass flow rate can be calculated:

$$\dot{m} = \rho u A = \left(973.9 \text{ kg/m}^3\right)(0.057 \text{ m/s})\left(2.83\text{E}{-}5 \text{ m}^2\right) = 0.0016 \text{ kg/s}.$$

Now that u and $\dot{m}$ are known, are we done? Not yet. It is necessary to check Re and Ma to verify that the correct equations were used. To this effect,

$$Re = \frac{x_{char} u_{char} \rho}{\mu} = \frac{D u \rho}{\mu} = \frac{(0.002 \text{ m})(0.057 \text{ m/s})(973.9 \text{ kg/m}^3)}{368.6\text{E}{-}6 \text{ Pa} \cdot \text{s}} = 301$$

and

$$Ma = \frac{u}{u_{sound}} = \frac{0.057 \text{ m/s}}{1555 \text{ m/s}} = 3.67\text{E}{-}05 \ll 0.3.$$

Therefore, the flow is laminar and incompressible, thereby corroborating that the equations used to solve the problem are valid in the applied domain.

2.3 Introduction to Conservation of Energy

As with conservation of mass and momentum, the key idea regarding conservation of energy is that it is neither created nor destroyed but is instead transformed from one type of energy into another. That is, in this non-quantum domain, energy does not appear from "nowhere," as some quantum mechanics processes do when nuclear particles and energies appear and disappear during a certain amount of time. Instead, the CFD point of view considers conservation of energy in the nano-, micro-, and macro-world. Furthermore, conservation of energy demands (proves) that there is no such thing as a "perpetual motion machine" with a limitless energy supply that appears out of thin air.

Conservation of energy is sufficiently general that it applies to both stationary and mobile fluids, as they undergo complex energy transfer mechanisms. Thus, it is expected that in its full form, the conservation of energy PDE is a formidable expression. On the other hand, the conservation of energy equation collapses to the familiar heat conduction equation for an immobile fluid, as will be shown later.

Simply stated, the energy time rate of change in a given body is the sum of the net heat fluxes plus the rate of work done on the body as a result of body and surface forces. That is, the net useful product from an energy system is a function of how much heat it generates minus how much work was required to extract the useful energy minus any heat losses/irreversible processes. *In other words, the governing equation for the conservation of energy is the first law of thermodynamics, but in a highly-generalized format.*

$$\dot{E} = \sum_i \dot{Q}_i - \sum_i \dot{W}_i. \tag{2.19A}$$

Note that the traditional Rudolf Clausius sign convention is adopted in the above equation, with a negative sign in front of the work term; whichever form is used, the key is consistency in the application of the terms.

The heat rates are represented as

$$\dot{Q}_i = \frac{\partial Q_i}{\partial t} [=] \frac{\text{Nm}}{\text{s}} = \frac{\text{J}}{\text{s}} \left(\begin{array}{l} \text{heat generation, conduction,} \\ \text{heat transferred by the fluid, etc.} \end{array} \right). \tag{2.19B}$$

The rate of work W done by a force F on a fluid moving at velocity V is

$$\dot{W} = \frac{\partial W}{\partial t} = \vec{F} \cdot \vec{V} [=] \text{N} \frac{\text{m}}{\text{s}} = \frac{\text{J}}{\text{s}} \left(\begin{array}{l} \text{pressure, viscous, body,} \\ \text{pumping, etc.} \end{array} \right) \tag{2.19C}$$

Note that a "motionless" body has internal energy (e). For a gas, e arises from the random movement of the atoms/molecules via their translational, rotational, vibrational, and electronic modes. On the other hand, a fluid in motion has the following *specific* kinetic energy K due to its bulk movement,

$$K = \frac{V^2}{2}. \tag{2.20}$$

Therefore, a fluid has a total energy based on its internal and kinetic energy:

$$E_{\text{tot}} = e + \frac{V^2}{2}. \tag{2.21}$$

In its glorified, full-expression, differential Eulerian form,

$$\begin{aligned}
\frac{\partial(\rho E_{\text{tot}})}{\partial t} + \vec{\nabla} \cdot \left(\rho E_{\text{tot}} \vec{V} \right) &= \rho \dot{Q} + \left[\frac{\partial}{\partial x}\left(k\frac{\partial T}{\partial x} \right) + \frac{\partial}{\partial y}\left(k\frac{\partial T}{\partial y} \right) + \frac{\partial}{\partial z}\left(k\frac{\partial T}{\partial z} \right) \right] \\
&- \left[\frac{\partial(uP)}{\partial x} + \frac{\partial(vP)}{\partial y} + \frac{\partial(wP)}{\partial z} \right] + \left\{ \begin{array}{l} \left[\frac{\partial(u\tau_{xx})}{\partial x} + \frac{\partial(v\tau_{xy})}{\partial x} + \frac{\partial(w\tau_{xz})}{\partial x} \right] \\ + \left[\frac{\partial(u\tau_{yx})}{\partial y} + \frac{\partial(v\tau_{yy})}{\partial y} + \frac{\partial(w\tau_{yz})}{\partial y} \right] \\ + \left[\frac{\partial(u\tau_{zx})}{\partial z} + \frac{\partial(v\tau_{zy})}{\partial z} + \frac{\partial(w\tau_{zz})}{\partial z} \right] \end{array} \right\} + \rho \vec{F} \cdot \vec{V}
\end{aligned} \tag{2.22}$$

where k = thermal conductivity.

[Note: the "modern" convention (perhaps "goal" is more appropriate) is to have the thermal conductivity k be replaced with λ. As to how successful that has been is debatable. But, in the turbulence literature, λ is traditionally reserved for the Taylor length scale, and k is reserved for the turbulence kinetic energy. Because the thermal conductivity will be used sparingly in this book, k and λ will be reserved primarily for the traditional turbulence variables. As always, the definition of a term is based on context!]

Each term in the energy PDE has the following interpretation:

$\frac{\partial(\rho E_{\text{tot}})}{\partial t}$ = accumulation of total energy; rate change of total energy in the fluid,

$\vec{\nabla} \cdot \left(\rho E_{\text{tot}} \vec{V}\right)$ = total energy change due to fluid convection,

$\rho \dot{Q}$ = volumetric heat source (e.g., point source, nuclear, chemical, heated surface),

$\left[\frac{\partial}{\partial x}\left(k \frac{\partial T}{\partial x}\right) + \frac{\partial}{\partial y}\left(k \frac{\partial T}{\partial y}\right) + \frac{\partial}{\partial z}\left(k \frac{\partial T}{\partial z}\right)\right]$ = rate of energy change due to conduction,

$\left[\frac{\partial(uP)}{\partial x} + \frac{\partial(vP)}{\partial y} + \frac{\partial(wP)}{\partial z}\right]$ = work done on the fluid by the pressure forces,

$\left\{ \begin{array}{l} \left[\frac{\partial(u\tau_{xx})}{\partial x} + \frac{\partial(v\tau_{xy})}{\partial x} + \frac{\partial(w\tau_{xz})}{\partial x}\right] \\[2mm] + \left[\frac{\partial(u\tau_{yx})}{\partial y} + \frac{\partial(v\tau_{yy})}{\partial y} + \frac{\partial(w\tau_{yz})}{\partial y}\right] \\[2mm] + \left[\frac{\partial(u\tau_{zx})}{\partial z} + \frac{\partial(v\tau_{zy})}{\partial z} + \frac{\partial(w\tau_{zz})}{\partial z}\right] \end{array} \right\}$ = rate of work done on the fluid by the

viscous forces, and

$\rho \vec{F} \cdot \vec{V}$ = rate of work done on the fluid by the body forces (note: $\vec{V}$ = velocity, not volume!).

At this point, constitutive equations in the form of thermodynamic relationships are used to provide closure. For instance, the internal energy e can be a function solely of pressure and temperature,

$$e = e(T, P). \tag{2.23}$$

As a simple application, a calorically perfect gas has constant specific heat and no pressure dependency:

$$e = C_v T. \tag{2.24}$$

Of course, C_v need not be constant and in fact is often a function of P and T.

As mentioned earlier, Eq. 2.22 includes fluid motion in various terms, thereby coupling energy and momentum transfer. However, should the fluid become motionless (and with certain assumed conditions, such as $k \neq k(x,y,z)$, $\rho \neq \rho(t)$, no heat sources, no body forces, and a calorically perfect gas), the generalized energy equation collapses onto the well-known transient, 3D heat conduction equation, as will be shown next. First, expand E_{tot} into its two components (internal and kinetic energy),

$$E_{\text{tot}} = C_v T + \frac{V^2}{2}.$$ (2.25)

Next, substitute Eq. 2.25 into Eq. 2.22, and delete the terms discussed above based on those assumptions. Then,

$$\frac{\partial\left[\rho\left(C_v T + \frac{V^2}{2}\right)\right]}{\partial t} + \vec{\nabla} \cdot \left[\rho\left(C_v T + \frac{V^2}{2}\right)\right] \vec{V} = \rho\dot{Q} + k\left(\frac{\partial^2 T}{\partial x^2} + \frac{\partial^2 T}{\partial y^2} + \frac{\partial^2 T}{\partial z^2}\right)$$

$$-\left[\frac{\partial(uP)}{\partial x} + \frac{\partial(vP)}{\partial y} + \frac{\partial(wP)}{\partial z}\right] + \left\{ + \begin{aligned} &\left[\frac{\partial(u\tau_{xx})}{\partial x} + \frac{\partial(v\tau_{xy})}{\partial x} + \frac{\partial(w\tau_{xz})}{\partial x}\right] \\ &\left[\frac{\partial(u\tau_{yx})}{\partial y} + \frac{\partial(v\tau_{yy})}{\partial y} + \frac{\partial(w\tau_{yz})}{\partial y}\right] \\ &\left[\frac{\partial(u\tau_{zx})}{\partial z} + \frac{\partial(v\tau_{zy})}{\partial z} + \frac{\partial(w\tau_{zz})}{\partial z}\right] \end{aligned} \right\} + \rho\vec{F}\cdot\vec{V}$$ (2.26A)

which reduces to

$$\rho C_v \frac{\partial T}{\partial t} = k\left(\frac{\partial^2 T}{\partial x^2} + \frac{\partial^2 T}{\partial y^2} + \frac{\partial^2 T}{\partial z^2}\right).$$ (2.26B)

At this point, it is convenient to define the thermal diffusivity as

$$\alpha = \frac{k}{\rho C_v}$$ (2.26C)

and insert it into Eq. 2.26B, yielding

$$\frac{\partial T}{\partial t} = \alpha\left(\frac{\partial^2 T}{\partial x^2} + \frac{\partial^2 T}{\partial y^2} + \frac{\partial^2 T}{\partial z^2}\right) = \alpha\nabla^2 T.$$ (2.26D)

This completes the derivation of the well-known 3D transient conduction equation. Though fairly easy to derive from the generalized energy conservation equation, the point is just how physics-rich the generalized energy equation is and how similar it is to the viscous flow of fluids in PDE:

$$\frac{\partial u}{\partial t} = \nu\left(\frac{\partial^2 u}{\partial x^2} + \frac{\partial^2 v}{\partial y^2} + \frac{\partial^2 w}{\partial z^2}\right) = \nu\nabla^2 V.$$ (2.26E)

2.4   Introduction to Turbulence Theory

Consider the Navier-Stokes equation in 1D, and assume the pressure and gravity terms are negligible. WLOG, choose flow in the x coordinate, with v and w being zero. Of course, turbulence is a 3D phenomenon, but the 1D case being considered at this point describes a crucial transition from stability to instability—that is, from laminar to turbulent flow:

$$
\rho \frac{\partial u}{\partial t} + \rho \left(u \frac{\partial u}{\partial x} + v \frac{\partial u}{\partial y} + w \frac{\partial u}{\partial z} \right) = \mu \left(\frac{\partial^2 u}{\partial x^2} + \frac{\partial^2 u}{\partial y^2} + \frac{\partial^2 u}{\partial z^2} \right) - \frac{\partial p}{\partial x}
$$
$$
+ \rho g_x. \tag{2.11A$'$}
$$

Collecting the remaining terms,

$$
\rho \frac{\partial u}{\partial t} = \mu \frac{\partial^2 u}{\partial x^2} - \rho u \frac{\partial u}{\partial x}, \tag{2.11A$''$}
$$

which is the transient Burgers' Equation. As discussed earlier, the second-order derivative acts as a damping force, driven by viscosity. Thus, the higher the viscosity, the higher the damping. This is confirmed when the above equation is nondimensionalized (represented with the "~" symbol), whereby the viscous PDE term now shows a crucial inverse Reynolds number (Re) dependency,

$$
\frac{\partial \tilde{u}}{\partial \tilde{t}} = \frac{1}{Re} \frac{\partial^2 \tilde{u}}{\partial \tilde{x}^2} - \tilde{u} \frac{\partial \tilde{u}}{\partial \tilde{x}} \tag{2.11A$'''$}
$$

where

$$
Re \equiv \frac{x_{char} u_{char} \rho}{\mu} = \frac{\text{Inertial (convective) force}}{\text{Viscous force}}. \tag{2.27}
$$

At steady state, the Burgers' Equation shows the following momentum balance, $\rho u \frac{\partial u}{\partial x} = \mu \frac{\partial^2 u}{\partial x^2}$; that is, the convective term equals the viscous term. Using some scaling analysis (refer to Sect. 6.4.2) allows the following interpretation for Re:

$$
Re = \frac{\text{Convective force term}}{\text{Viscous force term}} = \frac{nonlinear}{linear} = \frac{\rho u \frac{\partial u}{\partial x}}{\mu \frac{\partial^2 u}{\partial x^2}} \approx \frac{\rho u \frac{u}{x}}{\mu \frac{u}{x^2}} = \frac{x u \rho}{\mu}. \tag{2.28}
$$

Re is therefore the ratio of the inertial and viscous force, and the higher the Re is, the stronger the convective force is compared to the viscous force. For example, as the dynamic viscosity μ increases, viscous damping increases, so Re decreases. Conversely, Re increases as the viscous term decreases or the velocity increases.

Note that the inertial (convective) term (the second term on the RHS of Eq. 2.11A''') increases rapidly as a function of velocity times the gradient of the velocity. As the magnitude of the convective term continues to increase and eventually becomes much larger than the viscous term, there results a "runaway" flow instability whereby the damping effect is marginalized. Therefore, 3D instabilities form and become rampant as Re continues to increase. From this critical point on, the laminar flow becomes turbulent and evermore chaotic as Re continues to increase. That is, in nondimensionalized form,

$$\tilde{u}\frac{\partial \tilde{u}}{\partial \tilde{x}} \ll \frac{1}{Re}\frac{\partial^2 \tilde{u}}{\partial \tilde{x}^2} \text{ (laminar)} \tag{2.29A}$$

and

$$\tilde{u}\frac{\partial \tilde{u}}{\partial \tilde{x}} \gg \frac{1}{Re}\frac{\partial^2 \tilde{u}}{\partial \tilde{x}^2} \text{ (turbulent).} \tag{2.29B}$$

Thus, when the damping force is much larger than the inertial (convective) force, the flow is stable (laminar). By contrast, the damping force is much smaller than the inertial force, and the flow becomes unstable (turbulent). Again, this is not surprising, as Re is the ratio of the inertial and viscous damping forces.

For convenience, let the momentum equation be described in its nonconservative form, using the material derivative (defined later Eq. 2.42):

$$\frac{D\vec{V}}{Dt} = \frac{\partial \vec{V}}{\partial t} + \vec{\nabla} \cdot \vec{V}\vec{V} = \frac{\mu}{\rho}\nabla^2 \vec{V} - \frac{1}{\rho}\vec{\nabla}P + \vec{g}. \tag{2.30A}$$

After non-dimensionalizing the primitive variables and some algebra, Re appears again:

$$\frac{D\vec{\tilde{V}}}{D\tilde{t}} = \frac{\partial \vec{\tilde{V}}}{\partial \tilde{t}} + \vec{\tilde{\nabla}} \cdot \vec{\tilde{V}}\vec{\tilde{V}} = \frac{1}{Re}\tilde{\nabla}^2\vec{\tilde{V}} - \vec{\tilde{\nabla}}\tilde{P} + \frac{1}{Fr}\vec{\tilde{g}}, \tag{2.30B}$$

where the Froude number, Fr, is defined as (White 1991)

$$Fr \equiv \frac{u^2}{g_x x}. \tag{2.30C}$$

Because there is a $1/Re$ coefficient in front of the normalized viscous damping term, a decreased Re implies the second-order differential increases in magnitude. That is, its "braking (damping) power" increases, thereby mitigating potential flow instabilities. Thus, if any instabilities attempt to form turbulent coherent structures, the damping force is sufficiently large to mitigate eddy formation, thereby allowing the flow to continue in a stable, laminar fashion.

The converse is true: decreasing μ decreases the viscous damping, and Re increases. A larger Re means that the nonlinear inertial term increases. Because of the inverse Re relationship, the second-order differential decreases in magnitude. That is, its "braking power" decreases, thereby rendering it unable to mitigate flow instabilities, inevitably causing a transition from stable laminar flow to unstable turbulent flow. Thus, the turbulent instabilities arise from the nonlinear convective term.

Thus, turbulence arises over the excess of the nonlinear convective term compared with the linear viscous momentum term; the *linear* viscous term tends to produce laminar sheets, while the *nonlinear* convective term tends to produce sinuous coherent structures. Basically, laminar flow can be visualized as sheets of flow moving past each other, much like poker cards being pushed gently, with the cards sliding past each other, as shown at the top of Fig. 2.5. Laminar flows can also be viewed as consisting of straight, coherent filament motion that does not crisscross. As Osborne Reynolds noticed in 1883, when ink was injected into a horizontal transparent tube with water flowing internally, the ink basically followed straight lines when the flow was laminar, without deviating from their path (Reynolds 1883). However, as the water velocity increased in "small stages," the ink "would all at once mix up with the surrounding water and fill the rest of the tube with a mass of colored water. . ." Then, as Reynolds illuminated the flow with an electric spark, he noticed that it was characterized by "distinct curls, showing eddies," which are shown conceptually at the bottom of Fig. 2.5. As the convective term increased, the instabilities increased, producing an *eddying motion* that caused energetic local swirling of the fluid—turbulent eddies. The vorticity in eddies is extremely high as a result of their rotational nature, as the coherent structures stretch. Energy from the main flow is nearly instantaneously transferred to eddies via mechanisms that are not well-known (at least as of 2019). It is this energy that forms and feeds the chaotic, large eddies. Because turbulent mixing generates hundreds to thousands of times (or more) mixing than laminar flow, the net benefits are much higher heat transfer, augmented species mixing, more efficient combustion, and so forth.

More often than not, unstable states eventually tend to transition into more stable states. Paradoxically, turbulent flow is not a "stable" state onto which nature transitions in order to achieve additional stability as the flow leaves the laminar regime—quite the opposite! In general, the higher the Re, the higher the randomness

Fig. 2.5 Conceptual view of flow if ink were placed in the fluid: laminar vs. turbulent

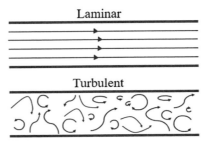

in the flow. More paradoxically yet, the flow drag coefficient for certain curved surfaces can decrease precipitously as *Re* increases (Eiffel paradox; discussed in Sect. 3.7).

Figure 2.6 shows the velocity profiles in a pipe for the laminar and turbulent cases. Notice that turbulence flattens the velocity profile in the center and increases the velocity distribution near the wall; otherwise, the profiles would appear remarkably similar. Nevertheless, the flow behaves remarkably different! In reality, the turbulent velocity profile shown in Fig. 2.6 is actually a *time-averaged* distribution. As a result, the turbulent profile looks quite continuous. However, if instantaneous snapshots were taken, the profile would be quite irregular (wavy), and because turbulent flow is random, no two snapshots would ever look exactly the same! However, in the limit as more and more snapshots are taken and averaged, the profile approaches the shape shown in Fig. 2.6. The chaotic, non-repetitive behavior is shown conceptually in Fig. 2.7.

As the flow transitions into turbulence, rapid pressure and velocity fluctuations (also known as pulsations) occur in both compressible and incompressible flows. Here, "fluctuations" refer to random, fast changes in a given variable, having dynamic spatial regions that are generated as molecular motion swirls collectively, forming "coherent structures." Coherent structures are "grouped" fluid particles that move in unison as they rotate, stretch, translate, and decay. These are the "distinct curls" envisioned by Reynolds—turbulent eddies. As the flow becomes turbulent, it forms 3D coherent structures that are unstable WRT time and space. This is what is best described classically by Hinze as turbulent flow with "irregular" (nondeterministic) fluid motion (Hinze 1987):

> Turbulent fluid motion is an irregular condition of flow in which the various quantities show a random variation with time and space coordinates, so that statistically distinct average values can be discerned.

That is, random variations in pressure and velocity occur in space and time for incompressible flows. And these random variations (fluctuations) conform to predictable statistics. *Paradoxically, turbulent flow involves a high degree of randomness that generates rather predictable, coherent patterns!* Besides velocity and pressure fluctuations, compressible flows include density variations as well.

Fig. 2.6 Laminar vs. turbulent velocity distribution in a pipe

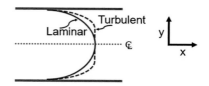

Turbulence flattens the velocity profile in the center and increases the velocity distribution near the wall. Otherwise, the profiles are remarkably similar, but the flow behaves remarkably different!

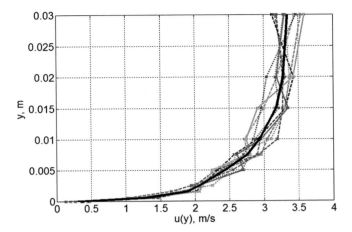

Fig. 2.7 Time-averaged distribution (thick black line) vs. instantaneous velocity distributions (all others)

There exist a wide range of spatial and temporal scales for eddies. Their spatial range is as small as the Kolmogorov scale and about as large as the characteristic flow length scale—the integral eddies (this subject and related definitions are covered in detail in Sects. 3.1, 3.2, and 3.3). The time scales are associated with a range of eddy sizes, with the largest eddies having the largest time scales and the smallest eddies having the smallest time scales (they have shorter lives before they decay unto laminarity). As a result of the different time scales, some fluctuations seemingly appear as near-instantaneous manifestations, while others change at a much slower pace. The fluctuations are much faster when compared to the time-averaged flow properties. The time and space scales can be modeled mathematically via Fourier analysis, with the time scale being associated with a frequency, while the space scale is associated with a wavelength.

Turbulent flows require an energy source to feed its turbulent behavior, as eddies decay rapidly WRT time; that is, the coherent structures diminish in size and fluctuating magnitude, until they transform onto a laminar region in space having no fluctuations and no randomness.

2.5   Introduction to Turbulence Modeling

Most engineering and industrial flows are turbulent (White 1991). One exception is blood flow in a healthy person. However, as fatty deposits and cholesterol block the veins and arteries, the flow area becomes constricted, so the flow becomes turbulent. Another exception is a reentry vehicle as it pierces through the rarefied atmosphere. This is a low *Re* problem because the kinematic viscosity is extremely large (density is very small, so the kinematic viscosity becomes very large). The vast majority of nano- and micro-devices exhibit laminar flow, but there are exceptions (Groisman and Steinberg 2000; Wang et al. 2014). For fluids with a relatively low viscosity, such as water, if the fluid is flowing internally in a pipe or duct with a characteristic length on the order of a cm or more, and at a speed of 1 m/s or higher, then the flow will inevitably be turbulent. On the one hand, turbulence requires energy (e.g., a pump). Generated at the cost of some pumping force or mechanism, turbulent flows are highly diffusive, meaning they have a strong mixing capacity. Therefore, momentum, mass, chemical species, energy, and so forth are rapidly "mixed," thereby generating countless practical engineering applications. So long as the cost to generate turbulence is repaid via its mixing capacity, turbulence is desirable, but the converse is also true. Consequently, it is crucial to understand how to model turbulent flows for more efficient industrial processes, increased safety margins, and higher profitability.

As noted in Fig. 2.5, a turbulent flow consists of the superposition of the continuous spectrum of small to large eddies, as well as discrete regions where the flow is laminar (e.g., the viscous sublayer, regions in the intermittent boundary, and numerous tiny regions where the Kolmogorov eddies have decayed back to laminar flow). Each eddy represents a coherent cluster of fluid atoms or molecules that move in unison, hence the name, "coherent structure." Because all turbulent flows involve randomness, if an "instantaneous velocity snapshot" were taken of the fluid, the velocity distribution would appear discontinuous, as shown in Fig. 2.7. For all curves except the black, thick solid line shown in the center, the velocity distributions are similar to the random walk (path) of a dizzy person—they are rather discontinuous and erratic. This is the case because the eddies move fairly independently of each other, and it is their superposition over time that results in the more familiar, time-averaged continuous distribution shown by the thick, black curve in the central region of Fig. 2.7. The black curve was generated by summing 10 random instances—the sum of the random instantaneous velocities converges onto a statistically predictable, continuous, time-averaged velocity distribution. *Thus, when it comes to turbulence, there is order in disorder and determinism in randomness!*

As noted already, the instantaneous velocity distribution is not smooth but is instead composed of a multitude of irregular velocity distributions shaped by the superposition of the eddies: any eddy moving forward generates a positive bump in the velocity distribution, while those eddies that are circulating in a backward motion generate backward bumps. In other words, the shape of the time-averaged velocity is forged by the sum of the eddies, and the small, mid-sized, and large wiggles are caused by the small, mid-sized, and large eddies, respectively; see Fig. 2.8. Note that eddies typically have velocities that are about 1/100th to 1/10th of the mean velocity; a methodology for calculating eddy velocity is presented in Chap. 3.

In summary, turbulent flow is the dynamic superposition of an extremely large number of eddies with random (irregular) but continuous spectrum of sizes and velocities that are interspersed with small, discrete pockets of laminar flow (as a result of the Kolmogorov eddies that decayed, as well as in the viscous laminar sublayer and in the intermittent boundary). In this sense, turbulent flow is intractable

Fig. 2.8 The relationship
between eddies,
instantaneous velocity
distribution, and time-
averaged distribution

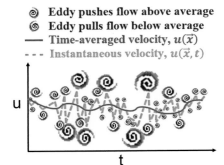

in its fullest manifestation; this is where good, engineering common sense and
approximations can deliver reasonable solutions, albeit approximate.

The first attempt to tract a seemingly intractable problem consisted of considering
eddy behavior over a sufficiently long period of time. This allows their behavior to
be time-averaged, thereby greatly simplifying its mathematical behavior (Reynolds
1895). Following Reynolds' original notation (Reynolds 1895) (as well as others,
e.g., (Bird et al. 2007; CFD-online 2017a)),

$$u\left(\vec{x},t\right) = \bar{u}\left(\vec{x}\right) + u'\left(\vec{x},t\right) \tag{2.31A}$$

or

$$u = \bar{u} + u' \text{ (short-hand)} \tag{2.31B}$$

where

u = instantaneous velocity $\left(\text{has } \vec{x}, t \text{ dependence}\right)$

$\bar{u}$ = mean velocity $\left(\text{only has } \vec{x} \text{ dependence because it is integrated over a time}\right)$

u' = instantaneous velocity fluctuation $\left(\text{has } \vec{x}, t \text{ dependence}\right)$

The above splitting (decomposition) of instantaneous flow into an average quantity and its associated fluctuating quantity is called Reynolds decomposition, in
honor of its originator, Osborne Reynolds. The fluctuating quantities, e.g., u', are
mathematical abstracts intended to quantify the eddy fluctuations. This reasoning is
similar to the stock market, where the minute-by-minute changes in stock prices are
analogous to the instantaneous changes in fluid velocity that are generated by the
eddy fluctuations. But, as stock prices are smoothed over a longer period, say daily,

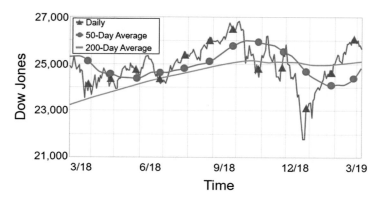

Fig. 2.9 Stock market instantaneous and averaged prices have a similar nature as turbulent flow

50-day average, and a 200-day average, the fluctuating price curve becomes smoother and smoother, as shown in Fig. 2.9 (the longer the time-averaging, the flatter the curve). Thus, the time-averaged price curves are analogous to the time-averaged velocity, so economists ought to be good turbulence modelers, and perhaps good turbulence modelers ought to make great stock market profits!

Note that some authors use a different notation than the original Reynolds' formulation (e.g., Wilcox):

$$u\left(\vec{x}, t\right) = U\left(\vec{x}\right) + u'\left(\vec{x}, t\right) \tag{2.32A}$$

or

$$u = U + u' \text{ (short hand).} \tag{2.32B}$$

Clearly, $\bar{u} \equiv U$. The Reynolds notation offered by Eqs. 2.31A and 2.31B will be used throughout this text because:

- Reynolds was the first to define the variable notation.
- A bar gives the sense of an operation (usually an average), while an uppercase letter provides no mathematical guidance.
- The Reynolds notation is the most prevalent in the literature.

That said, serious turbulence students should always follow the context, as well as be familiar with both notations.

Now define arbitrary variables ϕ and $\bar{\phi}$ to represent the following primitive variables: ρ, T, P, u, v, and w. More precisely, let $\phi = f\left(\vec{x}, t\right)$ indicate a spatial and temporal dependency. Further, $\bar{\phi} = F\left(\vec{x}\right)$ represents the time-averaged primitive variable obtained by integration WRT time,

$$\overline{\phi}\left(\vec{x}\right) \equiv \lim_{T \to \infty} \frac{1}{T} \int\limits_{t}^{t+T} \phi\left(\vec{x}, t\right) dt. \qquad (2.33)$$

$T \to \infty$ refers to a sufficiently long time period, at least sufficiently long to capture a reasonable number of fluctuations that yield a stationary domain. Equation 2.33 defines how a time-averaged ("mean") turbulence quantity is obtained by integrating all the instantaneous changes over a period of time. It is referred here as the "bar operation," and shortly, its utility will be demonstrated in the derivation of the turbulent mass and momentum equations.

Example 2.12 Express the three instantaneous turbulence velocity equations in terms of their mean and fluctuating velocity terms. For simplicity, consider a Cartesian system.

Solution Let $\phi\left(\vec{x}, t\right) = \overline{\phi}\left(\vec{x}\right) + \phi'\left(\vec{x}, t\right)$ in vector notation, or

$\phi_i(x_i, t) = \overline{\phi}_i(x_i) + \phi_i'(x_i, t)$ in index notation where $i = 1, 2, 3 = x, y, z$.
This means that:

$$x_1 = x, \quad x_2 = y, \text{and } x_3 = z.$$
$$\phi_1 = u_1 = u, \phi_2 = u_2 = v, \text{and } \phi_3 = u_3 = w.$$
$$\text{Then, } u(x, y, z, t) = \overline{u}(x, y, z) + u'(x, y, z, t),$$

where

$u = \overline{u} + u'$ (same as above, but in simpler notation)
$v = \overline{v} + v'$
$w = \overline{w} + w'$

Example 2.13 What is the time-averaged x-direction velocity in Cartesian coordinates?

Solution Recall that

$$\overline{\phi}\left(\vec{x}\right) = \lim_{T \to \infty} \frac{1}{T} \int\limits_{t}^{t+T} \phi\left(\vec{x}, t\right) dt$$

and

$$\phi\left(\vec{x}, t\right) = \overline{\phi}\left(\vec{x}\right) + \phi'\left(\vec{x}, t\right).$$

For this situation, $\overline{\phi} = \overline{u}$ and $\phi = u$. From $\phi = \overline{\phi} + \phi' \Rightarrow u = \overline{u} + u'$. Now substitute $\overline{\phi} = \overline{u}$ and $\phi = u$ onto the above integral:

$$\bar{u} = \lim_{T \to \infty} \frac{1}{T} \int\limits_{t}^{t+T} u \, dt = \lim_{T \to \infty} \frac{1}{T} \int\limits_{t}^{t+T} (\bar{u} + u') \, dt.$$

The above expression is simply the definition of $\bar{u}$.

Example 2.14 What is the time-average of a time-averaged value?

Solution Consider $\bar{u}$ and perform the bar operation on it:

$$\bar{\bar{u}} = \lim_{T \to \infty} \frac{1}{T} \int\limits_{t}^{t+T} \bar{u} \, dt = \bar{u}.$$

So, the mean of a mean is still a mean.

Example 2.15 What is the mean of a fluctuating quantity?

Solution Apply the averaging operation onto u' (i.e., the bar operation):

$$\bar{u'} = \lim_{T \to \infty} \frac{1}{T} \int\limits_{t}^{t+T} u' \, dt = \lim_{T \to \infty} \frac{1}{T} \int\limits_{t}^{t+T} (u - \bar{u}) \, dt = \bar{u} - \bar{\bar{u}} = \bar{u} - \bar{u} = 0.$$

Therefore, the time-averaged value for the instantaneous fluctuations is zero, because in due time, the fluctuations should average out.

The equations listed below summarize many useful time averages:

$$\bar{u'} = 0, \text{(ups and downs average out to zero; equal probability)} \qquad (2.34\text{A})$$

$$\bar{\bar{u}} = \bar{u}, \text{(the mean of a mean is a mean \ldots)} \qquad (2.34\text{B})$$

$$\overline{\bar{u}u'} = 0, \text{(taking the mean of a mean quantity times its fluctuation is 0)} \qquad (2.34\text{C})$$

$$\overline{u'\bar{u}} = 0, \text{(the mean of a fluctuation times its mean is 0)} \qquad (2.34\text{D})$$

$$\overline{\frac{\partial u}{\partial x}} = \frac{\partial \bar{u}}{\partial x}, \text{(differential is insensitive in space; commutation)} \qquad (2.34\text{E})$$

$$\overline{\frac{\partial u}{\partial t}} = \frac{\partial \bar{u}}{\partial t}, \text{(differential is insensitive in time)} \qquad (2.34\text{F})$$

$$\overline{u'_i u'_j} \neq 0; \text{where } i \neq j, \qquad (2.34\text{G})$$

$$\overline{u'_i u'_i} = \overline{u'^2_i} \neq 0, \qquad (2.34\text{H})$$

$$I = I_t = \text{turbulence intensity} = \frac{\sqrt{\overline{u'^2}}}{\overline{u}} \sim \frac{u'}{\overline{u}}, \tag{2.34I}$$

$$\overline{\overline{ab}c'} = 0, \tag{2.34J}$$

$$\overline{\overline{ab'}c'} \neq 0, \tag{2.34K}$$

and

$$\overline{a+b} = \overline{a} + \overline{b} \quad \text{(linearity)}. \tag{2.34L}$$

The turbulence intensity is a very useful metric for characterizing turbulence. Of course, the fluctuating velocity is difficult to estimate, though several very useful approximations exist. For example, the turbulence intensity can be approximated as a function of Re, showing that the turbulence fluctuations decrease as Re increases (CFD-online 2017b).

$$I \approx \frac{u'}{\overline{u}} = 0.16\, Re_h^{-1/8}. \tag{2.35A}$$

Equation 2.35A is suitable for the core of a fully developed internal pipe/duct flow. Re uses the hydraulic diameter and hence the lower case "h." From Eq. 2.35A, it follows that

$$u' \approx 0.16\overline{u}\, Re_h^{-1/8}. \tag{2.35B}$$

Of course, usage of Eq. 2.35B to estimate u' alludes to a fluctuating quantity that is somehow "characteristic" of a given turbulent flow and must be solely construed as such. In reality, u' will occupy a significant velocity range. Furthermore, Eq. 2.35B shows a smooth, continuous Re dependency, whereas u' is more akin to a random compilation of discontinuous velocity spikes with an irregular pattern. Therefore, how can u' possibly relate to a continuous function of Re? Because turbulence averages are routinely used to characterize flow, the above Re relationship is fair game, especially if the eddy distribution covers a narrow wavenumber range that contains a significant fraction of the eddy population—it therefore characterizes the collective performance of a group of fairly similar eddies, whose behavior reflects a reasonable segment of the turbulence behavior in question.

Note that when performing Reynolds decomposition, it is not a good idea to ignore the instantaneous (primed) quantities. For example, ignoring u' because $|u'| \ll |\overline{u}|$ is not a good approximation, as fluctuations can be much larger than $0.1\overline{u}$. That is, the turbulence intensity can often exceed 10% or more, as those who have flown can certify when they felt a turbulence bump in midair. Note that absolute quantities were considered because fluctuations can be negative.

Large fluctuations can be expected in the following situations:

- Flow near the wall (y^+ in the range of 7–30)
- Free shear flow
- Atmospheric boundary layer flow
- High speed flows
- Complex geometries (e.g., heat exchangers, turbomachinery)

Now, if instead of a time average, a spatial average is desired, then

$$\overline{\phi}(t) = \lim_{V \to \infty} \frac{1}{V} \iiint\limits_{V} \phi\left(\vec{x}, t\right) dx dy dz. \text{(Note that } V \text{ is volume, not velocity.)}$$

$$(2.36)$$

Because the definite integral is computed over a specific range in space, the resultant is solely a function of time. Spatial averages are suitable for *homogeneous* turbulence, i.e., for turbulent flows that, on an average basis, are uniform in any direction. Other averages exist, such as the ensemble and Fabre averages; refer to Wilcox (2006) for additional information. Ensemble averages are useful for experimental flow measurements that decay as a function of time, while Fabre averages are suitable for compressible fluids, where the density now requires an additional Reynolds decomposition, $\rho = \overline{\rho} + \rho'$.

For the purpose of this text, only time averages as applied to Reynolds-averaged Navier-Stokes (RANS) turbulence models will be considered. Thus, the flows considered herein are incompressible and therefore do not require Fabre averages. An "average redefinition" suitable for large eddy simulation (LES) will be considered in Sect. 5.1.1.

At this point, the turbulent conservation of mass and momentum equations can be derived by using Reynolds decomposition and integrating the equations over time (the bar operation). Assume the fluid is incompressible (divergence of the flow field is zero), SS, there are no heat sources, and the system is Cartesian (WLOG). Then, the conservation of mass equation becomes

$$\frac{\partial u}{\partial x} + \frac{\partial v}{\partial y} + \frac{\partial w}{\partial z} = 0. \qquad (2.37A)$$

Using Reynolds' decomposition, let

$$\begin{cases} u = \overline{u} + u' \\ v = \overline{v} + v' \\ w = \overline{w} + w' \end{cases} \qquad (2.38)$$

Now substitute the decomposed velocities into the conservation of mass PDE,

$$\frac{\partial}{\partial x}(\bar{u} + u') + \frac{\partial}{\partial y}(\bar{v} + v') + \frac{\partial}{\partial z}(\bar{w} + w') = 0,$$

which expands to

$$\frac{\partial \bar{u}}{\partial x} + \frac{\partial u'}{\partial x} + \frac{\partial \bar{v}}{\partial y} + \frac{\partial v'}{\partial y} + \frac{\partial \bar{w}}{\partial z} + \frac{\partial w'}{\partial z} = 0.$$

Next, perform time-averaging on the conservation of mass PDE, i.e., integrate over time via the bar operation. Recall that

$$\overline{u'} = 0 \tag{2.39A}$$

and

$$\bar{\bar{u}} = \bar{u}. \tag{2.39B}$$

Therefore,

$$\frac{\partial \bar{\bar{u}}}{\partial x} + \frac{\partial \overline{u'}}{\partial x} + \frac{\partial \bar{\bar{v}}}{\partial y} + \frac{\partial \overline{v'}}{\partial y} + \frac{\partial \bar{\bar{w}}}{\partial z} + \frac{\partial \overline{w'}}{\partial z} = 0,$$

which results in the mass conservation PDE for turbulent flow,

$$\frac{\partial \bar{u}}{\partial x} + \frac{\partial \bar{v}}{\partial y} + \frac{\partial \bar{w}}{\partial z} = 0. \tag{2.37B}$$

Thus, turbulent mass conservation is solely a function of the time-averaged velocities. Next, a similar approach can be applied to the momentum equations, as follows. WLOG, consider the x-direction in a Cartesian system with a turbulent flow that is incompressible, unheated, and Newtonian. Because the flow is incompressible and unheated, the density fluctuations can be ignored, and the gravitational term is immediately neglected.

$$\rho \frac{\partial u}{\partial t} + \rho\left(u\frac{\partial u}{\partial x} + v\frac{\partial u}{\partial y} + w\frac{\partial u}{\partial z}\right) = \rho\frac{\partial u}{\partial t} + \rho\left(\frac{\partial uu}{\partial x} + \frac{\partial uv}{\partial y} + \frac{\partial uw}{\partial z}\right)$$
$$= -\frac{\partial P}{\partial x} + \mu\nabla^2 u + \rho\bcancel{g_x}. \tag{2.40A}$$

Now apply Reynolds' decomposition, where

$$u = \bar{u} + u'; \ v = \bar{v} + v'; \ w = \bar{w} + w'; \ P = \bar{P} + P'. \tag{2.41}$$

Then,

$$\rho\frac{\partial(\bar{u}+u')}{\partial t} + \rho\left[\frac{\partial(\bar{u}+u')(\bar{u}+u')}{\partial x} + \frac{\partial(\bar{u}+u')(\bar{v}+v')}{\partial y} + \frac{\partial(\bar{u}+u')(\bar{w}+w')}{\partial z}\right]$$

$$= -\frac{\partial(\bar{P}+P')}{\partial x} + \mu\nabla^2(\bar{u}+u') + \rho g_x$$

Again, recall that $\overline{u'}=0$ and $\bar{\bar{u}}=\bar{u}$, etc. After some straightforward simplification,

$$\rho\frac{\partial\bar{u}}{\partial t} + \rho\left(\bar{u}\frac{\partial\bar{u}}{\partial x} + \bar{v}\frac{\partial\bar{u}}{\partial y} + \bar{w}\frac{\partial\bar{u}}{\partial z}\right) + \rho\left(\frac{\partial\overline{u'u'}}{\partial x} + \frac{\partial\overline{u'v'}}{\partial y} + \frac{\partial\overline{u'w'}}{\partial z}\right) = -\frac{\partial\bar{P}}{\partial x} + \mu\nabla^2\bar{u}.$$

For comparison with other texts, the averaged velocity in the second term on the left-hand side (LHS) can be included into the partial differential terms by noting that the system is incompressible. In such case, then

$$u\frac{\partial u}{\partial x} + v\frac{\partial u}{\partial y} + w\frac{\partial u}{\partial z} = \frac{\partial uu}{\partial x} + \frac{\partial uv}{\partial y} + \frac{\partial uw}{\partial z}.$$

Therefore, the turbulent x-momentum becomes

$$\rho\frac{\partial\bar{u}}{\partial t} + \rho\left(\frac{\partial\overline{uu}}{\partial x} + \frac{\partial\overline{uv}}{\partial y} + \frac{\partial\overline{uw}}{\partial z}\right) + \rho\left(\frac{\partial\overline{u'u'}}{\partial x} + \frac{\partial\overline{u'v'}}{\partial y} + \frac{\partial\overline{u'w'}}{\partial z}\right)$$

$$= -\frac{\partial\bar{P}}{\partial x} + \mu\nabla^2\bar{u}. \tag{2.40B}$$

Applying the same operations to the y- and z-momentum equations,

$$\rho\frac{\partial\bar{v}}{\partial t} + \rho\left(\frac{\partial\overline{vu}}{\partial x} + \frac{\partial\overline{vv}}{\partial y} + \frac{\partial\overline{vw}}{\partial z}\right) + \rho\left(\frac{\partial\overline{v'u'}}{\partial x} + \frac{\partial\overline{v'v'}}{\partial y} + \frac{\partial\overline{v'w'}}{\partial z}\right)$$

$$= -\frac{\partial\bar{P}}{\partial y} + \mu\nabla^2\bar{v} \tag{2.40C}$$

and

$$\rho\frac{\partial\bar{w}}{\partial t} + \rho\left(\frac{\partial\overline{wu}}{\partial x} + \frac{\partial\overline{wv}}{\partial y} + \frac{\partial\overline{ww}}{\partial z}\right) + \rho\left(\frac{\partial\overline{w'u'}}{\partial x} + \frac{\partial\overline{w'v'}}{\partial y} + \frac{\partial\overline{w'w'}}{\partial z}\right)$$

$$= -\frac{\partial\bar{P}}{\partial z} + \mu\nabla^2\bar{w}. \tag{2.40D}$$

The material derivative operator, D, is defined as the superposition of the following temporal and spatial derivatives:

$$\frac{D}{Dt} \equiv \frac{\partial}{\partial t} + \left(\vec{V} \cdot \vec{\nabla} \right). \tag{2.42}$$

Now, Eqs. 2.40B, 2.40C, and 2.40D can be written succinctly in vector-tensor format, along with the material derivative, such that

$$\rho \frac{D\vec{V}}{Dt} = \rho \frac{\partial \vec{V}}{\partial t} + \rho \vec{V} \vec{\nabla} \cdot \vec{V} = -\vec{\nabla} P - \vec{\nabla} \cdot \left(\vec{\tau} + \vec{R} \right) \tag{2.43}$$

where

$$\vec{\tau}_{ij} = \begin{vmatrix} \vec{\tau}_{xx} & \vec{\tau}_{xy} & \vec{\tau}_{xz} \\ \vec{\tau}_{yx} & \vec{\tau}_{yy} & \vec{\tau}_{yz} \\ \vec{\tau}_{zx} & \vec{\tau}_{zy} & \vec{\tau}_{zz} \end{vmatrix} \tag{2.44A}$$

with $i = 1{:}3$ for x, y, and z and $j = 1{:}3$ for x, y, and z in Cartesian form. In its full expression, the shear tensor $\vec{\tau}$ is

$$\vec{\tau} = -\mu \begin{bmatrix} \left(\frac{\partial \overline{u}}{\partial x} + \frac{\partial \overline{u}}{\partial x} \right) - \frac{2}{3} \vec{\nabla} \cdot \vec{V} & \left(\frac{\partial \overline{u}}{\partial y} + \frac{\partial \overline{v}}{\partial x} \right) & \left(\frac{\partial \overline{u}}{\partial z} + \frac{\partial \overline{w}}{\partial x} \right) \\ \left(\frac{\partial \overline{v}}{\partial x} + \frac{\partial \overline{u}}{\partial y} \right) & \left(\frac{\partial \overline{v}}{\partial y} + \frac{\partial \overline{v}}{\partial y} \right) - \frac{2}{3} \vec{\nabla} \cdot \vec{V} & \left(\frac{\partial \overline{v}}{\partial z} + \frac{\partial \overline{w}}{\partial y} \right) \\ \left(\frac{\partial \overline{w}}{\partial x} + \frac{\partial \overline{u}}{\partial z} \right) & \left(\frac{\partial \overline{w}}{\partial y} + \frac{\partial \overline{v}}{\partial z} \right) & \left(\frac{\partial \overline{w}}{\partial z} + \frac{\partial \overline{w}}{\partial z} \right) - \frac{2}{3} \vec{\nabla} \cdot \vec{V} \end{bmatrix}. \tag{2.44B}$$

Note that $\vec{\nabla} \cdot \vec{V} = 0$ for incompressible fluids. The Reynolds stress tensor is defined as

$$\vec{R} \equiv -\overline{u_i' u_j'} = - \begin{bmatrix} \overline{u'u'} & \overline{u'v'} & \overline{u'w'} \\ \overline{v'u'} & \overline{v'v'} & \overline{v'w'} \\ \overline{w'u'} & \overline{w'v'} & \overline{w'w'} \end{bmatrix}. \tag{2.45}$$

Note that R is a tensor formed by each of the three terms in the three momentum equations, for a total of nine terms. Like τ, R is symmetric, meaning that only six of the nine terms are independent. That is, $R_{ij} = R_{ji}$.

Because it appears ubiquitously in turbulence literature, the mean strain rate tensor S is conveniently defined here as

$$\vec{S} = \frac{1}{2\mu} \vec{\tau}. \tag{2.46}$$

At this point, either the three expanded momentum equations (Eqs. 2.40B, 2.40C, and 2.40D) or the momentum vector-tensor representation (Eq. 2.43) can be used in conjunction with Eqs. 2.44A, 2.44B, and 2.45 to calculate the RANS velocity equation, *if the expressions for the primed quantities for the six independent Reynolds stresses were known!* Because there are six unknown quantities, an additional six independent quantities are still needed! This is the famous "closure problem" that has plagued turbulence modeling since 1895 to the present and is not going away any time soon. The Reynolds stresses are therefore *approximated* (modeled) experimentally, empirically, or numerically (e.g., DNS). Bold engineers, in a desire to reach closure, began attempting ways to relate the fluctuating properties onto the averaged quantities, based on dimensional arguments and a gradient transport analogy. So, for better or for worse, these arguments are employed fairly commonly in turbulence research. In this case,

$$-\overline{u_i'\phi'} \sim \frac{\partial\overline{\phi}}{\partial x_i}, \qquad (2.47A)$$

where ϕ is any primitive variable, e.g., u, v, w, T, P, ρ, etc. An inspection of Eq. 2.47A shows that it can be made fully dimensionless by including a quantity that has units of m^2/s. Out of incredible inspiration (or desperation!), a new quantity was coined, "turbulent kinematic viscosity," ν_t, which is analogous to the kinematic viscosity. Hence,

$$-\overline{u_i'\phi'} = \nu_t \frac{\partial\overline{\phi}}{\partial x_i}. \qquad (2.47B)$$

The turbulent kinematic viscosity is typically orders of magnitude higher than the fluid viscosity, thus providing a reasonable way to gauge the degree of turbulence in a flow:

$$\frac{\nu_t}{\nu} \gg 1. \qquad (2.48)$$

- *Some Notes Regarding Turbulent Viscosity*

That $\nu_t \gg \nu$ is consistent with the expectation that the more turbulent the flow is, the larger ν_t will be. At this point, it is noteworthy to point out that both the kinematic viscosity and the turbulence viscosity have the same units and similar names. But, that is where the similarities end. The kinematic viscosity is based on the fluid type, pressure, and temperature and is a measure of damping, with higher kinematic viscosity resulting in higher damping. On the other hand, the turbulent kinematic viscosity is based on *Re*, where the higher *Re* is, the higher the turbulent kinematic viscosity. Thus, the turbulent viscosity is a measure of the degree of turbulence being experienced by the fluid, with higher turbulent viscosities implying a higher degree of turbulence. Conversely, the smaller ν_t is, the lower the degree of turbulence in the local region. In fact, if $\nu_t \to \nu$, then it is a safe bet that the flow is laminar. When turbulence codes under predict ν_t, they are said to be "overdamping" (e.g., the SKE has a tendency to overdamp near the wall) (Zhao et al. 2017).

Note that whereas the dynamic and kinematic viscosities are always positive, the turbulent dynamic and turbulent kinematic viscosities can become negative if the eddies are violent enough to impart energy onto the mean flow. Though not common, it is an issue associated with large-scale flows with high anisotropy such as violent weather patterns (Sivashinsky and Yakhot 1985; Dubrulle and Frisch 1991; Sivashinsky and Frenkel 1992) and ferrofluids under alternating magnetic fields at high frequencies (Shliomis and Morozov 1994).

Continuing on, each of the nine Reynolds stresses can finally be approximated using Eq. 2.47B and letting ϕ' refer to u', v', or w',

$$
R \equiv - \begin{vmatrix} \overline{u'u'} & \overline{u'v'} & \overline{u'w'} \\ \overline{v'u'} & \overline{v'v'} & \overline{v'w'} \\ \overline{w'u'} & \overline{w'v'} & \overline{w'w'} \end{vmatrix} \approx \nu_t \begin{vmatrix} \dfrac{\partial \overline{u}}{\partial x} & \dfrac{\partial \overline{u}}{\partial y} & \dfrac{\partial \overline{u}}{\partial z} \\ \dfrac{\partial \overline{v}}{\partial x} & \dfrac{\partial \overline{v}}{\partial y} & \dfrac{\partial \overline{v}}{\partial z} \\ \dfrac{\partial \overline{w}}{\partial x} & \dfrac{\partial \overline{w}}{\partial y} & \dfrac{\partial \overline{w}}{\partial z} \end{vmatrix} \tag{2.49A}
$$

where

$$
\nu_t = \frac{\mu_t}{\rho}. \tag{2.49B}
$$

However, a new unknown was introduced, ν_t, that now requires an additional equation to solve.

The approximation from Eq. 2.49A takes inspiration from the Newtonian stress tensor,

$$
\overrightarrow{\tau} = 2\mu \overrightarrow{s} + \left(\frac{2}{3}\mu - \lambda \right) \left(\overrightarrow{\nabla} \cdot \overrightarrow{V} \right) \overrightarrow{I} \quad \left(\text{compressible, } \overrightarrow{\nabla} \cdot \overrightarrow{V} \neq 0 \right) \tag{2.50A}
$$

$$
\overrightarrow{\tau} = 2\mu \overrightarrow{s} \quad \left(\text{incompressible, } \overrightarrow{\nabla} \cdot \overrightarrow{V} = 0 \right) \tag{2.50B}
$$

where

$$
s_{ij} = \frac{1}{2} \left(\frac{\partial u_i}{\partial x_j} + \frac{\partial u_j}{\partial x_i} \right) \quad \text{(Laminar strain rate tensor.)} \tag{2.50C}
$$

The Boussinesq turbulence approximation (not to be confused with Boussinesq buoyancy) assumes an analogous expression to the laminar Newtonian stress tensor, thereby asserting that

$$\vec{R} \equiv 2\nu_t \, \vec{\vec{S}} - \frac{2}{3} k \, \vec{\vec{I}} = \nu_t \left(\frac{\partial \bar{u}_i}{\partial x_j} + \frac{\partial \bar{u}_j}{\partial x_i} \right) - \frac{2}{3} k \delta_{ij} \qquad (2.51A)$$

where

$$I = \delta_{ij} = \begin{vmatrix} 1 & 0 & 0 \\ 0 & 1 & 0 \\ 0 & 0 & 1 \end{vmatrix}, \qquad (2.51B)$$

$$\bar{S}_{ij} = \frac{1}{2} \left(\frac{\partial \bar{u}_i}{\partial x_j} + \frac{\partial \bar{u}_j}{\partial x_i} \right) \quad \text{(turbulent strain rate tensor)}, \qquad (2.51C)$$

and k is the turbulence kinetic energy.

The approximation asserts that the Reynolds stress tensor R is proportional to the mean strain rate tensor S, with the proportionality being twice the turbulent kinematic viscosity. Thus, R is "aligned" with S along the principal axes (Hinze 1987; Peng and Davidson 1999; Wilcox 2006). The vast majority of the RANS-based turbulence models use this linear constitutive relationship for closure to estimate the Reynolds stresses; this is at the core of many modern one- and two-equation RANS models. The approximation is sometimes referred to as the Boussinesq hypothesis, but because it is more akin to an "assumption," Wilcox referred to it as the Boussinesq assumption (Wilcox 2006). The "assumption," of course, lies in the belief that the Reynolds stresses behave in a similar fashion as the Newtonian stress tensor; refer to the latter part of Sect. 2.5 for more information. Though greatly successful, the Boussinesq approximation has some shortcomings, including the following situations:

- Secondary motion in ducts (rectangular pipe flow in the corner region, semi-truck vortex)
- Rapid changes in the mean strain rate tensor (rapid dilatation resulting in large volume change, high Mach)
- Curved surfaces (concave, convex, large swirl angle, airfoils)
- Rotating fluids (turbomachinery, wind turbines)
- Nonhomogeneous turbulent flows

Under such situations, it is advisable to consider extended, nonlinear versions of the Boussinesq approximation. There are many nonlinear constitutive equations in the literature, some of which include the curl operator to evaluate the mean vorticity. Such models are suitable for modeling curved surfaces, high swirl, secondary flows, flow separation, recirculation, highly anisotropic flows, and so forth (Lumley 1970; Bakker 2005; Alfonsi 2009; Wilcox 2006). Recent DNS and experimental data investigations continue to show issues with the Boussinesq approximation; some of these can be resolved using nonlocal, nonequilibrium approaches (Speziale and Eringen 1981; Hamba 2005; Schmitt 2007; Hamlington and Dahm 2009; Wilcox 2006; Spalart 2015).

In any case, from dimensional analysis and the Newtonian analogy, Boussinesq related the nine Reynolds stresses (with their primed quantities) into averaged quantities. A few examples show how these quantities are derived.

Example 2.16 Find R_{xx}, R_{xy}, and R_{xz}.

Solution For $R_{ij} = R_{xx}$, it is clear that $i = x$ and $j = x$, while $u'_x = u'$.

$$R_{xx} = -\overline{u'_x u'_x} = -\overline{u'u'} = \nu_t \left(\frac{\partial \overline{u}}{\partial x} + \frac{\partial \overline{u}}{\partial x} \right) - \frac{2}{3} k \delta_{xx} = 2\nu_t \frac{\partial \overline{u}}{\partial x} - \frac{2}{3} k.$$

For R_{xy}, it is clear that $i = x$ and $j = y$, while $u'_x = u'$ and $u'_y = v'$.

$$R_{xy} = -\overline{u'_x u'_y} = -\overline{u'v'} = \nu_t \left(\frac{\partial \overline{u}}{\partial y} + \frac{\partial \overline{v}}{\partial x} \right) - \frac{2}{3} k \delta_{xy} = \nu_t \left(\frac{\partial \overline{u}}{\partial y} + \frac{\partial \overline{v}}{\partial x} \right).$$

For R_{xz}, $i = x$ and $j = z$; $u'_x = u'$ and $u'_z = w'$.

$$R_{xz} = -\overline{u'_x u'_z} = -\overline{u'w'} = \nu_t \left(\frac{\partial \overline{u}}{\partial z} + \frac{\partial \overline{w}}{\partial x} \right) - \frac{2}{3} k \delta_{xz} = \nu_t \left(\frac{\partial \overline{u}}{\partial z} + \frac{\partial \overline{w}}{\partial x} \right).$$

The remaining Reynolds stresses can be approximated in the same fashion.

At this point, the averaged-velocity Newtonian stress tensor and the Reynolds stress tensor can be incorporated into the momentum equation(s) to solve turbulent flows, *provided a closure relationship for the kinematic turbulence viscosity is properly defined*. It is at this point that many RANS turbulence relationships come into play, e.g., Prandtl's one-equation model, the two-equation models (e.g., k-ε, k-ω), and so forth. Once the Reynolds stress terms are determined, the momentum RANS equations can be further rearranged and simplified. For example, the x-direction in Cartesian coordinates for a Newtonian fluid yields

$$\frac{\partial \overline{u}}{\partial t} + \left(\overline{u} \frac{\partial \overline{u}}{\partial x} + \overline{v} \frac{\partial \overline{u}}{\partial y} + \overline{w} \frac{\partial \overline{u}}{\partial z} \right) = -\frac{1}{\rho} \frac{\partial \overline{P}}{\partial x} + \frac{\partial \left[\nu_t \left(\frac{\partial \overline{u}}{\partial x} + \frac{\partial \overline{u}}{\partial x} \right) - \frac{2}{3} k \right]}{\partial x}$$

$$+ \frac{\partial \left[\nu_t \left(\frac{\partial \overline{u}}{\partial y} + \frac{\partial \overline{v}}{\partial x} \right) \right]}{\partial y} + \frac{\partial \left[\nu_t \left(\frac{\partial \overline{u}}{\partial z} + \frac{\partial \overline{w}}{\partial x} \right) \right]}{\partial z} + \nu \left(\frac{\partial^2 \overline{u}}{\partial x^2} + \frac{\partial^2 \overline{u}}{\partial y^2} + \frac{\partial^2 \overline{u}}{\partial z^2} \right).$$

$$(2.40B')$$

2.6   The Importance of the Reynolds Number and How to Calculate It

When Osborne Reynolds considered internal pipe flow, he discovered that beyond a certain criterion, flow became turbulent (unstable). He defined a dimensionless quantity, based on three different types of fundamental system characteristics, that is now called the Reynolds number (*Re*) in his honor. He also defined two "*K*" numerical constants that bound the "critical (transition)" range in *Re* space. More specifically, he showed via experiments that *K* ranged from 1900 to 2000 and summarized his results as follows (Reynolds 1895):

$$\frac{\rho x u}{\mu} < 1900 \text{ (stable; laminar)} \tag{2.52A}$$

and

$$\frac{\rho x u}{\mu} > 2000 \text{ (unstable; turbulent)}. \tag{2.52B}$$

As a side note, it is pointed out that Reynolds' original formulation in the numerator has density first, followed by a length scale, and then a velocity. This formulation is very popular and is the first time this dimensionless ratio was written explicitly in such detail and particular order. However, other important literature uses density first but swaps Reynolds' ordering of the length scale with the velocity scale (Schlichting 1979; Hinze 1987; Holman 1990). And, in yet another popular permutation found in an esteemed classic, the order is length scale, followed by the velocity scale, and finally density (Bird et al. 1960) (BSL). Wilcox probably subscribed to the ordering of BSL, with the velocity being first, and then followed by the length scale. However, Wilcox used the kinematic viscosity, so it is not clear in which order he would have placed the density. Other fluid dynamics books follow BSL (Bennet and Myers 1982), while the classic for flow drag follows yet a fourth permutation (Hoerner 1992). Because there are three variables, there are six total permutations, and at least four have been identified here. In any case, the present work follows the ordering of BSL, as a way of honoring their outstanding contributions in many fluid fields. Of course, the order is immaterial in the sense that *Re* is a product of three scalar quantities and scalar multiplication is associative. Thus, the order is not important in terms of the final result. Furthermore, it is pointed out that many popular permutations already exist in numerous serious tomes. That said, who would feel comfortable using $E = c^2 m$?

It is also noted for historical purposes that the oldest reference to the usage of dimensionless dynamic similarity to solve flows occurred several decades before Osborne Reynolds conducted his experiments (Reynolds 1883) and before he did his theoretical work (Reynolds 1895). Indeed, the first reference related to the usage of *dimensionless dynamic similarity* is attributed to George Gabriel Stokes (Stokes 1850; Rayleigh 1892, 1913; Rott 1990), who considered ratios that are remarkably *Re*-like, such as (Stokes 1850):

$$\frac{\mu u}{x P} \approx \text{constant} \tag{2.52C}$$

where P is the system pressure. *Nevertheless, the earliest evidence of Re as used in modern times is that of Osborne Reynolds* (Reynolds 1883, 1895).

Reynolds' "aha moment" came when he substituted Stokes' pressure P with the pressure drop from the Darcy experiments, which closely follows Poiseuille's law for laminar flow (Reynolds 1883). Then, he compared how the velocity dependence changed as it transitioned from linear (1.0) to a power of 1.723 (Reynolds 1883; Rott 1990).

What is outstanding about Re is that it requires a simple set of just three—but very distinct in essence—system parameters to characterize flow: one velocity, one length, and one fluid physical property (if kinematic viscosity is chosen, which is the ratio of the dynamic viscosity and density). This unique set describes whether the system undergoes flow via contiguous, stable sheets or is instead comprised of the superposition of unstable, multi-scaled eddies having random coherent structures. Therefore, before any analysis is begun, wise analysts will *first* calculate Re (and perhaps Ma and so forth), to determine the type of flow being analyzed. Though simple enough to understand why, it is shocking to see how many systems are modeled with the incorrect fluid dynamics. In other words, *never* use a turbulence model if the flow is laminar and vice versa!

The kinematic viscosity ν is typically a function of temperature and pressure and is well-defined for thousands of fluids. But, what are reasonable values for x_{char} and u_{char}? In this context, "characteristic" refers to a system's fundamental length and velocity that best describe the system flow. The flow geometry dictates which characteristic dimension x_{char} ought to be used to calculate Re. As might be expected, the critical Re is geometry-dependent. Hence, criteria are summarized in the section that follows for many important engineering applications.

2.6.1   Calculating Re for Diverse Geometries

As mentioned previously, different geometries involve different Re_{crit}, where the critical Re is defined as the point at which the flow has transitioned into turbulence. Note that the ranges for a given geometry can vary substantially, depending on fluid purity, experimental care, and other factors, including external vibrations induced by pumps, surface roughness, and so forth. For example, under extremely careful conditions, an internal pipe flow can reach Re up to 10,000 and still exhibit laminar behavior. Although this is an extreme example, variations of plus or minus 10% are not uncommon in Re_{crit} experiments. Thus, an internal pipe flow could be turbulent at 1980 or possibly be laminar at 2420. Therefore, care should be taken when interpreting the value of Re.

Here, some semantics are in order. It is common to find "critical" and "transition" associated with Re, and this can cause confusion. In particular, as a flow *transitions* from laminar to turbulent, some people refer to this as a "*transition* based on a *critical Re*," so the symbol Re_{crit} *is used*. Then, for certain external flows, another type of *transition* occurs, which is the *transition* whereby the drag coefficient drops precipitously by a factor of about four or so; unfortunately, this "transition" is also referred to as a "critical" transition and is oftentimes labelled as Re_{crit} as well (Hoerner 1992). Thus, Re_{crit} can be based on the laminar to turbulent transition, or it can be based on the abrupt drop in the drag coefficient (e.g., the Eiffel paradox). In order to reduce ambiguity, the type of transition is specified for the geometries listed below.

- *Internal Pipe Flow*

For a pipe with inner diameter D and length L, $x_{char} = D$. u_{char} = average velocity, which can be easily obtained from the mass flow rate,

$$\dot{m} = \rho \overline{V} A$$

or

$$\overline{V} = \frac{\dot{m}}{\rho A} = u_{char}.$$

The laminar to turbulent transition occurs at $Re_{crit} = 2200\text{--}2400$. Wilcox uses 2300 (Wilcox 2006).

- *Internal Duct Flow*

For ducts (e.g., rectangular pipes; non-circular internal flow), x_{char} = hydraulic diameter (D_h),

$$D_h = \frac{4A}{\text{WP}} \tag{2.53}$$

where

A = flow area
WP = wetted perimeter.

u_{char} = average velocity and can be obtained from the mass flow rate. The laminar to turbulent transition occurs at $Re_{crit} = 2200\text{--}2400$.

- *Flow External to a Cylinder*

For flow around (external to) a vertical cylinder, x_{char} = outer diameter D. $u_{char} = U_\infty$ = approach velocity. The laminar to turbulent transition occurs at $Re_{crit} \approx 600\text{--}1000$ (Schlichting 1979); drag transition $Re_{crit} > 2 \times 10^5$ to 5×10^5 (Schlichting 1979; Hoerner 1992).

- *Flow External to a Sphere*

For flow around a sphere, x_{char} = outer diameter D. $u_{char} = U_\infty$ = approach velocity. The laminar to turbulent transition occurs at $Re_{crit} > 300$ (Blevins 1992); drag transition $Re_{crit} > 1 \times 10^5$ to 3×10^5 (Schlichting 1979; Moradian et al. 2008).

- *Flow External to a Cube or "Square Cylinder"*

For flow external to a cube with side S, $x_{char} = S$. $u_{char} = U_\infty$ = approach velocity. The laminar to turbulent transition is in the range of $200 \le Re_{crit} \le 260$ (Sohankar 2006; Bai and Alam 2018); drag transition Re_{crit}: none has been observed experimentally up to $Re = 10^7$ (Sohankar 2006; Bai and Alam 2018).

- *Jet Flow*

For jets, x_{char} = jet orifice diameter D. u_{char} = average velocity, which can be obtained from the jet mass flow rate. The jet is fully turbulent when $Re_{crit} > 3000$ (McNaughton and Sinclair 1966).

- *Couette Flow*

For Couette flow (specifically, flow between rigid parallel plates separated by a gap H), $x_{char} = H$. u_{char} = plate velocity [note: this type of flow is always fully developed (FD)]. The laminar to turbulent transition is at $Re_{crit} = 1500$ (Wilcox 2006).

- *Flow Over a Flat Plate*

For external flow in the x-direction (WLOG) parallel to a flat plate of length L, the laminar boundary layer $\delta(x)$ grows in the x-direction as follows (Blasius 1908):

$$\delta(x)_{lam} \approx \frac{5x}{Re_x^{1/2}}, \tag{2.54A}$$

while the turbulent boundary layer grows as

$$\delta(x)_{turb} \approx \frac{0.036x}{Re_x^{1/5}}. \tag{2.54B}$$

However, the characteristic length x_{char} for a flat plate is x, so Re reaches a maximum at $x = L$; this type of flow never becomes FD. $u_{char} = U_\infty$ = approach velocity. The laminar to turbulent transition occurs when $Re_{crit} \sim 5 \times 10^5$ to 10^6 (White 1991; Schlichting and Gersten 2000).

- *Flow Over an Airfoil or Wing*

For airfoils, x_{char} = chord width (distance from the leading edge to the trailing edge); for wings, use the mean aerodynamic chord, which can be approximated as the average of the tip chord and the root chord (note that more complex approximations include integrals and the center of gravity). $u_{char} = U_\infty$ = relative approach velocity; wind tunnel velocity. The laminar to turbulent transition occurs at $Re_{\text{crit}} > 5 \times 10^5$ (approximated as flow over a flat plate).

2.6.2 Fully Developed Laminar and Turbulent Flow

If a flow is fully developed (FD), the velocity distribution does not change WRT position; the partial of the velocity component for such direction is zero. Use the following expression to determine if a *laminar* flow is FD:

$$\frac{L_e}{D} = C_1 + C_2\, Re_{\text{h}}, \tag{2.55A}$$

where

$L_e =$ entrance length (i.e., the length at which the flow becomes fully developed)
$D =$ hydraulic diameter
$Re_{\text{h}} =$ Reynolds number based on the hydraulic diameter

$$C_1 = 0.5, \tag{2.55B}$$

$$C_2 = 0.05. \tag{2.55C}$$

For turbulent flows, a common expression is

$$\frac{L_e}{D} = C_1\, Re_{\text{h}}^{\,n}, \tag{2.55D}$$

where

$$C_1 = 1.36, \tag{2.55E}$$

and

$$n = 1/4. \tag{2.55F}$$

Note that the literature has several variants for the FD laminar and turbulent expressions. In addition, various "rules of thumb" exist, including some for turbulent flows, such as

$$\frac{L_e}{D} \geq 10, \tag{2.55G}$$

while Nikuradse was more conservative, with

$$\frac{L_e}{D} \geq 40. \tag{2.55H}$$

Example 2.17   Find the entrance length for ammonia at 400 K and 1 MPa. The mass flow rate is 0.001 kg/s in one case and 0.01 kg/s in the second case. The pipe in question has a diameter $D = 0.3$ m. What are the entrance lengths?

Solution   Start by obtaining the necessary physical properties, $\rho = 5.31$ kg/m^3 and $\nu = 2.62 \times 10^{-6}$ m^2/s. Use the density to back track the average fluid velocity for the first case,

$$u = \frac{\dot{m}}{\rho A} = \frac{0.001 \text{ kg/s}}{(5.31 \text{ kg/m}^3)(0.0707 \text{ m}^2)} = 0.00266 \text{ m/s}$$

$$Re = \frac{(0.3 \text{ m})(0.00266 \text{ m/s})}{2.62 \times 10^{-6} \text{ m}^2/\text{s}} = 304.$$

For this case, the flow is laminar, so

$$L_e = D(C_1 + C_2 \, Re_h) = (0.3 \text{ m})[0.5 + 0.05(304)] = 4.7 \text{ m}.$$

For the second case, the flow rate is 10 times larger, so the fluid velocity is 10 times larger, 0.0266 m/s. In this case Re is also 10 times larger, so $Re = 3040$, which is clearly turbulent. Thus,

$$L_e = C_1 D \, Re_h^n = 1.36 D \, Re_h^{1/4} = 1.36(0.3 \text{ m})(3040)^{1/4} = 3.03 \text{ m}.$$

Notice that the entrance length for turbulent flow is always smaller than that of laminar flows. Also, note that the rule of thumb *for turbulent flows can be significantly different than the curve-fit expression based on Re*, as clearly shown in Fig. 2.10.

2.7   Isotropic Turbulence Decay, Taylor Eddies, and Some Applications

The word "isotropic" is derived from two Greek words, "isos" and "tropos," which mean "equal" and "way," respectively. Whereas its root words might convey the connotation that an isotropic flow is the "same way" throughout, a less restrictive condition is applied in the turbulence world. Instead, for a turbulence flow to be isotropic, *its statistical properties must be the same in all directions, implying a mathematical order.* However, because no direction has a preferred *statistical* turbulence distribution, a perfect isotropic flow is in perfect disorder, in the sense

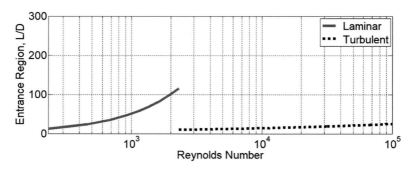

Fig. 2.10 Laminar vs. turbulent entrance length

that all directions are very well mixed (redistributed) within the same magnitude; hence, any initial molecular order is randomized equally in all directions. Thus, the statistical properties of an isotropic turbulent flow will be the same if the coordinate system is rotated or if the coordinate planes are reflected and translated. Because isotropy is a mathematical idealization for a random process, *perfect isotropy* cannot be generated experimentally nor does it exist in nature. Just like no two snowflakes are ever the same, no two eddies can ever be the same, let alone an agglomeration of millions of eddies. Hence, isotropy considers a *statistical* equivalence for eddies—a sort of "composite behavior." Fortunately, there are flows that reasonably *approximate* isotropic conditions, including atmospheric flows, the core region of a square duct, grid flow, high *Re* flows, and flows whose eddies are much smaller than integral eddies (e.g., the flow core region at very high *Re*).

So, for example, in a Cartesian system, a homogeneous turbulence flow would be indistinguishable whether an observer looked in the *x, y*, or *z* directions. WLOG, this concept can be extended to non-Cartesian systems.

Much useful research has been conducted in the area of isotropic flow over the past century (Taylor 1935a, b; von Karman and Howarth 1938; Millionshtchikov 1941; Hinze 1987), and there are literally thousands of references. Hinze provides an excellent review for this topic; the interested reader is encouraged to consult it. A rather small sampling of recent advances in isotropic flows includes (Nakano 1972; Sivashinsky and Frenkel 1992; Gryanik et al. 2005; Holm and Kerr 2007; Mazumdar and Mamaloukas 2012; Boschung et al. 2016; Boffeta and Musacchio 2017; Hirota et al. 2017; Stepanov 2018). A good review of recent advances in homogeneous isotropic flow is found in Benzi and Biferale (2015). Despite these advances, researchers continue investigating and debating key aspects of isotropic turbulence, albeit that this is a *very simplified* turbulence state (CFD-online 2017b)! Nearly a century after Sir Horace Lamb's jovial pessimism regarding turbulent flows, his concerns regarding turbulence continue to be validated, even for the "simplest" of turbulence flows—isotropic turbulence.

The *k* PDE is derived in Sect. 4.4. For convenience, it is expressed here to show how substantially simplified isotropic homogeneous flows are in the mathematical sense. In its full form for Cartesian coordinates, the *k* PDE is expressed formidably as:

$$\frac{\partial k}{\partial t} + \overline{u}\frac{\partial k}{\partial x} + \overline{v}\frac{\partial k}{\partial y} + \overline{w}\frac{\partial k}{\partial z} =$$

$$\left\{ \left[\nu_t\left(\frac{\partial \overline{u}}{\partial x} + \frac{\partial \overline{u}}{\partial x}\right) - \frac{2}{3}k \right]\frac{\partial \overline{u}}{\partial x} + \nu_t\left(\frac{\partial \overline{u}}{\partial y} + \frac{\partial \overline{v}}{\partial x}\right)\frac{\partial \overline{u}}{\partial y} + \nu_t\left(\frac{\partial \overline{u}}{\partial z} + \frac{\partial \overline{w}}{\partial x}\right)\frac{\partial \overline{u}}{\partial z} \right\}$$

$$+\left\{ \nu_t\left(\frac{\partial \overline{v}}{\partial x} + \frac{\partial \overline{u}}{\partial y}\right)\frac{\partial \overline{v}}{\partial x} + \left[\nu_t\left(\frac{\partial \overline{v}}{\partial y} + \frac{\partial \overline{v}}{\partial y}\right) - \frac{2}{3}k \right]\frac{\partial \overline{v}}{\partial y} + \nu_t\left(\frac{\partial \overline{v}}{\partial z} + \frac{\partial \overline{w}}{\partial y}\right)\frac{\partial \overline{v}}{\partial z} \right\}$$

$$+\left\{ \nu_t\left(\frac{\partial \overline{w}}{\partial x} + \frac{\partial \overline{u}}{\partial z}\right)\frac{\partial \overline{w}}{\partial x} + \nu_t\left(\frac{\partial \overline{w}}{\partial y} + \frac{\partial \overline{v}}{\partial z}\right)\frac{\partial \overline{w}}{\partial y} + \left[\nu_t\left(\frac{\partial \overline{w}}{\partial z} + \frac{\partial \overline{w}}{\partial z}\right) - \frac{2}{3}k \right]\frac{\partial \overline{w}}{\partial z} \right\}$$

$$+\frac{\partial}{\partial x}\left[\left(\nu + \frac{\nu_t}{\sigma_k}\right)\frac{\partial k}{\partial x} \right] + \frac{\partial}{\partial y}\left[\left(\nu + \frac{\nu_t}{\sigma_k}\right)\frac{\partial k}{\partial y} \right] + \frac{\partial}{\partial z}\left[\left(\nu + \frac{\nu_t}{\sigma_k}\right)\frac{\partial k}{\partial z} \right] - \varepsilon.$$

$$(2.56A)$$

In this context, k is the specific turbulence kinetic energy, in units of length squared per time squared. Without any doubt, the 3D k PDE as expressed in Eq. 2.56A presents a formidable challenge that has yet to be solved analytically!

Certainly, turbulent flow is always strictly a 3D phenomenon. Eddies are 3D coherent structures shaped as curved sheets, ovaloids, and so forth that stretch asymmetrically and chaotically in space and time. These 3D structures break up into smaller structures as they stretch, until the eddies eventually become so small that the viscous force overcomes them, causing these tiny coherent-motion clusters to decay into laminar sheets. *In other words, eddies are irregular 3D structures that move irregularly through 3D space and time.* Nevertheless, consider a drastic simplification for turbulence flow, where the turbulent flow is assumed 1D, and WLOG, choose the x-direction. Therefore, any derivatives WRT y or z are zero, and so are the v and w velocity components. This results in the following significant simplification of the k PDE expressed in Eq. 2.56A, which is now reduced to

$$\frac{\partial k}{\partial t} + \overline{u}\frac{\partial k}{\partial x} = \left[\nu_t\left(\frac{\partial \overline{u}}{\partial x} + \frac{\partial \overline{u}}{\partial x}\right) - \frac{2}{3}k \right]\frac{\partial \overline{u}}{\partial x} - \varepsilon + \frac{\partial}{\partial x}\left[\left(\nu + \frac{\nu_t}{\sigma_k}\right)\frac{\partial k}{\partial x} \right]. \quad (2.56B)$$

Despite its simplicity, the k PDE is still not solvable in an analytic fashion. The above exercise shows just how significant and complex turbulence effects are, even if just 1D is considered.

Now assume further still that the turbulence flow is far away from the wall, such that there are neither large velocity gradients nor buoyancy effects that generate significant turbulence. Assume further that there is no eddy convection or diffusion. The turbulence phenomenon that is being considered at this point consists solely of the transient decay behavior of eddies, with no explicit spatial dependency. Then, with all these assumptions, such turbulent flow is often described as isotropic homogeneous turbulence decay. Thus, the PDE equation is reduced such that it only describes the eddy decay for homogeneous isotropic flows (i.e., the dynamic eddy behavior as the eddies collapse onto laminar sheets):

$$\frac{\partial k}{\partial t} + \bar{u}\frac{\partial k}{\partial x} = \left[\nu_t\left(\frac{\partial \bar{u}}{\partial x} + \frac{\partial \bar{u}}{\partial x}\right) - \frac{2}{3}k\right]\frac{\partial \bar{u}}{\partial x} - \varepsilon + \frac{\partial}{\partial x}\left[\left(\nu + \frac{\nu_t}{\sigma_k}\right)\frac{\partial k}{\partial x}\right]. \quad (2.56C)$$

Thus, the formidable Eq. 2.56A is reduced to

$$\frac{\partial k}{\partial t} = \frac{dk}{dt} = -\varepsilon \quad\quad\quad (2.57)$$

where ε = turbulent eddy dissipation (decay).

Notice that the partial derivative becomes an ordinary derivative because k and ε are not functions of position for this extremely simplified situation. Therefore, the ordinary differential equation expresses that as time increases, the eddy kinetic energy decreases rapidly per unit time at a rate ε, until the flow becomes laminar; at that point, k is zero, and there are no remaining eddies. This phenomenon can be seen in a hot tub with jets—turn the switch off, and the eddies decay within a few seconds into laminar flow. The same happens in a blender. As the blender's blades move rapidly through the liquid, they induce turbulent flow with perhaps hundreds of thousands to millions of eddies. But, once the blender is turned off, the eddies decay rapidly, and motion quickly ceases—all the eddies have "died off." The laminarization observed in the hot tub and blender confirms that the decay mechanism for water-based fluids typically works within a time frame of approximately 1 s or so, depending on the situation. Indeed, for isotropic homogeneous flow, theoretical and experimental results indicate a fairly strong exponential decay of the turbulent kinetic energy WRT time,

$$k \sim t^{-1.25}. \quad\quad\quad (2.58)$$

Example 2.18 Use Eq. 2.58 to derive an expression for the isotropic homogeneous decay (dissipation), ε.

Solution Begin by noting that $\varepsilon = -\frac{dk}{dt}$.

Therefore, $\varepsilon \sim -\frac{d\left(t^{-1.25}\right)}{dt} = 1.25t^{-2.25}$.

For isotropic turbulence, the time average of the square of the velocity fluctuations is equal, indicating that

$$\overline{u'^2} = \overline{v'^2} = \overline{w'^2}, \quad\quad\quad (2.59)$$

where k is defined for a *nonisotropic* turbulent flow as (Prandtl 1945)

$$k = \frac{1}{2}\left(\overline{u'^2} + \overline{v'^2} + \overline{w'^2}\right). \quad\quad\quad (2.60)$$

Therefore, for an *isotropic* flow,

$$k = \frac{1}{2}\left(\overline{u'^2} + \overline{v'^2} + \overline{w'^2}\right) = \frac{1}{2}\left(\overline{u'^2} + \overline{u'^2} + \overline{u'^2}\right) = \frac{3}{2}\overline{u'^2}. \qquad (2.61)$$

The dissipation ε for an *anisotropic* turbulent flow is rather complex (Taylor 1935a):

$$\varepsilon = \mu \left[\begin{array}{c} 2\left(\frac{\partial \overline{u'}}{\partial x}\right)^2 + 2\left(\frac{\partial \overline{v'}}{\partial y}\right)^2 + 2\left(\frac{\partial \overline{w'}}{\partial z}\right)^2 \\ \overline{+\left(\frac{\partial v'}{\partial x} + \frac{\partial u'}{\partial y}\right)^2} + \overline{\left(\frac{\partial w'}{\partial y} + \frac{\partial v'}{\partial z}\right)^2} + \overline{\left(\frac{\partial u'}{\partial z} + \frac{\partial w'}{\partial x}\right)^2} \end{array} \right]. \qquad (2.62)$$

But, once isotropy is invoked, the decay simplifies significantly (Taylor 1935a),

$$\varepsilon = 6\mu \left[\left(\frac{\partial \overline{u'}}{\partial x}\right)^2 + \left(\frac{\partial \overline{u'}}{\partial y}\right)^2 + \frac{\partial \overline{v'}}{\partial x} \frac{\partial \overline{u'}}{\partial y} \right]. \qquad (2.63)$$

It is insightful that one of the pillars of contemporary turbulence modeling, David Wilcox, referred to Taylor eddies in the following manner (Wilcox 2006), "Because the Taylor microscale is generally too small to characterize large eddies and too large to characterize small eddies, it has generally been ignored in most turbulence-modeling research."

Not surprisingly, as noted by Wilcox, the integral and Kolmogorov eddies are most often than not readily discussed in the literature, but the Taylor eddies tend to be ignored (Andersson et al. 2012). In what follows, some ideas from Taylor are expanded, for the purpose of promoting Taylor eddies. One of Taylor's goals was to derive a "characteristic eddy" size described through various definitions. This includes Taylor's assertion that certain mid-sized eddies represent (Taylor 1935a), "'the average size of the smallest eddies', which are responsible for the dissipation of energy by viscosity."

However, it can be inferred, that in Taylor's context, "smallest eddies" do not actually refer to the Kolmogorov eddies in this case but rather onto the eddies associated with dissipation that are *larger* than the Kolmogorov scale. This was clearly his intent, considering that he assumed a turbulent isotropic flow that can be characterized using k, which is associated with *energetic eddies*. By contrast, Kolmogorov eddies are characterized with some form of decay, say ε. Furthermore, Taylor's "characteristic eddies" are somewhat smaller than integral eddies because they arise after breaking up from relatively larger eddies. Thus, using Taylor's own lingo, he referred to his "characteristic eddies" as some sort of mid-sized eddies whose (Taylor 1935a) "length may be taken to represent roughly the diameter of the smallest eddies into which the eddies defined by the scales ℓ_1 or ℓ_2 break up."

The ℓ_1 and ℓ_2 eddies discussed by Taylor apply to a wind tunnel, and their length scales are Prandtl's mixing length and $0.2S$, respectively. S is the grid size used to generate the isotropic flow (Taylor 1935a), and therefore, $0.2S$ represents the length of the integral eddies generated by the grid. Consequently, these "characteristic eddies" in Taylor's research are most certainly eddies that are much larger than the Kolmogorov eddies but are smaller than the integral eddies. *Roughly speaking, these are mid-sized, isotropic-like eddies. The eddies Taylor defined in this context are now referred as "Taylor eddies" in his honor.*

Wilcox and others also defined a mid-sized turbulence correlation scale λ such that (Lesieur et al. 2005; Wilcox 2006)

$$\lambda \equiv \sqrt{\frac{-2}{\left(\frac{\partial^2 f}{\partial y^2}\right)_{y=0}}} \tag{2.64A}$$

where

$$f = f(x;r) = \frac{R_{11}(x;r)}{\overline{u'^2}(x)}. \tag{2.64B}$$

After a rather extensive derivation, Taylor developed an expression for a "characteristic length" λ based on turbulence correlation arguments for isotropic flows,

$$\frac{1}{\lambda^2} \equiv \lim_{y \to 0} \frac{1 - R_y}{y^2}, \tag{2.65}$$

where R_y is a complicated expression that describes the eddy curvature based on a geometric series that is based on the mean of u' derivatives (Taylor 1922, 1935a, b; Hinze 1987). For the sake of brevity, Taylor's key result indicates that

$$\varepsilon = 15\mu\overline{u'^2} \lim_{y \to \lambda} \frac{1 - R_y}{y^2}. \tag{2.66}$$

Now, once the R_y function is substituted into Eq. 2.66 and its limit taken, Taylor's classic turbulence equation for his "characteristic eddy" size becomes

$$\varepsilon = 15\frac{\mu\overline{u'^2}}{\rho\lambda^2} = 15\frac{\nu\overline{u'^2}}{\lambda^2}. \tag{2.67}$$

Solving for $\overline{u'^2}$ from Eq. 2.61 and substituting into Eq. 2.67 results in the familiar dissipation equation for Taylor eddies, which is clearly only applicable under *isotropic* conditions (Taylor 1935a; von Karman and Howarth 1938; Hinze 1987; Wilcox 2006),

$$\varepsilon = 15\frac{\nu\left(\overline{u'^2}\right)}{\lambda^2} = 15\frac{\nu\left(\frac{2}{3}k\right)}{\lambda^2} = 10\frac{\nu k}{\lambda^2}. \tag{2.68}$$

Upon solving for the "characteristic eddy" size λ, the classic Taylor length scale is readily obtained.

$$\lambda = \sqrt{10\frac{\nu k}{\varepsilon}}. \tag{2.69}$$

For large Re, the decaying Taylor length can be estimated as a function of time as (von Karman and Howarth 1938)

$$\lambda(t) = \sqrt{\frac{\nu t}{\alpha}}, \tag{2.70}$$

where $\alpha = 1/5$ is ideal for situations where the larger eddies remain fairly constant in size (von Karman and Howarth 1938).

As noted by Wilcox, Eq. 2.69 uses both k and ε simultaneously, where k is typically associated with the larger eddies that contain the highest share of kinetic energy (on the order of 80%), while ε is associated with decay, where the eddies are at the opposite spectrum—their length scale is much smaller and have much less kinetic energy. In fact, Taylor eddies are on the order of at least 70 times larger than Kolmogorov eddies. Thus, as "hybrid" eddies, *Taylor eddies are sufficiently large to merit an energy definition that includes the turbulent kinetic energy k and sufficiently small to merit approximation with dissipation ε.*

It is therefore enticing that a single eddy characteristic scale can adequately describe a very large number of eddies that span a large section in the continuous, eddy-length range (e.g., Fig. 3.1). This implies that to some degree, there is a "characteristic eddy" size that adequately models the behavior for an eddy size range that can potentially do useful engineering work. This was no doubt a key reason why Taylor expended so much effort to find such eddy scale. Furthermore, as noted by Kolmogorov, the "most characteristic" turbulence-behavior motion properties are associated with eddies that are much smaller than the integral eddies (Kolmogorov 1942); and no, he was not referring to the Kolmogorov dissipation eddies. Instead, Kolmogorov was referring to the local eddy structure that can be viewed as the energetic, "spatially homogeneous and isotropic structures"—the Taylor eddies. *Furthermore, the importance of Taylor eddies in the dissipative process was advanced by Chou, and this notion guided part of the development of both the k-ε and k-ω RANS models* (Chou 1940, 1945); refer to Chap. 4. In addition, researchers have recently demonstrated the benefits of incorporating features of the behavior of Taylor eddies onto k-ε models (Myong and Kasagi 1990; Bae et al. 2016); see Chap. 4. Moreover, a relationship for eddy dissipation ε based on Taylor's λ eddy length scale was derived as well (Chou 1945).

$$\varepsilon_{\text{diss,Chou}} = 2\nu g^{mn}\overline{\frac{\partial u'_i}{\partial x_m}\frac{\partial u'_k}{\partial x_n}} = -\frac{2\nu}{3\lambda^2}(C-5)q^2 g^{ik} + \frac{2C\nu}{\lambda^2}\overline{u'_i u'_k} \tag{2.71A}$$

where

$$g^{mn} = 1, \tag{2.71B}$$

$$q^2 = \overline{u_i' u_k'}, \tag{2.71C}$$

C = unitless constant

Thus, the larger the Taylor eddies, the smaller their impact on turbulent decay and vice versa.

Expanding Eq. 2.71A and substituting the q^2 relationship onto the RHS term,

$$\varepsilon_{\text{diss,Chou}} = \left(-\frac{2\nu\cancel{C}}{3\cancel{\lambda}^2} + \frac{10\nu}{3\lambda^2} + \frac{2\nu\cancel{C}}{3\cancel{\lambda}^2} \right) q^2. \tag{2.72A}$$

If $k = q^2/3$, then the above expression collapses onto Taylor's expression:

$$\varepsilon_{\text{diss,Chou}} = \frac{10\nu k}{\lambda^2}. \tag{2.72B}$$

So, not only is Taylor's eddy length scale (Eq. 2.69) confirmed by von Karman (von Karman and Howarth 1938) but by Chou as well (Chou 1945). *This further reinforces the notion presented in Chap. 4 that Chou's original k-ε work was associated with the Taylor eddies and not the Kolmogorov eddies.* Nevertheless, it is clear that within some constant c,

$$k_{\text{Chou}} = cq^2. \tag{2.73}$$

It is also noteworthy that various double and triple correlations developed by Chou involve the Taylor length scale (Chou 1945). It is therefore not surprising that the Taylor eddy length scale has been included in the standard k-ε turbulence model for improved turbulence modeling (Myong and Kasagi 1990).

So, what do Taylor eddies do for a living? Being much smaller than integral eddies, Taylor eddies carry about 1/5th of the total turbulent kinetic energy. On the other hand, they have a much higher count and are more ubiquitous than integral eddies. In addition, integral eddies are anisotropic, whereas Taylor eddies are isotropic. Thus, Taylor eddies exhibit a more nuanced disorderly behavior, having more uniform randomness than integral eddies; that is, because Taylor eddies are isotropic, this implies that the fluid is so well-mixed that it looks mathematically similar in all directions; the same amount of k is expended in all directions, as opposed to the preferred direction of integral eddies (which is along the main flow direction). The higher degree of more uniform random fluctuations associated with Taylor eddies inevitably contributes toward more thorough and uniform mixing of the fluid in the flow core and in the range of $7 \leq y^+ \leq 30$. Furthermore, because Taylor eddies have shorter lives than integral eddies, Taylor eddies tend to be in local equilibrium and thereby react much faster under external stimulus.

Therefore, because Taylor eddies are numerous and ubiquitous, carry a reason-able amount of energy, have a uniformly randomizing isotropic nature, and react rapidly to external flow conditions, their impact on turbulent flows cannot be ignored. For example, isotropy can be engineered in compact heat exchangers, resulting in enhanced heat transfer (Elyyan et al. 2008). Isotropy can also be engineered onto solar energy collectors and other heat transfer surfaces, for the purpose of increased mixing and heat transfer near walls (Rodriguez 2016).

2.8   Problems

2.1 A round pipe with a contraction reduces in diameter from 0.5 to 0.25 m. The fluid is helium at 350 K and 2.0 atmospheres and is flowing into the pipe at 1.5 kg/s. The system is isothermal and incompressible. What is the exit velocity?

2.2 A square pipe with a linear expansion increases from $S = 0.2$ to 1.0 m. The fluid is liquid lead at 700 K and 1 atmosphere and is flowing into the pipe at 10 kg/s. The system is isothermal and incompressible. What is the exit velocity?

2.3 Helium at 950 K and 4.9 MPa flows through a nuclear reactor. A positive displacement pump increases the gas's velocity until it becomes compressible. At what velocity will the helium become compressible?

2.4 If   a   system   has   the   following   velocity   distribution, $\vec{V} = 4x^2 y\,\vec{i} + 4xy^2\,\vec{j} - 60\vec{k}$, is the flow compressible or incompressible? Why?

2.5 Derive the transient equation for 1D momentum in the x direction for a viscous fluid, with an external pressure force, and negligible gravitational effect. Now assume the system approaches steady state and solve the PDE by letting the pressure gradient be constant.

2.6 Explain why $\overline{u'} = 0$ but $\overline{u'^2} \neq 0$ (at least most of the time).

2.7 Find expressions for the following Cartesian Reynolds stresses: R_{yx}, R_{yy}, and R_{yz}. Hint: use the Boussinesq approximation.

2.8 A jet at Re_{crit} has an orifice at $D = 0.055$ m, a mass flow rate of 1.9 kg/s, and $\mu = 1.55 \times 10^{-6}$ kg/m-s. What is the average velocity at the orifice?

2.9 Water at 300 K and 200,000 Pa flows through a pipe. At what velocity will it become compressible? Redo the problem for nitrogen gas at the same pressure and temperature.

2.10 Assume a fluid is non-Newtonian, such that the Ostwald-de Waele model applies, $\tau_{yx} = -m\left|\frac{du}{dy}\right|^{n-1}\frac{du}{dy}$.

Let $n = 3$ (so the fluid is dilatant), $m = \mu$, and $u(y) = u_0\left[\frac{y}{y_0} - \left(\frac{y}{y_0}\right)^4\right]$.

Find $d\tau_{yx}/dy$ and obtain the maximum value as a function of y. Hint: set the derivative to zero and solve for y.

2.11 Your new boss, Mr. Minimalist, wants you to use the CFD tool of your choice to simulate a 2D rectangular flat-plate flow with the following characteristics. The fluid velocity is uniform on the LHS at $U_\infty = 100$ m/s. The RHS is open to the atmosphere. The bottom is a smooth, horizontal, motionless wall. The top boundary is open to ambient. The system length is 5.0 m and is 0.25 m wide. The air is at 500 °C and atmospheric pressure. Describe and justify the following key problem characteristics: compressible vs. incompressible; Newtonian vs. non-Newtonian; and laminar vs. turbulent. Which fluid models would you use?

2.12 A grad student had too much coffee and accidentally broke the only viscosity meter. Unfortunately, the viscosity measurements were not completed. Consider the student's lab experiment, where flow over a long, horizontal flat plate is initially laminar at $U_\infty = 55$ m/s. Is it possible to obtain one more variable by observing the flow, such that a kinematic viscosity can be backtracked, and if so, which variable would that be? Assume that you now have the value of such variable. Now calculate the fluid's kinematic viscosity.

References

Alfonsi, G. (2009). Reynolds-averaged Navier-Stokes equations for turbulence modeling. *Applied Mechanics Reviews, 62*, 1–20.

Anderson, J. (1995). *Computational fluid dynamics: The basics with applications.* New York: McGraw-Hill International Editions.

Andersson, B., et al. (2012). *Computational fluid dynamic for engineers.* Cambridge, UK: Cambridge University Press.

Bae, Y. Y., Kim, E. S., & Kim, M. (2016). Numerical simulation of supercritical fluids with property-dependent turbulent Prandtl number and variable damping function. *International Journal of Heat and Mass Transfer, 101*, 488–501.

Bai, H., & Alam, M. M. (2018). Dependence of square cylinder wake on Reynolds number. *Physics of Fluids, 30*, 1–20.

Bakker, A. (2005). Turbulence models. Applied computational fluid dynamics, Lecture 10, www. bakker.org

Bennet, C. O., & Myers, J. E. (1982). *Momentum, heat, and mass transfer* (3rd ed.). New York: McGraw-Hill.

Benzi, R., & Biferale, L. (2015). Homogeneous and isotropic turbulence: A short survey on recent developments. *Physics Fluid Dynamics, 161*, 1351–1365.

Bird, R., Stewart, W., & Lightfoot, E. (1960). *Transport phenomena* (1st ed.). New York: Wiley.

Bird, R., Stewart, W., & Lightfoot, E. (2007). *Transport phenomena* (2nd ed.). New York: Wiley.

Blasius, H. (1908). Grenzschichten in Flussigkeiten mit kleiner Reibung. *Zeitschrift für Mathematik und Physik, 56*(1). Also available in English as, The boundary layers in fluids with little friction. National Advisory Committee for Aeronautics, Technical Memorandum 1256, 1950.

Blevins, R. D. (1992). *Applied fluid dynamics handbook.* Malabar: Kreiger Publishing Company.

Boffeta, G., & Musacchio, S. (2017). Chaos and predictability of homogeneous-isotropic turbulence. *Physical Review Letters, 119*, 054102.

Boschung, J., et al. (2016). Streamlines in stationary homogeneous isotropic turbulence and fractal-generated turbulence. *Fluid Dynamics Research, 48*, 021403.

CFD-online.   https://www.cfd-online.com/Wiki/Introduction_to_turbulence/Statistical_analysis/ Ensemble_average#Mean_or_ensemble_average. Accessed on 24 June 2017a.

CFD-online. https://www.cfd-online.com/Wiki/Turbulence_intensity. Accessed on 4 July 2017b.

Chou, P. Y. (1940). On an extension of Reynolds' method of finding apparent stress and the nature of turbulence. *Chinese Journal of Physics, 4*, 1–33.

Chou, P. Y. (1945). On velocity correlations and the solutions of the equations of turbulent fluctuation. *Quarterly of Applied Mathematics, 3*, 38–54.

CoolProp. Welcome to CoolProp. http://coolprop.sourceforge.net. Accessed on 4 July 2017.

Dubrulle, B., & Frisch, U. (1991). Eddy viscosity of parity-invariant flow. *Physical Review A, 43* (10), 5355–5364.

Elyyan, M. A., Rozati, A., & Tafti, D. K. (2008). Investigation of dimpled fins for heat transfer enhancement in compact heat exchangers. *International Journal of Heat and Mass Transfer, 51*, 2950–2966.

Engineering ToolBox. http://www.engineeringtoolbox.com. Accessed on 22 Aug 2017.

Groisman, A., & Steinberg, V. (2000). Elastic turbulence in a polymer solution flow. *Nature, 405*, 53–55.

Gryanik, V. M., et al. (2005). A refinement of the Millionshtchikov quasi-normality hypothesis for convective boundary layer turbulence. *Journal of the Atmospheric Sciences, 62*, 2632–2638.

Hamba, F. (2005). Nonlocal analysis of the Reynolds stress in turbulent shear flow. *Physics of Fluids, 17*, 1–10.

Hamlington, P. E., & Dahm W. J. A. (2009). Reynolds stress closure including nonlocal and nonequilibrium effects in turbulent flows. 39th AIAA fluid dynamics conference, AIAA 2009–4162.

Hinze, J. (1987). *Turbulence* (McGraw-Hill classic textbook reissue series). New York: McGraw-Hill.

Hirota, M., et al. (2017). Vortex stretching in a homogeneous isotropic turbulence. *Journal of Physics: Conference Series, 822*, 012041.

Hoerner, S. F. (1992). *Fluid-dynamic drag*. Bakersfield: Hoerner Fluid Dynamics.

Holm, D. D., & Kerr, R. M. (2007). Helicity in the formation of turbulence. *Physics of Fluids, 19*, 025101.

Holman, J. P. (1990). *Heat transfer* (7th ed.). New York: McGraw-Hill.

Kolmogorov, A. N. (1942). Equations of turbulent motion of an incompressible fluid. Izvestiya Akademii Nauk SSSR, Seria fizicheska VI(1–2). (Translated by D. B. Spalding.)

Lemmon, E., Huber M., McLinden M., & REFPROP (2010). Reference fluid thermodynamic and transport properties, NIST Standard Reference Database 23, Version 9.0.

Lesieur, M., Metais, O., & Comte, P. (2005). *Large-eddy simulations of turbulence*. Cambridge: Cambridge University Press.

Lumley, J. L. (1970). Toward a turbulent constitutive relation. *Journal of Fluid Mechanics, 41*(Part 2), 413–434.

Mazumdar, H. P., & Mamaloukas C. (2012). On Millionshtchikov's zero-fourth cumulant hypothesis applied to turbulence. ATINER's conference paper series, MAT2012-0103.

McNaughton, K., & Sinclair, C. (1966). Submerged jets in short cylindrical flow vessels. *Journal of Fluid Mechanics, 25*(Part 2), 367–375.

Millionshtchikov, M. (1941). On the theory of homogeneous isotropic turbulence. *Doklady Akademii Nauk SSSR, 32*(9), 615–618.

Moradian, N., King, D., & Cheng, S. (2008). The effects of freestream turbulence on the drag coefficient of a sphere. *Experimental Thermal and Fluid Science, 33*(3), 460–471.

Myong, H. K., & Kasagi, N. (1990). A new approach to the improvement of k-ε turbulence model for wall-bounded shear flows. *JSME International Journal, 33*(1), 63–72.

Nakano, T. (1972). A theory of homogeneous, isotropic turbulence of incompressible fluids. *Annals of Physics, 73*(2), 326–371; Online 2004.

Peng, S. H., & Davidson, L. (1999). Computation of turbulent buoyant flows in enclosures with low-Reynolds-number k-ω models. *International Journal of Heat and Fluid Flow, 20*, 172–184.

Prandtl, L. (1945). Uber ein neues Formelsystem fur die ausgebildete Turbulenz, (On a new formulation for the fully developed turbulence). Nachrichten der Akademie der Wissenschaften in Gottingen, Mathematisch-Physikalische Klasse.

Rayleigh, L. (1892). VIII. on the question of the stability of fluids. The London, Edinburgh, and Dublin Philosophical Magazine and Journal of Science.

Rayleigh, L. (1913). Sur la Resistance des Spheres dans l'air en Mouvement", Comptes Rendus.

Reynolds, O. (1883). An experimental investigation of the circumstances which determine whether the motion of water shall be direct or sinuous, and of the law of resistance in parallel channels. Philosophical Transactions of the Royal Society of London, 174. (Can also be obtained through JSTOR, http://www.jstor.org)

Reynolds, O. (1895). On the dynamical theory of incompressible viscous fluids and the determination of the criterion. *Philosophical Transactions of the Royal Society of London A, 186,* 123–164. (Can also be obtained at http://www.jstor.org/stable/90643).

Rodriguez, S. (2016, July). Next-generation heat transfer surfaces. Sandia National Laboratories, Tracking No. 476209.

Rodriguez, S., & Turner D. Z. (2012). Assessment of existing Sierra/Fuego capabilities related to grid-to-rod-fretting (GTRF). Sandia National Laboratories, SAND2012-0530.

Rott, N. (1990). Note on the history of the Reynolds number. *Annual Review of Fluid Mechanics, 22*(1), 1–12.

Schlichting, H. (1979). *Boundary layer theory* (7th ed.). New York: McGraw-Hill Book Company.

Schlichting, H., & Gersten, K. (2000). *Boundary layer theory* (8th ed.). New York: Springer.

Schmitt, F. G. (2007). About Boussinesq's turbulent viscosity hypothesis: Historical remarks and a direct evaluation of its validity. *Comptes Rendus Mecanique, 335,* 617.

Shliomis, M. I., & Morozov, K. I. (1994). Negative viscosity of ferrofluid under alternating magnetic field. *Physics of Fluids, 6*(8), 2855–2861.

Sivashinsky, G. I., & Frenkel, A. L. (1992). On negative eddy viscosity under conditions of isotropy. *Physics of Fluids A, 4*(8), 1608–1610.

Sivashinsky, G., & Yakhot, V. (1985). Negative viscosity effect in large-scale flows. *Physics of Fluids, 28*(4), 1040.

Sohankar, A. (2006). Flow over a bluff body from moderate to high Reynolds numbers using large eddy simulation. *Computers & Fluids, 35,* 1154–1168.

Spalart, P. R. (2015). Philosophies and fallacies in turbulence modeling. *Progress in Aerospace Sciences, 74,* 1–15.

Speziale, C. G., & Eringen, A. C. (1981). Nonlocal fluid mechanics description of wall turbulence. *Computers and Mathematics with Applications, 7,* 27–41.

Stepanov, R. (2018). Helical bottleneck effect in 3D homogeneous isotropic turbulence. *Fluid Dynamics Research, 50,* 011412.

Stokes, G. G. (1850). On the effect of the internal friction of fluids on the motion of pendulums. *Transactions of the Cambridge Philosophical Society, 9,* 1–99.

Taylor, G. (1922). Diffusion by continuous movements. *Proceedings of the London Mathematical Society, s2–s20*(1), 196–212.

Taylor, G. (1935a). Statistical theory of turbulence. Royal Society of London, Series A, Mathematical and Physical Sciences. Can also be downloaded from rspa.royalsocietypublishing.org

Taylor, G. (1935b). Statistical theory of turbulence II. *Proceedings of the Royal Society of London, Series A, Mathematical and Physical Sciences, 151*(873), 444–454.

von Karman, T., & Howarth L. (1938). On the statistical theory of isotropic turbulence. Proceedings of the Royal Society of London. Series A, 164. (The title page has "de" Karman, which also means "von" Karman in German, which means "of").

Wang, G., Yang, F., & Zhao, W. (2014). There can be turbulence in microfluidics at low Reynolds number, doi:https://doi.org/10.1039/C3LC51403J, Lab on a Chip.

White, F. (1991). *Viscous fluid flow* (2nd ed.). New York: McGraw-Hill.

Wilcox, D. C. (2006 k-ω model). *Turbulence modeling for CFD* (3rd ed.). La Cañada: DCW Industries, printed on 2006 and 2010.

Zhao, C. R., et al. (2017). Influence of various aspects of low Reynolds number k-ε turbulence models on predicting in-tube buoyancy affected heat transfer to supercritical pressure fluids. *Nuclear Engineering and Design, 313,* 401–413.

Chapter 3
Applied Theory: Practical Turbulence Estimates

Big whorls have little whorls that feed on their velocity, and little whorls have lesser whorls and so on to viscosity.

—L. F. Richardson (1922)

We really don't know a whole lot for sure about turbulence. And worse, we even disagree about what we think we know! ... Thus it might be wise to view most 'established' laws and theories of turbulence as more like religious creeds than matters of fact.

—W. K. George (2013)

Abstract The three key eddies are described in detail, with practical equations to calculate their length, life time, and velocity. The LIKE algorithm is presented to compute key turbulence parameters. Flow regions are defined and represented mathematically for the viscous laminar sublayer and the log laws. Detailed descriptions for calculating y^+ are presented. The chapter concludes with a detailed description for dimpling and surface engineering applicable for drag reduction and heat transfer enhancements. Both compressible and incompressible applications are discussed; examples include golf balls, shark skin, vehicle surfaces, heat transfer systems, wings and airfoils, turbulators, and diverse dimple geometries.

Without doubt, good simulations involving Reynolds-averaged Navier Stokes (RANS), large eddy simulation (LES), or direct numerical simulation (DNS) provide valuable knowledge that enable useful engineering designs, analysis, safety, optimization, and so forth. But how much information can be gleaned from back-of-the-envelope (BOTE) calculations? *The answer is that an astonishing amount of information can be obtained quickly and inexpensively, well before any CFD or experiments are conducted.* These turbulence BOTE metric estimates can yield much useful information, including:

- Eddy velocity, size, and time scales
- Number of elements needed
- Guidance for turbulence model selection and input

© Springer Nature Switzerland AG 2019
S. Rodriguez, *Applied Computational Fluid Dynamics and Turbulence Modeling*,
https://doi.org/10.1007/978-3-030-28691-0_3

- Dissipation ε, turbulent kinetic energy k, and turbulent kinematic viscosity ν_t
- Peak turbulence velocity
- Fluctuation metrics
- And much more

In what follows, the equations needed to calculate three key characteristic metrics (eddy size, lifetime, and velocity) are described, starting with integral eddies, continuing on to Taylor eddies, and finally, the Kolmogorov eddies. In these sections, it will be noted that two key variables, k and ε, are needed to calculate many of the eddy scales. These two turbulence variables can be obtained from experimental data, CFD analysis, or through the LIKE algorithm, which is discussed in Sect. 3.4.

To estimate this insightful information, it is necessary to have a feel for how eddies (coherent structures) characterize turbulence. In particular, turbulent flows can be understood as the superposition of small, medium, and large eddies that are interspersed and dynamically interacting throughout the flow. The large eddies correspond to the "integral eddies," the medium eddies are the "Taylor eddies," and the smallest of the small eddies correspond to the "Kolmogorov eddies."

Eddies span the continuum from small to large, with no single eddy size having a preferential distribution; their inverse length distribution as a function of turbulent kinetic energy k resembles a somewhat compressed Gaussian curve, with each curve being lower as the Reynolds number (Re) decreases; see Fig. 3.1. The figure shows that the eddy spectrum is continuous, with no single discrete length being unique

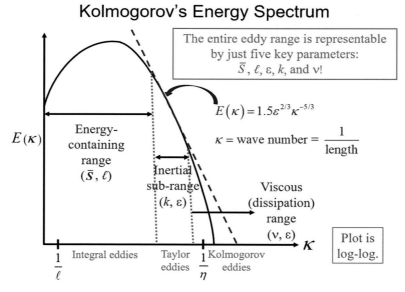

Fig. 3.1 Turbulent energy distribution as a function of eddy size—the Kolmogorov energy spectrum

within the energy-containing and inertial subrange. On the other hand, the Kolmogorov scale is uniquely determined for any system.

Whereas eddies span a wide length and turbulent energy range as shown conceptually in Fig. 3.1, *their collective behavior can be visualized as having "characteristic" scales (e.g., length, time, and velocity) that provide reasonable engineering estimates.*

As Fig. 3.1 suggests, certain domains based on length can be observed, such that eddies can be *grouped* into integral and Taylor eddy regions, as well as the Kolmogorov eddy limit. *But most importantly, the grouped performance of eddies in each domain displays a behavior that allows analysts to reasonably estimate their key turbulence metrics.* These metrics include a characteristic eddy size, lifetime, and velocity for all three eddy types. Then, these quantities can be used to compute valuable properties of the turbulent flow, as will be shown in the sections that follow.

3.1 Integral Eddies

Large eddies are generated as the flow becomes unstable, as it becomes nonlinear. The flow instabilities generate large clusters of fluid that are born as highly energetic coherent structures that nearly instantaneously obtain k from the bulk fluid energy. These eddies are very vigorous, as they carry approximately 80% of the total k. They are therefore responsible for most of the diffusive processes that result in the well-known energetic mixing that occurs in turbulence, including mass, momentum, and energy. That is, integral eddies provide the highest degree of mixing within the fluid, the greatest degree of heat transfer due to the mixing of hot and cold fluid clusters, and so forth. Because of their relatively large size and k compared to the Taylor and Kolmogorov eddies, integral eddies provide the most mixing, thereby easily cornering this most important turbulence role. This situation is akin to comparing bowling balls (integral eddies) moving at large velocities vs. golf balls and pebbles (Taylor and Kolmogorov eddies, respectively) moving at slower velocities.

Integral eddies can be as large as the characteristic length of a system, and even larger on occasion, up to about twice as large. How can this be? Consider internal flow in a pipe where the characteristic length scale is diameter D. But, because integral eddies stretch as they travel within the fluid, they can elongate like a 3D oval, into large coherent structures with sizes up to twice the pipe diameter. Another feature of large eddies is that they can flow up to about 30 pipe diameters before decaying.

Integral eddies have a large degree of vorticity, akin to sheets of paper being rolled, with the roll's diameter becoming smaller and smaller as the large velocity gradients continue curling the sheets. This is called vortex stretching and is the primary process that breaks up the large eddies. Thus, as the large eddies stretch, they reach a critical point as they break up into smaller eddies, which in turn continue stretching, thereby generating even smaller eddies. This process is called cascading, whereby integral eddies break up into smaller and smaller eddies. However, this

Fig. 3.2 Energy transfer through cascading, from large to small

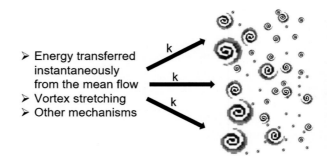

> Energy transferred instantaneously from the mean flow
> Vortex stretching
> Other mechanisms

process cannot go on forever; it is limited by the Kolmogorov eddy size. The cascading process is shown schematically in Fig. 3.2.

Eddy cascading has been studied for many decades but is still not fully understood. Researchers recently obtained images of a large eddy as it broke up into two smaller eddies (Shesterikov et al. 2012). The turbulent flow occurred in a tokomak plasma field, and five images taken every six microseconds show the large eddy as it split into two smaller eddies.

As noted earlier, eddies in turbulent flows have a continuous-length spectrum that is bounded from the integral to the Kolmogorov scale. In this context, an eddy can be viewed as a local swirling motion associated with a length scale. Hence, the large eddies are unstable due to various factors, which includes their highly energetic motion, large structure, and by being surrounded by other eddies that stretch and push the larger eddies in many directions. This inevitably causes the larger eddies to stretch and break into smaller eddies, and so forth.

As can be seen, integral eddies play a substantial role in turbulence, so finding their characteristic scales helps analysts estimate many fundamental properties of the turbulent flow in question.

The integral eddy size can be characterized by ℓ_o, which represents the *larger* of the large eddies (i.e., those for which the wavenumber $\kappa \to 0$; i.e., the eddies that reside on the extreme LHS of Fig. 3.1). On the other hand, ℓ represents the "average" size of the integral eddies. Fortunately, the two integral eddy length scales form a linear relationship that can be expressed as (Hinze 1987)

$$\ell_o \approx C_1 \ell, \tag{3.1A}$$

where C_1 is a constant.

In any case, ℓ is used throughout the book, as it is representative of the "averaged behavior" of the integral eddies and because most relationships in the literature use the scale directly.

Early researchers deduced a relationship for ℓ based on dimensional arguments (Taylor 1935):

$$\ell \approx \frac{k^{3/2}}{\varepsilon}. \tag{3.1B}$$

However, Eq. 3.1B can be made more precise by fitting it to experimental data (Wilcox 2006),

$$\ell = C_\mu \frac{k^{3/2}}{\varepsilon}, \tag{3.1C}$$

where a reasonable value for C_μ is 0.09.

However, the values for k and ε are usually not known a priori. If such is the case, the integral eddy size can be approximated as follows:

$$\ell = C_1 x_{\text{char}}, \tag{3.1D}$$

where x_{char} and C_1 are defined by the system geometry. Section 3.4, under "Calculation of ℓ," lists the relevant parameters for pipes, ducts (noncircular internal flow), external surfaces, and wind tunnels; refer to Eqs. 3.17A, 3.17B, 3.18, 3.19 and 3.20.

Note that Eqs. 3.1B, 3.1C, and 3.1D do not correspond to an average integral eddy size based on a rigorous mathematical definition nor do they represent a maximum eddy size. Instead, the equations provide a practical metric for a "characteristic" length that embodies a "representative" integral eddy size that the system can support, a sort of eddy length that provides a reasonable scale for the *composite* behavior of the larger eddies.

The integral eddy has a lifetime that corresponds to a time scale that represents how long the eddy will exist before it decays away. If Eq. 3.1C is used, then the appropriate (self-consistent) time scale is

$$\tau_\ell = C_\mu \frac{k}{\varepsilon}. \tag{3.2A}$$

However, if an equation analogous to Eq. 3.1D is used instead (refer to Sect. 3.4 for these geometry-specific expressions that are based on x_{char}), then the corresponding self-consistent equation is as follows:

$$\tau_\ell = \frac{C_1 x_{\text{char}}}{k^{1/2}}. \tag{3.2B}$$

The integral eddy velocity is the ratio of its characteristic length divided by its time scale, which is appropriate (self-consistent) for both Eqs. 3.1C and 3.1D,

$$u_\ell = \frac{\ell}{\tau_\ell} = k^{1/2}. \tag{3.3}$$

It is noted that some researchers include a factor of the square root of 2/3 in the integral eddy velocity (Andersson et al. 2012).

Note that the integral eddy length and velocity scales can be used along with the kinematic viscosity to obtain the integral eddy turbulent Reynolds number, or more specifically, Re_T or Re_ℓ. If Eq. 3.1C is used, then

$$Re_\ell = \frac{x_{\text{char}} u_{\text{char}}}{\nu} = \frac{\ell u_\ell}{\nu} = \frac{\left(C_\mu \frac{k^{3/2}}{\varepsilon}\right)\left(k^{1/2}\right)}{\nu} = C_\mu \frac{k^2}{\varepsilon \nu}, \tag{3.4A}$$

where the kinematic viscosity is $\nu = \frac{\mu}{\rho}$.

But, if an expression based on x_{char} in the form of Eq. 3.1D is used, then the self-consistent Re_ℓ is

$$Re_\ell = \frac{\ell u_\ell}{\nu} = \frac{\left(C_1 x_{\text{char}}\right)\left(k^{1/2}\right)}{\nu} = C_1 \frac{x_{\text{char}} k^{1/2}}{\nu}. \tag{3.4B}$$

An inspection of Eqs. 3.4A and 3.4B shows that $Re_\ell = C_\mu \frac{k^2}{\varepsilon \nu} = C_1 \frac{x_{\text{char}} k^{1/2}}{\nu}$, and when this is divided by $k^{1/2}$, implies that $C_\mu \frac{k^{3/2}}{\varepsilon} = C_1 x_{\text{char}} \equiv \ell$, which comes as no surprise.

The integral eddy Reynolds number is uniquely specified for the integral eddies as Re_ℓ, which is not to be confused with the system's hydraulic Reynolds number, Re_h. The magnitude for Re_ℓ is typically lower than Re_h, and the same trend can be observed for the Taylor and Kolmogorov Reynolds numbers.

Note that as the flow becomes more turbulent (i.e., as Re_h or Re_ℓ increases), the typical length of the eddy population decreases, while the number of eddies becomes more numerous as the larger eddies split. It is possible to associate the integral eddies with the magnitude of Re_ℓ and the Kolmogorov eddies (η), as follows:

$$\frac{\eta}{\ell} \approx Re_\ell^{-3/4}. \tag{3.5}$$

The rapid decrease of the smallest eddies compared with the integral eddies is shown in Fig. 3.3.

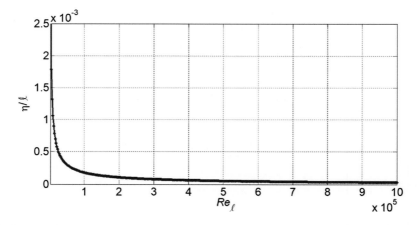

Fig. 3.3 Ratio of Kolmogorov vs. integral eddies as a function of the turbulent Reynolds number

3.2 Taylor Eddies

Taylor eddies are the "intermediate-sized eddies" that are too small to behave as integral eddies and too large to behave as Kolmogorov eddies. In more ways than one, Taylor eddies can be viewed as having a behavior that is a hybrid mixture of integral and Kolmogorov eddies. For example, both Taylor and integral eddies are characterized by turbulent kinetic energy k, which is attributable to their energetic nature. So, whereas large eddies may contain about 4/5 of the total turbulent energy, Taylor eddies have k approaching 1/5, which is still a significant quantity, engineering-wise. Then, on the opposite extreme, both Taylor and Kolmogorov eddies are characterized by dissipation ε, which is typically associated with eddy decay. Therefore, *Taylor eddies* are best considered as "hybrid" eddies because they contain features of both integral and Kolmogorov eddies.

The Taylor eddy characteristic length can be approximated for isotropic flows based on a couple of straightforward simplifications (Wilcox 2006):

$$\lambda \sim \left(\ell\eta^2\right)^{1/3} \tag{3.6A}$$

where η is the Kolmogorov length scale; refer to Sect. 3.3. Note that the above expression collapses to the Taylor's length shown in Eq. 3.6B. However, because Taylor's expression is more familiar, is easier for obtaining the Taylor eddy characteristic length for isotropic flows, and was derived from basic principles (Taylor 1935), it is therefore the preferred expression:

$$\lambda = \left(\frac{10k\nu}{\varepsilon}\right)^{1/2}. \tag{3.6B}$$

The eddy lifetime before it decays is

$$\tau_\lambda = \left(\frac{15\nu}{\varepsilon}\right)^{1/2}. \tag{3.7}$$

The Taylor eddy velocity is the ratio of its characteristic length divided by its time scale:

$$u_\lambda = \frac{\lambda}{\tau_\lambda} = \left[\left(\frac{10k\nu}{\varepsilon}\right)\left(\frac{\varepsilon}{15\nu}\right)\right]^{1/2} = \left(\frac{2k}{3}\right)^{1/2}. \tag{3.8}$$

Finally, the Taylor Re is the product of the characteristic length of the eddy times the eddy velocity, divided by the kinematic viscosity:

$$Re_\lambda = \frac{x_{char}u_{char}}{\nu} = \frac{\lambda u_\lambda}{\nu} = \frac{\left(\frac{10k\nu}{\varepsilon}\right)^{1/2}\left(\frac{2k}{3}\right)^{1/2}}{\nu} = \frac{\left(\frac{10k\nu}{\varepsilon}\frac{2k}{3}\right)^{1/2}}{\nu} = k\left(\frac{20}{3\varepsilon\nu}\right)^{1/2}. \tag{3.9}$$

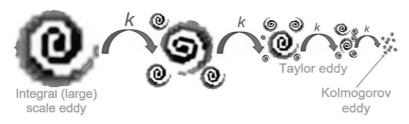

Fig. 3.4 The energy cascade

3.3 Kolmogorov Eddies

As eddies evolve in time, the energy cascade refers to the situation where the kinetic energy from the larger eddies is transferred onto smaller eddies, and from those smaller eddies, onto even smaller eddies, and so forth, until a physics-based minimum eddy length scale is reached, the Kolmogorov eddy. Thus, the length scale limit occurs because the ever-smaller eddy k eventually becomes less than or equal to the viscous term that dampens (overcomes) their small, rotational, eddying motions. Thus, the Kolmogorov eddies are at the tail end of the energy cascade. The cascade is a consequence of a break-up process as the eddy sheets rotate into finer and finer rolls having higher and higher vorticity, which eventually breaks up the larger eddies into the smallest eddies, as shown in Fig. 3.4. Another distinguishing feature of the Kolmogorov eddy is that they have far less k than any other type of eddy.

Thus, the smallest eddies that a turbulent flow can physically sustain are the Kolmogorov eddies; these are the eddies that decay onto very tiny laminar spots within the fluid. As this happens, the Kolmogorov eddy energy is transferred back to the fluid through viscous shear, from whence it initially came. Because Kolmogorov eddies continue rotating similarly to scroll sheets that have smaller radii, these eddies have the highest level of vorticity.

Kolmogorov eddies are very small and can be as small as a human hair, in the range of 17–180 microns. Of course, this value depends heavily on the fluid type and other flow characteristics, as will be shown later in this section. Notwithstanding their small characteristic size, Kolmogorov eddies tend to be larger than the viscous boundary layer near the wall. Because of their small size and energy, they have a much shorter life span compared to integral and Taylor eddies. Further, the Kolmogorov eddy time scale is much smaller than the mean-flow time scale (based on a characteristic mean-flow velocity and length scale). *Therefore, Kolmogorov eddy motion is independent of the mean flow and the large eddies.* As a result, *energy received by the large eddies corresponds to the heat dissipated by the small eddies* (provided that turbulence is at equilibrium). This premise was first postulated by Kolmogorov in 1941 and is called "the universal equilibrium theory." Kolmogorov hypothesized that small eddy motion depends solely on two phenomena.

The first phenomenon is the rate of turbulent kinetic energy transfer from the large eddies:

$$\varepsilon \equiv -\frac{dk}{dt}. \tag{3.10}$$

Thus, turbulent dissipation is defined as the rate of change of the turbulent kinetic energy with respect to (WRT) time—it represents the conversion of the eddy turbulent kinetic energy back into internal energy. This represents how fast eddies lose their fluctuating energy. The second phenomenon is the viscous dissipation, for which ν characterizes how strong the damping viscous force is. Then, from the above two, it is straightforward using dimensional arguments to derive the Kolmogorov length, velocity, and time scales *for the smallest eddies*. Therefore, the Kolmogorov length scale is

$$\eta = \left(\frac{\nu^3}{\varepsilon}\right)^{1/4}. \tag{3.11}$$

This is the smallest eddy length scale that can be supported by a fluid before its viscous forces dampen the coherent structure back into laminarity.

The Kolmogorov eddy has a time scale before it decays that is approximated as

$$\tau_\eta = \left(\frac{\nu}{\varepsilon}\right)^{1/2}. \tag{3.12}$$

The Kolmogorov eddy velocity is

$$u_\eta = \frac{\eta}{\tau} = (\nu\varepsilon)^{1/4}. \tag{3.13}$$

Finally, the Kolmogorov *Re* is

$$Re_\eta = \frac{x_{char} u_{char}}{\nu} = 1. \tag{3.14}$$

Example 3.1 Prove that $Re_\eta = 1$. Hint: apply the basic definition for Re, using the Kolmogorov characteristic length and velocity.

Solution $x_{char} = \eta = \left(\frac{\nu^3}{\varepsilon}\right)^{1/4}$ and $u_{char} = u_\eta = (\nu\varepsilon)^{1/4}$. Therefore,

$$Re_\eta = \frac{x_{char} u_{char}}{\nu} = \frac{\left(\frac{\nu^3}{\varepsilon}\right)^{1/4}(\nu\varepsilon)^{1/4}}{\nu} = \frac{\left(\frac{\nu^3}{\varepsilon}\nu\varepsilon\right)^{1/4}}{\nu} = \frac{(\nu^4)^{1/4}}{\nu} = 1.$$

3.4 The "LIKE" Algorithm: Computing Your Own Epsilon, Turbulent Kinetic Energy, Turbulent Kinematic Viscosity, and Reynolds Stresses

Oftentimes, beginners and perhaps even researchers who have studied turbulence for years realize that turbulence involves many equations that are useful. But, how are these equations related to each other? And perhaps most importantly, how can a manageable set of turbulence equations form a closed set of independent equations, with the goal of obtaining n equations that can be used to solve for an n number of useful turbulence variables? Here, the "LIKE" acronym is used to describe a closed set of four independent equations that can be used to estimate four key variables, and these in turn can be used to estimate a significant set of additional useful turbulence variables. Note that an acronym is typically a grouping of letters that is arranged to help people remember a set of words, such as a mnemonic. In this case, the following four letters represent four key turbulence variables that are "conveniently" arranged into an acronym, that with a little imagination, forms the word "LIKE":

$L = \ell$ = integral eddy length scale
$I = I_t$ = turbulence intensity
$K = k$ = turbulent kinetic energy

and

$E = \varepsilon$ = turbulent dissipation

In what follows, guidelines are provided for calculating each of the four variables.

- *Calculation of ℓ*

Taylor deduced a relationship for eddy length based on dimensional arguments for isotropic flow (Taylor 1935; Wilcox 2006),

$$\ell \sim \frac{u'^3}{\varepsilon},$$

(3.15A)

which is akin to saying that

$$\varepsilon \sim \frac{k^{3/2}}{\ell}$$

(3.15B)

because

$$u'^3 = k^{3/2}.$$

(3.15C)

Therefore,

$$\varepsilon \equiv \frac{W}{\rho} = C\frac{u'^3}{\ell} = C\frac{k^{3/2}}{\ell}, \qquad (3.15\text{D})$$

where C is a constant; Taylor used W/ρ to refer to the modern ε. Prandtl let $C = C_\text{D}$ for his one-equation transport model (Prandtl 1945). However, based on an experimental data fit, Wilcox recommends $C = C_\mu$ (Wilcox 2006). Therefore,

$$\varepsilon = C_\mu \frac{k^{3/2}}{\ell} \qquad (3.15\text{E})$$

where

$$C_\mu = 0.09. \qquad (3.15\text{F})$$

To see how this relates to the integral eddy length, simply solve for ℓ, yielding

$$\ell = C_\mu \frac{k^{3/2}}{\varepsilon}, \qquad (3.15\text{G})$$

which is the same expression found in CFD-Online (2017a) and many other references. That the full expression for ℓ must include C_μ is clear because it is needed to satisfy the Prandtl-Kolmogorov relationship; refer to Problem 4.3 in Chap. 4.

Note that u' is the fluctuating velocity and ε is the turbulent dissipation. Generally, both variables are unknown. Fortunately, ℓ can be approximated because it tends to be linearly proportional to the *system characteristic length*. That is, the system characteristic length places a linear scale limit upon the maximum magnitude of the integral eddy length:

$$\ell \sim x_\text{char}. \qquad (3.16)$$

However, the maximum length is usually not reached nor is there a "typical value." Therefore, conservative approximations are used, so some fraction of x_char is considered, and this depends on the geometry (von Karman and Howarth 1938). For example, for internal flow in a pipe, the characteristic length is D because it forms a reasonable upper boundary for the length of the large eddies (though stretching will make a small fraction of the large eddies larger than D). In any case, some fraction of D is used to provide a reasonable estimate for ℓ (CFD-Online 2017a):

$$\ell \approx 0.07D. \qquad (3.17\text{A})$$

A similar expression is (Minin and Minin 2011)

$$\ell \approx 0.1D. \tag{3.17B}$$

Either 0.07 or 0.1 is reasonable, with 0.07 being the preferred coefficient because it is more limiting.

For noncircular internal flow, the hydraulic diameter D_h is factored in

$$\ell \approx 0.07D_\mathrm{h}. \tag{3.18}$$

For boundary layers generated by external flow over a surface, the hydraulic boundary layer thickness δ is the system characteristic length scale, and therefore

$$\ell \approx 0.4\delta. \tag{3.19}$$

Finally, for wind tunnels, the grid spacing S is the system characteristic length (Taylor 1935):

$$\ell \approx 0.2S. \tag{3.20}$$

The above characteristic length represents the average eddy size downstream of the grid, ℓ_2. This length is twice ℓ_1, the "Mischungsweg of Prandtl," as Taylor would fondly call it (Taylor 1935).

- *Calculation of I*

The turbulence intensity is often represented by the symbol I or I_T, which the literature defines in various ways, including (CFD-Online 2017b)

$$I = \sqrt{\frac{\left(\frac{2}{3}k\right)}{\overline{u^2}}}, \tag{3.21}$$

where the flow is assumed isotropic, thus imposing that

$$\overline{u_x'^2} = \overline{u_y'^2} = \overline{u_z'^2}. \tag{3.22}$$

However, a simpler relationship can be obtained by approximating the turbulence intensity as the ratio of the mathematical definition for root-mean-square (RMS) velocity fluctuations u' and the time-averaged velocity $\bar{u}$. Therefore, the degree of turbulence fluctuation in a generic system can be expressed as the ratio of the "averaged velocity fluctuation" and the flow's time-averaged velocity,

$$I = I_t \equiv \frac{\sqrt{\frac{1}{3}\left(\overline{u'^2} + \overline{v'^2} + \overline{w'^2}\right)}}{\overline{u}} = \frac{u'}{\overline{u}}. \tag{3.23}$$

If the flow is isotropic or nearly isotropic, then Eq. 3.22 can be substituted into the above expression, thereby simplifying as follows:

$$I \approx \frac{\sqrt{\overline{u'^2_x}}}{\overline{u}}. \tag{3.24}$$

Moreover, if the following nonchalant approximation is made, $\sqrt{\overline{u'^2_x}} \approx \left|u'_x\right|$, then

$$u' \approx \left|u'_x\right| \approx \overline{u}I. \tag{3.25}$$

An absolute value is assigned to the above expression because fluctuations can be negative WRT the time-averaged mean velocity and hence can cause nearly-instantaneous decreases in the mean. Another way of visualizing negative fluctuations is that they *decrease* the mean velocity.

Thus, as turbulence intensity approaches 0, the flow becomes less and less turbulent, while a value approaching 1 implies huge velocity fluctuations; I can range to as low as 0.003 for highly aerodynamic airfoils, to as high as 0.3 for atmospheric boundary layer flows; this indicates that turbulence exhibits a large, two-decade range for I and should be approximated as carefully as possible (be weary of using the CFD code default without investigating your specific application).

Table 3.1 shows some estimates for I that can be used for various systems and general conditions. However, more precise approximations can be made, as will be shown later.

Generally speaking, as the hydraulic Re increases, I decreases, which implies that eddies become smaller. This behavior is reflected through analytic expressions for pipe and duct flows as power law functions of Re. These relationships have been

Table 3.1 Estimates for I

System or general condition	I
Aerodynamic airfoils	0.003
Core flows	0.02–0.05
Pipes, ducts, simple internal flows at intermediate to high Re	0.02–0.12
High-velocity flows in complex systems (e.g., turbomachinery, heat exchangers, baffles, high swirl devices, near the wall)	0.05–0.2
Atmospheric boundary layer flows, gusting winds, tornadoes, hurricanes, oscillating boundaries	0.3

derived for at least six decades and have the following form (Sandborn 1955; Minin and Minin 2011; Russo and Basse 2016; Basse 2017):

$$I = C_1 \, Re_{\mathrm{h}}^{-C_2} \tag{3.26A}$$

where Re_{h} refers to the hydraulic Re.

The earliest verifiable version of a power law expression dates to 1955 and is attributed to research conducted for the National Advisory Committee for Aeronautics (Sandborn 1955),

$$I_{\mathrm{meas,pipe\ center}} = 0.144 \, Re_{\mathrm{h}}^{-0.146}. \tag{3.26B}$$

Sandborn's equation was derived for the pipe center for fully developed flow and is based on data from several experiments.

Curiously, many unsuccessful attempts have been made by researchers (Russo and Basse 2016), etc., to find the original reference for the following power law equation:

$$I = 0.16 \, Re_{\mathrm{h}}^{-1/8}. \tag{3.26C}$$

An early citation for Eq. 3.26C is found in (Minin and Minin 2011), and perhaps there are earlier citations as well.

More recently, power law expressions were developed by using CFD and experimental data for fully developed pipe flow (Russo and Basse 2016). Russo and Basse considered many cases, including compressible and incompressible flows, as well as expressions for I that are based on the pipe axis and flow area. For I in the pipe axis, they found that C_1 is in the range of 0.0550 and 0.0947, while C_2 is in the range of 0.0407 and 0.0727. For I computed based on flow area, C_1 is in the range of 0.140 and 0.227, while C_2 is in the range of 0.0779 and 0.100. This wide range is shown in Fig. 3.5, which includes the Sandborn and 1/8th expressions as well. Figure 3.5 shows that I spans a range that differs by approximately 3.5 times between the peak and the minimum values.

It is therefore necessary to sort through the expressions to look for patterns. First, it is noted that compressible vs. incompressible flows did not result in significant differences when compared to other factors. Some expressions were derived solely on CFD, while others relied solely on experimental data. But without having error bars and detailed descriptions of the CFD calculations and the experiments, there is insufficient information to choose CFD over experiments, or vice versa. However, from Fig. 3.5, it is clear that the Russo and Basse equations for the pipe area have a higher I, which makes sense. In particular, turbulence fluctuations are larger near the wall (but not in the laminar sublayer, of course) than at the pipe centerline. So, it stands to reason that the area-based I is larger. As an aide to sort the expressions, Figs. 3.6 and 3.7 attempt to seek such pattern, where Fig. 3.6 shows the pipe axis expressions, and Fig. 3.7 shows the area-based expressions.

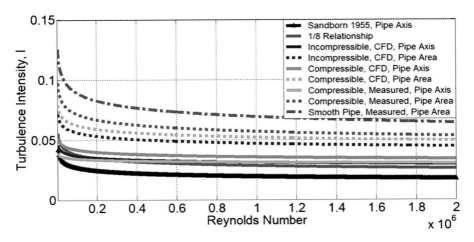

Fig. 3.5 The wide range in power law equations for turbulence intensity

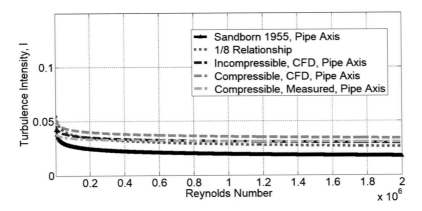

Fig. 3.6 Turbulence intensity based on pipe axis

Figure 3.6 indicates that the curves form a tighter fit; the compressible, the 1/8th relationship, the CFD-based expressions, and the measured experimental data expressions follow each other in a much closer pattern. If anything, the Sandborn Equation now appears as an outlier WRT the other expressions; but it is still in close range. Certainly, Eqs. 3.26B and 3.26C yield I values that can be sufficiently different; refer to Example 3.5 in this chapter for an instance where their difference was nearly 34%. However, the Sanborn Equation is the most limiting expression for I based on the pipe axis, followed by the 1/8th expression. Because of its close fit to the curves in Fig. 3.6, it is ventured that the 1/8th expression was derived for pipe axis I. However, because of its better documentation and its lower limit bound, the Sandborn Equation is preferred. On the other hand, if the reader seeks an upper limit, then the work of (Russo and Basse 2016) should be consulted.

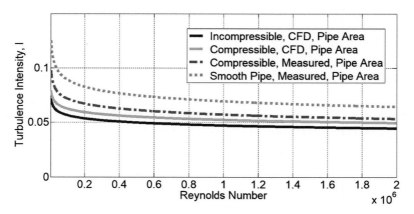

Fig. 3.7 Turbulence intensity based on pipe area

Figure 3.7 indicates essentially the same pattern found in Fig. 3.6, though the curves do not form as tight a pattern. Figure 3.7 indicates that the most limiting expression is Eq. 3.26F, which was formulated using CFD for an incompressible flow, with area-based I. The reader is encouraged to consult Russo and Basse (2016) for upper limits based on area.

Based on experimental data, Russo and Basse (2016) and Basse (2017) recommend that

$$I_{\text{meas,pipe axis}} = 0.055 \, Re_{\text{h}}^{-0.0407}. \tag{3.26D}$$

Equation 3.26D is also recommended for smooth pipes (Basse 2017).
Likewise, based on experimental data for a smooth pipe,

$$I_{\text{meas,pipe area}} = 0.227 \, Re_{\text{h}}^{-0.1}. \tag{3.26E}$$

Based on Fig. 3.7, the bounding limit expression for area-based incompressible turbulence intensity is

$$I_{\text{Inc,CFD,pipe area}} = 0.140 \, Re_{\text{h}}^{-0.079}. \tag{3.26F}$$

Finally, for smooth pipes, I based on the pipe area is

$$I_{\text{smooth,pipe area}} = 0.317 \, Re_{\text{h}}^{-0.11}. \tag{3.26G}$$

Quite recently, it was determined that there is a relationship between pipe surface roughness λ and turbulence intensity (Basse 2017), namely, that

$$I_{\text{pipe area}} \sim \frac{9}{10} \sqrt{\frac{d}{2}} \; (\text{Note: } d = \text{channel roughness}). \tag{3.26H}$$

To say the least, future research in the field of turbulence intensity should reconcile the differences in the multiple expressions for I, as evidenced in Fig. 3.5. An additional research area of strong merit is the development of I expressions for *external* flows, such as flow over flat plates and airfoils.

- *Calculation of k*

Once I or u' have been calculated using the previous expressions, it is straightforward to calculate k:

$$k = \frac{3}{2}(u')^2 = \frac{3}{2}(\bar{u}I)^2 \tag{3.27}$$

where $\bar{u}$ is the time-averaged velocity. If $\bar{u}$ is unknown, but the mass flow rate is given, then it is easy to obtain the velocity from $\dot{m} = \rho\bar{u}A$.

- *Calculation of ε*

Once k is known, and in conjunction with the expressions for ℓ shown earlier, the calculation of ε is straightforward:

$$\varepsilon = C_\mu \frac{k^{3/2}}{\ell}. \tag{3.28}$$

This completes the LIKE algorithm for estimating ℓ, I, k, and ε.

Example 3.2 Consider a pipe with $D = 0.025$ with at an average velocity of 0.02 m/s. The fluid consists of nitrogen gas at 600 K and 5.0×10^6 Pa. Use LIKE to determine the integral, Taylor, and Kolmogorov eddy length, time, and velocity.

Solution Actually, Example 3.2 was a "nefarious test question"; however, the "real world" offers no qualms in terms of such scenarios. It is preferable to spend a few minutes calculating Re than spending hours explaining why such calculation was not made. In any case, for this situation, $\nu = 1.09\text{E}{-}06$ m^2/s. Therefore,

$$Re_{\text{h}} = \frac{0.025 * 0.02}{1.09E - 06} = 459.$$

Clearly, the flow is laminar, so it cannot generate and sustain turbulent eddies! This example reinforces the absolute need to *always* calculate Re.

Example 3.3 Redo Example 3.2, except that now the average fluid velocity is 2.0 m/s.

Solution $$Re_h = \frac{0.025 * 2.0}{1.09E - 06} = 4.59E04, \therefore \text{turbulent.}$$

Usage of the equations in Sects 3.1, 3.2, 3.3 and 3.4 are sufficient to solve this problem. In particular, from Eq. 3.17A,

$$\ell \approx 0.07D = 0.07 * 0.025 = 0.0018 \text{ m.}$$

From Eq. 3.26C,

$$I = 0.16 \, Re_h^{-1/8} = 0.16(4.59E04)^{-1/8} = 0.0418.$$

From Eq. 3.27,

$$k = \frac{3}{2}(\bar{u}I)^2 = \frac{3}{2}(2.0 * 0.0418)^2 = 0.0105 \text{ m}^2/\text{s}^2.$$

Finally, from Eq. 3.28,

$$\varepsilon = C_\mu \frac{k^{3/2}}{\ell} = 0.09 \frac{(0.0105)^{3/2}}{0.0018} = 0.0538 \text{ m}^2/\text{s}^3.$$

Now that the four LIKE variables have been calculated, the entire set of eddy scales can be estimated. Namely, for the integral eddies, ℓ, τ_ℓ, and u_ℓ are

$\ell = 0.0018$ m (already calculated from Eq. 3.17A),

$\tau_\ell = \frac{C_l x_{\text{char}}}{k^{1/2}} = \frac{(0.07)(0.025)}{0.0105^{1/2}} = 0.0171$ s (Eq. 3.2B),

and

$u_\ell = k^{1/2} = 0.0105^{1/2} = 0.1025$ m/s (Eq. 3.3).

For the Taylor eddies, λ is found from Eq. 3.6B,

$$\lambda = \left(\frac{10k\nu}{\varepsilon}\right)^{1/2} = \left(\frac{10 * 0.0105 * 1.09E - 6}{0.0538}\right)^{1/2} = 0.0015 \text{ m,}$$

while τ_λ is obtained from Eq. 3.7,

$$\tau_\lambda = \left(\frac{15\nu}{\varepsilon}\right)^{1/2} = \left(\frac{15 * 1.09E - 6}{0.0538}\right)^{1/2} = 0.0174 \text{ s,}$$

and u_λ proceeds from Eq. 3.8,

$$u_\lambda = \left(\frac{2k}{3}\right)^{1/2} = \left(\frac{2 * 0.0105}{3}\right)^{1/2} = 0.0837 \text{ m/s.}$$

Finally, for the Kolmogorov eddies, Eq. 3.11 is used to obtain η,

$$\eta = \left(\frac{\nu^3}{\varepsilon}\right)^{1/4} = \left(\frac{1.09E-6^3}{0.0538}\right)^{1/4} = 6.99 \times 10^{-5} \text{ m},$$

while Eq. 3.12 provides τ_η,

$$\tau_\eta = \left(\frac{\nu}{\varepsilon}\right)^{1/2} = \left(\frac{1.09E-6}{0.0538}\right)^{1/2} = 0.0045 \text{ s},$$

and u_η is calculated from Eq. 3.13,

$$u_\eta = \frac{\eta}{\tau} = (\nu\varepsilon)^{1/4} = (1.09E-6*0.0538)^{1/4} = 0.0156 \text{ m/s}.$$

A check of the LIKE MATLAB script found in Chap. 7 confirms the above values to within round-off.

Example 3.4 Consider the same conditions as in Example 3.2, but now replace the fluid with molten liquid lead bismuth eutectic (LBE). Use the LIKE and liquid-metal physical properties calculator scripts found in Chap. 7.

Solution LBE melts at 397 K. Because the fluid is single-phase molten metal, the high pressure should not impact the kinematic viscosity. For $T = 600$ K, $\nu = 1.6873E{-}07$ m^2/s. For this situation,

$$Re_h = \frac{0.025*2.0}{1.687E-07} = 2.96E05, \therefore \text{turbulent.}$$

From the LIKE MATLAB script output:

Kolmogorov eddy size $= 2.04E{-}05$ m
Kolmogorov eddy velocity $= 8.25E{-}03$ m/s
Kolmogorov eddy time $= 2.48E{-}03$ s
Taylor eddy size $= 6.36E{-}04$ m
Taylor eddy velocity $= 6.62E{-}02$ m/s
Taylor eddy time $= 9.60E{-}03$ s
Integral eddy size $= 1.75E{-}03$ m
Integral eddy velocity $= 7.30E{-}03$ m/s
Integral eddy time $= 2.40E{-}01$ s

3.5   Using LIKE to Obtain Key Turbulence Variables

Once the LIKE equations in Sect. 3.4 are solved, estimating the characteristic length, velocity, and time for the integral, Taylor, and Kolmogorov eddies is trivial. The MATLAB-like script in Chap. 7 already has all the eddy scale equations found in

Sects. 3.1, 3.2 and 3.3. Furthermore, the four LIKE variables can provide additional useful information through BOTE calculations, such as valuable turbulence metrics and quantities required for input models. For example, the RANS k-ε model requires initial values for k and ε, which are already calculated by the LIKE algorithm. However, if the user is interested in the k-ω model, input values for ω can be estimated as follows:

$$\omega = \frac{k^{1/2}}{\ell} = \frac{\varepsilon}{\beta^* k} \tag{3.29A}$$

where

$$\beta^* = \frac{9}{100}. \tag{3.29B}$$

If the code requires input for the kinematic viscosity ν_t, then

$$\nu_t \equiv \frac{\mu_t}{\rho} = \frac{C_\mu k^2}{\varepsilon}, \tag{3.30}$$

which is the Prandtl-Kolmogorov relationship. Note that for turbulent flows, ν_t is generally 10 to 1000 times larger than the fluid kinematic viscosity, ν. Furthermore, the larger the ratio of ν_t vs. ν, the more turbulent the flow will be.

The turbulent Re, which is based on the integral eddies, can be estimated as

$$Re_T = \frac{\ell \sqrt{k}}{\nu} \approx \frac{\ell u'}{\nu}. \tag{3.31}$$

Re_T is typically about 10 to 100 times smaller than Re_h, appears in many turbulence models, and is useful for estimating the number of computational nodes N needed for DNS:

$$N_{1D} \approx (Re_T)^{3/4} = \text{nodes required in 1D DNS calculations} \tag{3.32}$$

$$N_{3D} \approx (Re_T)^{9/4} \text{to} (Re_T)^{11/4} = \text{nodes required in 3D DNS calculations} \tag{3.33}$$

A more precise relationship is as follows (Wilcox 2006; Sodja 2007):

$$N_{3D} = (110 Re_T)^{9/4}. \tag{3.34}$$

The total computational processing unit (CPU) needed to solve 3D DNS problems is proportional to the turbulent Re to the very costly third power, indicating an overwhelming increase in the required computational resources as Re increases:

$$\text{CPU} \quad \propto Re_{\text{T}}^{3}. \tag{3.35}$$

The Courant number for a DNS calculation can also be estimated:

$$\text{Courant number} = \frac{u'\Delta t}{\eta} < 1 \text{ (DNS)} \tag{3.36}$$

where η is the Kolmogorov length scale.

The key advantage of LES is that it calculates the dynamic behavior of the larger eddies, which generally includes the range from integral to Taylor eddies having the lowest κ (see Fig. 3.1). If faster calculations are desired at the expense of accuracy, then the LES calculation can capture the integral eddies and up to a scale that is five to ten times larger than the Taylor length. Therefore, as a useful guide for computational node spacing, the eddy lengths can be estimated from the LIKE algorithm, so that the mesh has node spacing that is suitable for LES simulations. Note that failure to estimate the eddy size to ensure that the mesh elements are sized to adequately model the larger eddies will inevitably result in significant errors, either the analyst risks having node spacing that is close to or even larger than the integral eddies or having elements that are at or smaller than the Kolmogorov length. In the first situation, the calculation will not resolve the larger eddies, and in the latter case, too many computational nodes will be generated, thereby unduly increasing computational time and calculating too much damping.

Likewise, when modeling with DNS, all the length scales ought to be considered (from integral to Kolmogorov). Thus, the mesh must be sufficiently discretized to permit reasonable calculation of the limiting Kolmogorov eddy. In this context, "reasonable" refers to the notion that there are sufficient calculational nodes to obtain spatial resolution for the dynamic behavior of eddies. (This applies to LES and DNS models, but not to RANS models because RANS does not calculate individual eddy behavior.) Therefore, in any given direction, no fewer than two or three computational nodes ought to discretize a single eddy. In two directions, this requires 2^2 to 3^2 computational nodes *per eddy* and up to 2^3 to 3^3 for three dimensions. If the total system volume is known, then it is just a matter of dividing its volume by the volume occupied by a single Kolmogorov eddy to estimate the total number of eddies and thus the total number of computational nodes required by the system. Considering that in 3D, each Kolmogorov eddy requires between 8 and 27 nodes, the total number of necessary computational nodes skyrockets very quickly. Fortunately, these conservative nodal constraints can be loosened a bit, especially as discretization studies are conducted. But clearly, LES and DNS calculations can be very computationally intensive!

Another useful approximation is obtained rather quickly once the average velocity is calculated, such as from the mass flow rate, $\dot{m} = \rho \bar{u} A$. Then, the peak turbulence velocity is estimated as a rule of thumb as (Bird et al. 2007)

$$u_{\text{turb,max}} \approx C\bar{u} = \frac{5}{4}\bar{u}. \tag{3.37}$$

This expression is somewhat conservative, as it assumes that the peak velocity fluctuation approaches 25%. Refer to Table 3.1 for more specific values.

Finally, if the turbulence is anisotropic, but flows over a flat plate, the root-mean-square (RMS) of the turbulent velocity fluctuations can be approximated as

$$\overline{u'^2} : \overline{v'^2} : \overline{w'^2} \approx 4 : 2 : 3. \tag{3.38}$$

This implies that

$$\frac{\overline{u'^2}}{\overline{v'^2}} \approx 2 \tag{3.39A}$$

and

$$\frac{\overline{u'^2}}{\overline{w'^2}} \approx \frac{4}{3}. \tag{3.39B}$$

Substitution of Eqs. 3.39A and 3.39B into Eq. 2.60 (nonisotropic k) implies

$$u' = \sqrt{\frac{1}{3}\left(\overline{u'^2} + \overline{v'^2} + \overline{w'^2}\right)} = \sqrt{\frac{1}{3}\left(\overline{u'^2} + \frac{\overline{u'^2}}{2} + \frac{3}{4}\overline{u'^2}\right)} = \sqrt{\frac{3}{4}\overline{u'^2}}. \tag{3.39C}$$

Wilcox briefly mentioned the following approximations, which are suitable for anisotropic turbulent flow over flat plates (Wilcox 2006):

$$\overline{u'^2} \approx k, \tag{3.40A}$$

$$\overline{v'^2} \approx \frac{2}{5}k, \tag{3.40B}$$

and

$$\overline{w'^2} \approx \frac{3}{5}k. \tag{3.40C}$$

If Eqs. 3.40A, 3.40B, and 3.40C are substituted into Eq. 2.60, then $\overline{u'^2} + \overline{v'^2} + \overline{w'^2} = k + \frac{2}{5}k + \frac{3}{5}k = 2k.$ Therefore,

$$k_{\text{Wilcox,anisotropic,flat plate}} = \frac{1}{2}\left(\overline{u'^2} + \overline{v'^2} + \overline{w'^2}\right). \tag{3.40D}$$

Thus, the turbulent kinetic energy split approximations proposed by Wilcox are not only reasonable and appropriately normalized but compare well with experimental data (Wilcox 2006).

Example 3.5 Consider water flowing in a pipe with $D = 0.1$ m, at an average velocity of 5.5 m/s. The water is at 350 K and 5.7×10^5 Pa. Suppose that you want to use various RANS turbulence models, for which you will need as input the following: k, ε, ω, and ν_t. Compute the four quantities and explain why this flow is highly turbulent. What is the expected peak turbulence velocity?

Solution The water has $\nu = 3.78 \times 10^{-07}$ m^2/s. So,

$$Re_h = \frac{0.1 * 5.5}{3.78E - 07} = 1.45E06, \therefore \text{turbulent.}$$

From Eq. 3.17A,

$$\ell \approx 0.07 D_h = 0.07 * 0.1 = 0.007 \text{ m.}$$

From Eq. 3.26C,

$$I = 0.16\,Re_h^{-1/8} = 0.16(1.45E06)^{-1/8} = 0.0272.$$

Note that if Eq. 3.26B were used instead,

$$I_{\text{meas,pipe center}} = 0.144\,Re_h^{-0.146} = 0.144(1.45E06)^{-0.146} = 0.0181,$$

which is considerably different, and thus somewhat disconcerting (hence the appeal for more research in this area). For now, proceed with the results from Eq. 3.26C. Then, from Eq. 3.27,

$$k = \frac{3}{2}(\overline{u}I)^2 = \frac{3}{2}(5.5 * 0.0272)^2 = 0.0336 \text{ m}^2/\text{s}^2.$$

From Eq. 3.28,

$$\varepsilon = C_\mu \frac{k^{3/2}}{\ell} = 0.09 * \frac{(0.0336)^{3/2}}{0.007} = 0.0787 \text{ m}^2/\text{s}^3.$$

Note that the above four turbulence variables (ℓ, I, k, and ε) were obtained with the four appropriate LIKE equations.
Then, from Eq. 3.29A,

$$\omega = \frac{k^{1/2}}{\ell} = \frac{(0.0335)^{1/2}}{0.007} = 26.1/\text{s}.$$

Next, the turbulent kinematic viscosity can be obtained from the Prandtl-Kolmogorov relationship, Eq. 3.30,

$$\nu_t = \frac{C_\mu k^2}{\varepsilon} = \frac{0.09 * (0.0335)^2}{0.0787} = 0.0013 \text{ m}^2/\text{s}.$$

Regarding degree of turbulence, it is known that for pipes, a flow becomes turbulent at a *Re* of about 2200. For this example, the flow is at $Re = 1.45 \times 10^6$, which indicates a high degree of turbulence. But furthermore, note that the ratio of ν_t vs. ν is extremely large:

$$\frac{\nu_t}{\nu} = \frac{0.0013}{3.78\text{E} - 07} = 3.38E3.$$

As noted in Chap. 2, the turbulent kinematic viscosity can be orders of magnitude higher than the fluid viscosity, thus providing a reasonable way to gauge the degree of turbulence, namely, that
$\frac{\nu_t}{\nu} \gg 1$ (with larger ratios implying a higher degree of turbulence; Eq. 2.48).
Finally, Eq. 3.37 can be used to estimate very quickly the peak turbulent velocity:

$$u_{\text{turb,max}} \approx \frac{5}{4}\bar{u} = \frac{5}{4} * 5.5 = 6.88 \text{ m/s}.$$

Example 3.6 Consider the same system as in Example 3.5. However, the analyst would now like to consider the feasibility of using LES and DNS models. What distance between the computational nodes will be sufficient to resolve the integral, Taylor, and Kolmogorov eddies?

Solution Use either the LIKE MATLAB script that is part of this book or the equations in Sect. 3.4 to obtain the following output:

Kolmogorov eddy size $= 2.88 \times 10^{-05}$ m
Taylor eddy size $= 1.27 \times 10^{-03}$ m
Integral eddy size $= 7.00 \times 10^{-03}$ m

So now, it is just a matter of dividing the length of each eddy by three along the coordinate of interest (for a very conservative approximation, or by two for a coarser estimate). For example, for the Kolmogorov eddy, the conservative approximation in 1D would be

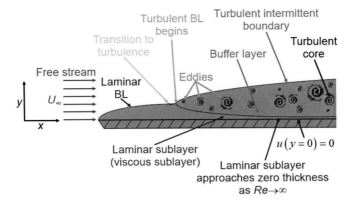

Fig. 3.8 Laminar and turbulent flow over a flat plate

$$\frac{\eta}{3} = \frac{2.88E - 05}{3} = 9.6E - 06 \text{ m.}$$

For the Taylor eddies, the desired nodal distance for this situation in 1D is

$$\frac{\lambda}{3} = \frac{7.00E - 03}{3} = 2.33E - 03 \text{ m.}$$

In summary, the LIKE algorithm is quite useful, so perhaps the user will truly "like" it!

3.6   The Viscous Sublayer, the Log Laws, and Calculating Y^+

Flow near a wall is complex, to say the least. To simplify the situation, imagine a laminar, uniform liquid flow at U_∞ approaching parallel to a flat plate, as shown in Fig. 3.8. The flow on the LHS is laminar; until at some point as the flow travels further to the right, it begins exhibiting instabilities and eventually becomes turbulent.

Proceeding perpendicularly from the wall ($y = 0$) to the turbulent core, the flow can be characterized into various regions that exhibit different behavior, starting with the laminar sublayer, on to the buffer layer, followed by the log layer, and finally, the defect layer. These regions are shown in Fig. 3.9, and are discussed next. Note that the layer regions are an engineering construct (a convenience), with the layers opportunely separated by dimensionless variable y^+ (to be defined later). Another dimensionless variable, u^+, forms theoretically and experimentally validated functions that are based on y^+ and are unique for each layer. Thus, $u^+ = u^+(y^+)$. u^+ will be defined shortly.

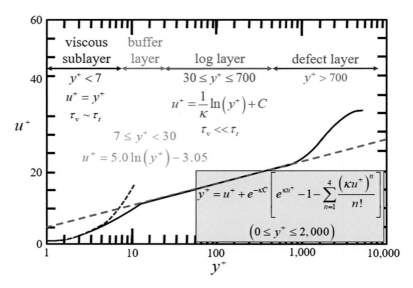

Fig. 3.9 Laminar (viscous), buffer, log, and defect layers

- *The viscous sublayer*

At the wall, a liquid has zero velocity due to "no slip." Thus, barring microbubbles formed by air or other miscellaneous gases, the liquid molecules will be attached to the wall, whether smooth or rough. Then, at some point, the viscous flow sheets begin to slip right by each other, without mixing, because they are laminar. This behavior normally occurs when y^+ is in the range of just slightly above 0, to as high as 5 to 8, with 7 being a more typical value. This is shown in the bottom LHS of Fig. 3.9. Thus, despite the fact that the flow in the sublayer occurs *beyond* the turbulence transition point and well into the turbulence region, the sublayer behavior consists of thin, laminar sheets without eddies. This is the case because near the wall, the viscous force dominates. It is rather straightforward to imagine that if the fluid velocity is zero at the wall and the velocity increases gradually away from the wall until turbulence is achieved, then there must be a laminar region where the flow is bounded by $0 \le Re < Re_{crit}$; such region is called the viscous sublayer or laminar sublayer.

That said, it is interesting that studies have shown velocity fluctuations in the viscous sublayer (Fage and Townend 1932; Chapman and Kuhn 1986; Hinze 1987; Mansour et al. 1988). These observations include experimental measurements of irregular fluctuations in w' (z direction velocity component) using dopants. Both experiments and DNS simulations show that k (the turbulence kinetic energy of eddies) is a function of y^{+2} in the range of $0.3 < y^+ < 3$. It is surmised here that such fluctuation observations could be due to the momentum effects arising from intermittency occurring at the interface between the viscous sublayer and the buffer layer. If so, the viscous sublayer is purely laminar in the narrow regions not affected by this particular type of intermittency. Intermittency has already been noted in the interface between the log layer and the free stream (Corrsin and Kistler 1954; Klebanoff 1954; Wilcox 2006). Another possibility for the experimental observation of fluctuations in the viscous sublayer may result from eddies in the buffer layer that are sufficiently large, energetic, and move toward the wall, such that they occasionally collide onto the laminar flow near the wall; it is noted that the buffer layer has a high degree of isotropy and turbulence intensity (Chapman and Kuhn 1986). Moreover, note that the relationship between the flow velocity in the viscous sublayer and distance from the wall is linear, thereby alluding to its laminar nature; turbulence is associated with nonlinear velocity. Another possibility is the existence of streamwise vortices, which may exhibit linear Rankine vortex-like motion in the domain of $0 < y^+ < 5$. Alternatively, vortices located at $y^+ = 20$ from the wall, but that are large enough to have a radius of $y^+ = 15$, could generate the spurious fluctuations (Kim et al. 1987; Rodriguez et al. 2012).

Finally, as will be shown later in this section, the viscous sublayer is thinner than Kolmogorov eddies, so the sublayer is not able to support eddies; they simply don't fit in this small region! In summary, the viscous sublayer, though crucial for turbulence CFD modeling, is not turbulent, despite existing in the "turbulence" region (e.g., at distance x, for flow in a flat plate where $Re_x > Re_{\text{crit}}$)! *However, intermittency and/or turbulence intensity may occasionally make the laminar viscous sublayer appear as turbulent.*

At last, the *dimensionless* variable y^+ is defined as follows:

$$y^+ = \frac{yu_*}{\nu} \tag{3.41}$$

where

$$u_* \equiv \sqrt{\frac{\tau_w}{\rho}}. \tag{3.42}$$

The variable u_* is commonly called the "friction velocity," "wall friction velocity," "wall shear stress velocity," and "shear velocity." It is also commonly represented as u_τ in the literature and has SI units of m/s. The variable y is the distance perpendicularly away from the wall, with $y = 0$ defined as the wall interface.

The dimensionless variable y^+ is a very interesting *physical* quantity, as it is analogous to the key fluid dynamics dimensionless quantity, Re; in particular, both are based on the product of length and velocity, which is then divided by kinematic viscosity:

$$\begin{cases} y^+ \leftrightarrow Re \\ \dfrac{yu_*}{\nu} \leftrightarrow \dfrac{x_{\text{char}} u_{\text{char}}}{\nu} \end{cases} \cdot \tag{3.43}$$

In this context, τ_w is the wall shear stress at $y = 0$, has SI units in kg/m-s^2, and is defined as

$$\tau_\mathrm{w} = \mu \frac{\partial u}{\partial y}\bigg|_{y=0}. \tag{3.44}$$

Unless an analytical solution is known for $u(y)$, Eq. 3.44 requires an experimental or theoretical velocity expression or input from a previous CFD simulation so that τ_w can be evaluated. Fortunately, the literature has many expressions for calculating τ_w directly, as will be noted later in this section.

Now that y^+ is defined and is calculable from Eq. 3.41, the dimensionless velocity distribution u^+ for the viscous sublayer can be calculated for a smooth wall as a simple linear function of y^+:

$$u^+ = y^+. \tag{3.45}$$

The above relationship is valid for the viscous sublayer, which is defined as the region where $y^+ < 7$.

The linear expression for the viscous sublayer has been corroborated extensively with experimental data. But most importantly, the viscous sublayer is a theoretical solution for the Navier-Stokes momentum equation (Hinze 1987; Oberlack 2001). Therefore, not surprisingly, the viscous sublayer is typically accurate to within 1% of the measured flow. Thus, its simplicity should not be underestimated.

If the wall is rough, then the surface roughness λ must be accounted as well (Hinze 1987; Wilcox 2006).

The dimensionless velocity u^+ is defined as follows:

$$u^+ \equiv \frac{\bar{u}}{u_*}. \tag{3.46}$$

At this point, y^+ is obtained from Eq. 3.41, u_* is known from Eq. 3.42, and u^+ is acquired from Eq. 3.45. Then, from Eq. 3.46, the sought-after turbulence time-averaged velocity $\bar{u}$ is computed. In this context, $\bar{u}$ is the viscous sublayer velocity in m/s if SI units are considered. That is, the ultimate goal is to calculate $\bar{u}$, which requires u_*, u^+, and y^+; a few examples will help clarify the process.

Note that each layer shown in Fig. 3.9 *requires its own u^+ formula as a function of y^+ because each layer exhibits different flow physics.* But u_* does not change, so once its value is known, it can be used in any of the layer formulas. Furthermore, Eq. 3.46 is appropriate for the entire y^+ domain.

Note that most engineering viscous sublayers are very thin, as they tend to be mostly on the order of micro-m to just a few mm in thickness; this is shown in various examples found later in this section.

- *The buffer layer*

The buffer layer is considered as the "intermediate layer" between the viscous sublayer and the log layer. The buffer layer therefore exhibits a hybrid behavior, having both laminar and turbulent characteristics. It is rich in coherent structures, such as low-velocity streaks, streamwise vortices, and hairpin vortices (Kawahara 2009). Its y^+ range is typically as low as 5 to 8 and as high as 30. One thing that makes the buffer layer unique and powerful is its sharp velocity gradient, which enables the turbulence production term to have the highest magnitude within the entire flow. Recall that the viscous sublayer is too small to sustain eddies, and it is noted that the log layer has a flatter velocity gradient than the buffer layer, so eddy production is the largest in the log layer. The turbulence intensity has the largest magnitude in the buffer region as well, thus implying that it generates the largest eddies. Therefore, the buffer layer is characterizable as the region with the highest degree of turbulence mixing.

A reasonable approximation for the buffer layer velocity distribution is (von Karman 1939, 1940)

$$u^+ = 5.0 \ln \left(y^+ \right) - 3.05, \tag{3.47}$$

which is valid for $5 \leq y^+ \leq 30$.

- *The log layer*

The "log layer" y^+ ranges from 30 to 700 and is a region where the viscous and inertia terms are much smaller than the turbulent shear terms. Like the buffer layer, this region is characterized by turbulence intensity and production, but to a lesser degree, because it has a smaller velocity gradient.

This layer is fondly referred to as the "log layer" because it includes a natural log expression for its velocity distribution,

$$u^+ = \frac{1}{\kappa} \ln \left(y^+ \right) + C, \tag{3.48A}$$

which is valid for $30 \leq y^+ \leq 700$. For smooth walls,

$$\kappa \approx 0.41 \tag{3.48B}$$

and

$$C \approx 5.0. \tag{3.48C}$$

Recent controversy regarding the log law has resulted in much research. That said, the log law has received confirmation using DNS and statistical methods, and this also applies to power laws (Buschmann and Gad-el-Hak 2003; Wilcox 2006). *The log equation is very accurate if the flow is not detached or has large pressure gradients.* Readers interested in flows with large pressure gradients are referred to (Shih et al. 1999) for a generalized form of the log wall. When used in the proper y^+

range, the log law equation typically compares to within $\pm 5\%$ of experimental values. *Fortunately, the log law is also applicable for both internal and external flows.* Most importantly, it is well-known that the log equation can be derived as an exact solution of the Navier-Stokes momentum PDE (Kolmogorov 1942; Nazarenko 2000; Oberlack 2001).

The log equation is sometimes referred to as the "law of the wall," and its most formal and extensive name is the "von Karman-Prandtl universal logarithmic velocity distribution." However, the inclusion of "wall" in its naming convention is unfortunate, as the flow in question is clearly not at the wall; other researchers have held the same misgivings for the naming convention (Schlichting and Gersten 2000). Nevertheless, because of the log equation's excellent correlation with experimental data for both internal and external flows, the term "universal" is appropriate. Because it is an analytical solution for Navier-Stokes, "law" is reasonable as well. Therefore, "universal log law velocity distribution" is recommended, and "log law" is acceptable for brevity.

- *The defect layer*

For y^+ greater than 700, the measured turbulent velocity exceeds the magnitude predicted by the log law and is therefore called the "defect layer." The Spalding log law formula (discussed next) can be used for the defect layer in the range of $700 < y^+ \leq 2000$. Various formulations exist for the calculation of the defect velocity; the interested reader is referred to Wilcox (2006).

- *The Spalding log law formula*

A single equation was proposed half a century ago that covers y^+ from 0 to 1000 (Spalding 1961; White 1991; Patankar and Minkowycz 2017). *Recent investigators have shown that the Spalding equation is suitable for y^+ up to 2000* (Osterlund 1999; McKeon et al. 2004; Mandal and Mazumdar 2015).

Spalding's equation has several amazing properties:

- A *single* equation covers the four layers (viscous, buffer, log, and defect) for $0 \leq y^+ \leq 2000$.
- Collapses to the linear viscous sublayer formulation as u^+ approaches 0.
- Collapses to the log law for large u^+.
- Subtracts the first five terms of a Taylor series expansion for $e^{\kappa u^+}$ and hence can be simplified by only considering the higher order terms, beginning with $n = 5$.
- Has excellent agreement with experimental data.

Spalding's equation is as follows:

$$y^+ = u^+ + e^{-\kappa C}\left[e^{\kappa u^+} - 1 - \sum_{n=1}^{4} \frac{(\kappa u^+)^n}{n!} \right] \quad \text{for } 0 \leq y^+ \leq 2,000. \quad (3.49A)$$

For this equation, it is recommended that the following κ and C values are used:

$$\kappa \approx 0.4 \qquad\qquad (3.49\text{B})$$

and

$$C \approx 5.5. \qquad\qquad (3.49\text{C})$$

A quick inspection shows these values are similar to those used in the log wall formula.

- *The turbulent core and intermittency*

Depending on the author, the turbulent core loosely includes the log and defect layers. Thus, the core represents the production region that generates mid to large eddies and whose span reaches the top of the turbulent boundary layer (recall that $y = 0$ is at the wall surface, the bottom).

Note that the turbulent boundary edge is typically shown as a smooth parabolic curve, indicating its "time-averaged" nature. However, if instantaneous snapshots were taken, the boundary would not be smooth but would instead have a discontinuous, jagged distribution as a result of the random velocity fluctuations that occur as eddies are generated, transported, and decayed. Thus, at the region where the turbulence boundary layer ends, there are myriads of superimposed dynamic subregions where the turbulent boundary layer edge not only oscillates randomly between laminar and turbulent flow, but its spatial distribution also changes WRT time with irregular, nondeterministic patterns and motions. This is referred to as "intermittency," and its effect has been incorporated successfully into many turbulence models (Corrsin and Kistler 1954; Klebanoff 1954; Wilcox 2006).

- *Calculation of y^+*

At this point, it should be very clear that the spatial distribution of computational nodes based on their y^+ location is crucial, particularly because of the different flow physics within turbulent flow. This assertion is important if accurate drag and lift calculations are desired, as well as accurate calculations for surfaces involving strong wall shear, swirl, rotational surfaces, turbomachinery, heat transfer, etc.

To find the value of y (in meters) for any desired dimensionless value of y^+, first calculate Re based on the characteristics of the flow. (Refer to Sect. 2.6.1 for recommendations for applying Re onto systems with internal and external flows.) At this point, it is necessary to obtain the appropriate characteristic length and velocity, x_{char} and u_{char}, so they properly reflect their impact on Re:

$$Re \equiv \frac{x_{\text{char}} u_{\text{char}} \rho}{\mu}. \qquad\qquad (3.50)$$

The next step is to calculate the skin friction, C_f, which is defined as

$$C_f \equiv \frac{\tau_w}{\frac{1}{2}\rho u_{char}^2}. \qquad (3.51A)$$

At this point, the skin friction is calculated based on whether the flow is internal or external. Traditionally, C_f is calculated for the skin friction of exterior flows over a wall, while a friction factor f is calculated for the skin friction of internal flows (Hoerner 1992). The boundary layer on an external flow can grow indefinitely, while for internal flows, the boundary layer grows until the duct becomes "full of boundary layer" if the duct is large enough; hence different x_{char} are used in Re, depending on the flow geometry. The two cases are considered next.

- *External flows*

  The Schlichting formula is valid up to $Re_x \leq 10^9$ for smooth *external* plate flow:

$$C_f = [2\log_{10}(Re_x) - 0.65]^{-2.3} \qquad (3.52A)$$

where x refers to the location along the flat plate (or curved surface), that is, it tracts how far the flow has traveled. Usually, $x = 0$ is at the beginning of the plate on the LHS, where the flow first comes into contact with the plate, as shown in Fig. 3.8, while $x = L$ refers to the RHS extreme (opposite end of the plate).

Prandtl developed a more simplified skin friction formula for a smooth, flat plate:

$$C_f = 0.074\, Re_x^{-1/5}. \qquad (3.52B)$$

Despite their formulation differences, an overlay of Eqs. 3.52A and 3.52B shows a remarkable similitude over a wide Re range.

- *Internal flows*

  Blasius determined a simple relationship for smooth walls with *internal* flow:

$$C_f = f_{Darcy} = 0.316\, Re_h^{-1/4} \qquad (3.53A)$$

in the domain of

$$2500 \leq Re_h \leq 1 \times 10^5,$$

where Re_h is based on $x_{char} =$ hydraulic diameter (refer to Eq. 3.50).

A more complex formulation was proposed by Moody, which is valid for internal flows with smooth to rough walls:

$$C_f = f_{\text{Darcy}} = 0.0055 \left[1 + \left(2 \times 10^4 \frac{d}{D} + \frac{1 \times 10^6}{Re_{\text{h}}} \right)^{1/3} \right], \qquad (3.53\text{B})$$

which is subject to

$$4000 \leq Re_{\text{h}} \leq 5 \times 10^8$$

and

$$0 \leq \frac{d}{D} \leq 1 \times 10^{-2}.$$

Note that the above Moody equation typically uses λ as the wall roughness, instead of d, as is done here. However, λ is the Taylor length. Furthermore, though ε is typically used in the literature as a metric for surface roughness, the turbulence literature uses ε for dissipation, hence the notation conundrum.

In any case, recall from Sect. 2.2 that the Darcy friction factor is four times the Fanning friction factor, so care must be taken to ensure that the proper friction factor is applied:

$$f_{\text{Darcy}} = 4 f_{\text{Fanning}}. \qquad (3.54\text{A})$$

Failure to use a consistent expression will result in the skin friction factor being off by four, causing u_* and the turbulent velocity to be incorrect. A clue as to which factor is being referred is that if Eq. 3.54B is present, then Darcy is being used; if Eq. 3.54C is used, then the Fanning friction factor is the appropriate friction factor. Another clue is that chemical engineers generally use Fanning, while most other engineering disciplines tend to use Darcy. In addition, the Moody chart is based on Eq. 3.54B and hence its name. Some clever engineers can tell the difference based on the relative magnitudes of the friction factors. These relationships are as follows:

$$\Delta P = f \frac{L}{D} \frac{\rho U^2}{2} \equiv f_{\text{Darcy}} \frac{L}{D} \frac{\rho U^2}{2}, \qquad (3.54\text{B})$$

and

$$\Delta P = \frac{f}{4} \frac{L}{D} \frac{\rho U^2}{2} \equiv f_{\text{Fanning}} \frac{L}{D} \frac{\rho U^2}{2}. \qquad (3.54\text{C})$$

An additional caution: many friction factor equations are commonly listed in the literature using both the Darcy and Fanning format, so it must never be assumed that a formula is associated with either. For example, the reader may encounter the following explicit formulation, $f = 0.316 Re_{\text{h}}^{-1/4}$; which version is this? The

Table 3.2 Calculation of y at the desired value for y^+

	Equation	Unknown
1	$Re = \frac{x_{char} u_{char}}{\nu}$	Re
2	*External flow:* $C_f = [2\log_{10}(Re_x) - 0.65]^{-2.3}$ *Internal flow (with Darcy formulation so that the shear equation for τ_w in row 3 below is applicable):* $C_f = f_{Darcy} = 0.0055\left[1 + \left(2 \times 10^4 \frac{d}{D} + \frac{1\times10^6}{Re_h}\right)^{1/3}\right]$	C_f, Re
3	$\tau_w = C_f \frac{\rho u_{char}^2}{2}$	τ_w, C_f
4	$u_* = \sqrt{\frac{\tau_w}{\rho}}$	u_*, τ_w
5	$y = \frac{y^+ \nu}{u_*}$ $\left(\text{Backtracked from } y^+ \equiv \frac{y u_*}{\nu}\right)$	y, u_*

Blasius equation is reported in the literature for $2000 \leq Re_h \leq 1 \times 10^5$ as the following two expressions:

$$f_{\text{Blasius,Darcy}} = 0.316\, Re_h^{-1/4}, \tag{3.55A}$$

$$f_{\text{Blasius,Fanning}} = 0.0791\, Re_h^{-1/4}. \tag{3.55B}$$

Of course, Darcy is four times larger than Fanning, but not all writers explicitly specify which equation is Darcy and which one is Fanning, caveat emptor.

In any case, once C_f is known, it can be used to solve for the wall shear τ_w from Eq. 3.51A, which, for convenience, is now rewritten as

$$\tau_w = C_f \frac{\rho u_{char}^2}{2}. \tag{3.51B}$$

Note that for *internal* flows, Eq. 3.51B must be used along with the Darcy friction factor (refer to the discussion in Rows 2 and 3 of Table 3.2).

Wrapping it up, the variable u_* can now be found from Eq. 3.42. Then, y^+ is obtained from Eq. 3.41, thereby allowing for the computation of u^+ from the appropriate layer equation; this is the case because $u^+ = u^+(y^+)$. And now that u^+ is known, Eq. 3.46 is easily solved, yielding $\bar{u}$. This multistep process can be a little daunting at first, so various examples are worked out later in this section (refer to Examples 3.7, 3.8, 3.9 and 3.10).

Because various equations are required in this process and it is important to show that n independent equations are used to solve for n unknown variables, for convenience, Table 3.2 lists all the equations and unknowns needed to calculate y for any desired value of y^+. Table 3.2 assumes that x_{char}, u_{char}, ρ, ν, d (wall roughness, if any), and flow type (internal or external) are known. In addition, the user decides a priori what value of y^+ is desired so that y can be calculated. Some guidelines are as follows:

- $y^+ = 0.05$ for the first computational node for DNS calculation of channel flow at $Re = 3300$.
- $y^+ = 1$ for the first computational node. This guideline is recommended by Fluent (Fluent 2012) and is typically used in most RANS and LES models.
- $y^+ = 5$ for the first computational node if the Wilcox 2006 k-ω model is used.
- $y^+ = 7$ to figure out where the buffer layer ends.
- $y^+ = 30$ to determine where the buffer layer ends.
- Peak y^+ (or y^+ of interest) to ensure that the appropriate layer equation is used.
- And so forth.

Summary for calculating y^+

As shown in Table 3.2, there are five unknowns for this process: Re, C_f, τ_w, u_*, and y. Therefore, five independent equations are required. For convenience, they are listed in the table in the order that they are typically solved. Thus, Table 3.2 serves as a roadmap for the calculation of y for any desired y^+. Where appropriate, the user may decide to employ a more relevant (system-specific) expression for C_f.

Example 3.7 Suppose an engineer wants to know how thick the viscous sublayer is for a fluid with a given kinematic viscosity ν and friction velocity u_*. Assume the sublayer extends up to $y^+ = 8$. How far away from the wall does the sublayer extend? (Refer to Fig. 3.8 for the orientation of y in terms of its relationship to the wall.)

Solution From Eq. 3.41,

$$y^+ = \frac{y u_*}{\nu}.$$

In this case, solving for y and inserting the value of interest for y^+ yields

$$y = \frac{y^+ \nu}{u_*} = \frac{8\nu}{u_*}.$$

Example 3.8 It is desired to conduct CFD for flow that is parallel over a flat plate that is 1.0 m long. The fluid is a gas with $\rho = 1.5$ kg/m^3 and $\mu = 2.4 \times 10^{-7}$ kg/m-s, and the flow has a free stream velocity of 5.0 m/s. If the first computational node is placed at $y^+ = 1$, how far away from the wall will it be?

Solution $\nu = \mu/\rho = 1.6 \times 10^{-7}$ m^2/s. u_{char} is 5.0 m/s and x_{char} is 1.0 m. This yields

$$Re_x = \frac{x_{char} u_{char}}{\nu} = \frac{(1.0 \text{ m})(5.0 \text{ m/s})}{1.6 \times 10^{-7} \text{ m}^2/\text{s}} = 3.13 \times 10^7.$$

So, the flow is clearly turbulent and the Schlichting formula (Eq. 3.52A) is valid. Note that to be conservative, Re is calculated at its maximum value, which is achieved at $x = L$; this ensures that the first computational node is as close to the wall as necessary. Once Re_x is known, the skin friction is calculated using Eq. 3.52A:

$$C_f = [2\log_{10}(3.13 \times 10^7) - 0.65]^{-2.3} = 2.19 \times 10^{-3}.$$

The wall shear can now be calculated from Eq. 3.51B:

$$\tau_w = C_f \frac{\rho u_{char}^2}{2} = 2.19 \times 10^{-3} \frac{(1.5 \text{ kg/m}^3)(5.0 \text{ m/s})^2}{2} = 4.10 \times 10^{-2} \text{ kg/m-s}^2.$$

At this point, u_* can now be found from Eq. 3.42:

$$u_* = \sqrt{\frac{\tau_w}{\rho}} = \sqrt{\frac{4.10 \times 10^{-2} \text{ kg/m-s}^2}{1.5 \text{ kg/m}^3}} = 0.165 \text{ m/s}.$$

Finally, y is solved from Eq. 3.41, yielding

$$y = \frac{y^+\nu}{u_*} = \frac{(1)(1.6 \times 10^{-7} \text{ m}^2/\text{s})}{0.165 \text{ m/s}} = 9.68 \times 10^{-7} \text{ m}.$$

Example 3.9 Using the input and results from Example 3.8, how large is the viscous sublayer, and how does it compare in this case with the Kolmogorov eddy size?

Solution Suppose the viscous sublayer extends to $y^+ = 8$. Then, $y_{(y^+=8)} = 8y_{(y^+=1)} = 8*9.68 \times 10^{-7}$ m $= 7.74 \times 10^{-6}$ m. Applying the LIKE algorithm (see Sect. 3.4), $\eta = 3.85 \times 10^{-5}$ m. Thus, for this example, the viscous sublayer is five times smaller than the Kolmogorov length scale and is therefore unable to accommodate Kolmogorov eddies—they simply don't fit! It might be noted that out of hundreds of cases considered to date by the present author, not a single case has been found where the viscous sublayer was larger than the Kolmogorov length scale. Furthermore, as shown in Fig. 3.10 for turbulent flow over a flat plate, the Kolmogorov scale is always larger than the viscous boundary layer for all Re.

Example 3.10 Consider a turbulent flow in a smooth pipe with a mass flow rate at 0.05 kg/s, $\rho = 15.0$ kg/m^3, $\mu = 3.89 \times 10^{-5}$ kg/m-s, and $D = 0.1$ m. Find the peak turbulent velocity.

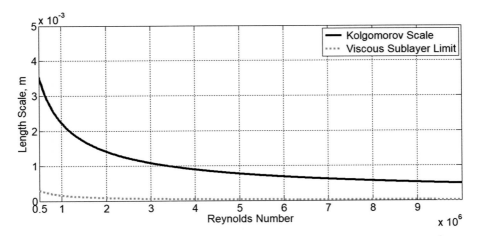

Fig. 3.10 Viscous sublayer vs. Kolmogorov eddy size

Solution $\nu = \mu/\rho = 2.59 \times 10^{-6}$ m^2/s. The flow area is

$$A = \pi \frac{D^2}{4} = \pi \frac{(0.1)^2}{4} = 7.85 \times 10^{-3} \text{ m}^2.$$

The average turbulent velocity can be obtained from the mass flow rate equation:

$$\dot{m} = \rho \bar{u} A$$

or

$$\bar{u} = \frac{\dot{m}}{\rho A} = \frac{0.05 \text{ kg/s}}{(15.0 \text{ kg/m}^3)(7.85 \times 10^{-3} \text{ m}^2)} = 0.425 \text{ m/s}.$$

This velocity represents the average velocity across the tube. *Re* can now be calculated so that y^+ can be obtained, and from it, the turbulent velocity at the pipe centerline:

$$Re = \frac{(0.1 \text{ m})(0.425 \text{ m/s})}{2.59 \times 10^{-6} \text{m}^2/s} = 1.64 \times 10^4.$$

Now calculate the friction factor based on *Re*:

$$C_f = f = 0.0055\left[1 + \left(2 \times 10^4 \frac{\lambda}{D} + \frac{1 \times 10^6}{Re_h}\right)^{1/3}\right]$$

$$= 0.0055\left[1 + \left(2 \times 10^4 \frac{0.0}{0.1} + \frac{1 \times 10^6}{1.64 \times 10^4}\right)^{1/3}\right] = 0.0068.$$

Note that the friction formula based on Darcy is consistent for use with the wall shear formula below:

$$\tau_w = C_f \frac{\rho u_{char}^2}{2} = 0.0068 \frac{(15.0 \text{ kg/m}^3)(0.425 \text{ m/s})^2}{2} = 0.0092 \text{ kg/m-s}^2.$$

Then, the friction velocity is

$$u_* = \sqrt{\frac{\tau_w}{\rho}} = \sqrt{\frac{0.0092 \text{ kg/m-s}^2}{15.0 \text{ kg/m}^3}} = 0.0247 \text{ m/s}.$$

The peak velocity is at the centerline, so $y = D/2 = 0.05$ m. At this point, y^+ can finally be calculated:

$$y^+ = \frac{yu_*}{\nu} = \frac{(0.05 \text{ m})(0.0247 \text{ m/s})}{2.59 \times 10^{-6} \text{m}^2/s} = 477.$$

The value of y^+ allows for the log wall equation to be used, in order to solve for u^+:

$$u^+ = \frac{1}{\kappa} \ln\left(y^+\right) + C = \frac{1}{0.41} \ln(477) + 5.0 = 20.0.$$

Now it is possible to solve for $\bar{u}$ from Eq. 3.46, so the turbulent velocity at the center of the pipe is

$$\bar{u} = u^+ u_* = 20.0(0.0247 \text{ m/s}) = 0.494 \text{ m/s}.$$

Alternatively, the turbulent velocity can be more easily approximated from Eq. 3.37:

$$u_{\text{turb,max}} \approx \frac{5}{4}\bar{u} = \frac{5}{4}(0.425 \text{ m/s}) = 0.531 \text{ m/s},$$

which shows that the agreement is heartwarming vs. the log wall.

Finally, as a couple of brief side notes in relationship to Fig. 3.8, it is pointed out that the approaching flow need not begin as laminar; a situation where a wind tunnel generates turbulent flow using grid spacers could have been considered, whereby the

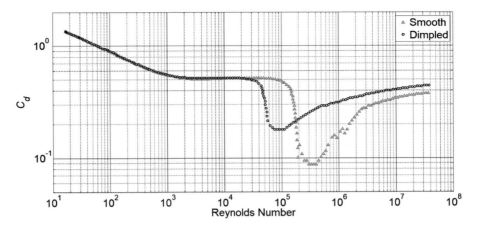

Fig. 3.11 Drag coefficient as a function of *Re* for smooth and dimpled golf balls (flow over a sphere)

flow over the plate starts with a *turbulent* boundary layer, instead of transitioning from laminar to turbulent flow. Furthermore, a liquid was chosen for this example, so the discussion of microbubbles could be introduced. Microbubbles attach to a wall and can act as a lubricant as the liquid slips past the microbubbles that are attached to the wall (Tabeling 2009; Bolaños and Vernescu 2017). This mechanism is a key factor in the usage of shark skin for the reduction of flow drag; refer to the next section for more discussion on this fascinating subject.

3.7   Dimples, Shark Skin, and Surface Engineering for Drag Reduction

It is well-known that as *Re* increases, the coefficient of friction decreases in the laminar region and generally decreases in the turbulent region, as shown in Fig. 3.11. But in some geometries (notably round surfaces, such as cylinders, spheres, and certain airfoils), the drag coefficient drops precipitously at a critical point under turbulent *Re*; this is known as the drag crisis (or the Eiffel paradox, after its discoverer (Eiffel 1913)). At this drag transition point, the wake behind a round object becomes smaller, thereby significantly reducing the frictional drag (Rodriguez et al. 2015). This is one of various key design parameters for golf balls, as they are intended to travel in *Re* space in the range of the critical drag domain, as shown in the following example.

Example 3.11   Consider a warm spring day at 70 °F (294.3 K) at a golf course at sea level. The official golf ball diameter is 0.0427 m. Suppose a lethargic amateur can hit the ball to 50 miles per hour (22.4 m/s). Will the ball be in the drag crisis domain?

Repeat the exercise for a professional golfer who can hit the golf ball to 180 miles per hour (80.5 m/s).

Solution The air kinematic viscosity is 1.52×10^{-5} m²/s. Getting Re is straightforward, and its magnitude is compared with the drag curve for the *dimpled* golf ball in Fig. 3.11.

For the listless amateur,

$$Re = \frac{x_{char} u_{char}}{\nu} = \frac{0.0427 * 22.4}{1.52 \times 10^{-5}} = 6.29 \times 10^4.$$

For the professional golfer,

$$Re = \frac{0.0427 * 80.5}{1.52 \times 10^{-5}} = 2.26 \times 10^5.$$

As one of life's "mysteries," the amateur and the professional golfers will indeed hit their dimpled golf balls such that they travel within the minimum and maximum range of the drag crisis regime, respectively (notice where Re for the two cases compares vs. Fig. 3.11). Therefore, both drastically different swings will benefit from significantly reduced air drag! Coincidence? Notice that if the golf ball were *smooth*, then all amateurs would never enter the drag reduction domain, because the minimum Re shifts at $\sim 1 \times 10^5$. Clearly, modern dimpled golf balls were designed with mass commercialization in mind!

Thus, the Eiffel paradox initiates the drag crisis, and the dimples precipitate (shift) the effect at a lower Re. But, that is not to say that dimpling cannot reduce drag as well, because it can, when done correctly.

In the golf world, people figured out centuries ago that a used golf ball travels farther than a new, smooth ball. Through the years, people have discovered that golf balls with dimples can travel nearly twice as far (all other things being equal) because the dimples reduce the surface drag (Baek and Kim 2013). Curiously, n engineers will provide at least $n + 1$ different explanations as to how this is accomplished. However, engineers generally agree that the boundary layer is reduced through dimpling, and the location at which detachment occurs on the golf ball is moved further downstream. Other effects may occur, including the rotation of the golf ball through the Magnus effect. Many researchers have recently used experiments and CFD to investigate drag reduction in golf balls, with just a few listed here for the interested reader (Ting 2003; Baek and Kim 2013; van Nesselrooij et al. 2016). However, the Magnus effect is not present in dimpled *stationary* fixed-wing airfoils, and yet they exhibit drag reduction, as will be discussed next. This indicates that various complex physics are involved in dimpled drag reduction.

Again, consider Fig. 3.11, which shows the drag coefficient as a function of Re for smooth and dimpled golf balls (or drag coefficient for flow over smooth and non-smooth spheres). For Re in the range of 1000 to about 100,000, the drag coefficient for a smooth sphere is about 0.47 to 0.5 (Hoerner 1992). But for Re

slightly higher than 100,000, a "critical point" is reached, whereby the drag drops precipitously by a factor of three for the dimpled ball and by a factor of five for the smooth ball. Note also that the dimples cause the golf ball to reach the critical point sooner in *Re* space, which essentially allows the golf ball to reach the highly reduced drag region at a lower velocity. (That is, the golf ball's characteristic length is constant and so are the physical properties of air at the moment the ball was struck, so lower *Re* implies lower velocity.)

That said, is it possible for a non-spherical geometry to benefit from dimpling? That is, can curved surfaces and even flat surfaces experience a drag decrease through judicious dimpling? The answer is yes, and many useful engineering benefits have been observed experimentally.

Dimpling has been applied to airfoils at subsonic to hypersonic Mach number (*Ma*), with many outstanding results. For example, in the subsonic range, wind tunnel experiments show drag reduction and increased angle of attack (Prasath and Angelin 2017). In the transonic range, with *Ma* between 0.8 and 1.4, decreased pressure was observed in dimpled vs. undimpled surfaces (Kontis and Lada 2005). Shock wave attenuation was observed at *Ma* = 1.4 because of dimpled surfaces (Kontis et al. 2008). In the supersonic range at *Ma* = 2.2, reduced lambda shock was observed in dimpled surfaces (Sekaran and Naik 2011). Finally, in the hypersonic range, experimenters noted a 20% skin friction reduction and reduced flow trail at the downstream surfaces at *Ma* = 5 for dimpled vs. undimpled surfaces (Babinsky and Edwards 1997). More recent experiments at *Ma* = 6 showed that dimpled surfaces had an impact on eddy spacing, increased flow symmetry, and increased heat transfer (Abney et al. 2013).

A theory-founded methodology ("right-sized dimpling," RSD) was developed recently to calculate dimple size and distribution based on system characteristics (Rodriguez 2017). RSD output was first compared with commercial golf ball dimpling and with a dimpled heat transfer experiment (Choi et al. 2013), to validate its predictive capability. Thereafter, RSD was applied to a CFD model of a Ford Mustang hood (Rodriguez et al. 2017). The simulations showed 25% less drag for the dimpled vs. the undimpled hood. In addition, it was shown that the flow pattern at the back (near the trunk) had a smaller wake pattern, as shown in Fig. 3.12. The simulations were based on an interesting, though informal experiment conducted by the "MythBusters," whereby a dimpled car showed an 11% increase in fuel efficiency (MythBusters 2009). Of course, even higher fuel savings can be achieved because the MythBusters dimpling pattern and geometry were ad hoc, as opposed to using theory-based engineered dimpling, such as RSD. Thus, the potential for reduced vehicle drag and fuel savings is significant. However, it is likely that most car buyers will not be attracted by the aesthetics of a dimpled vehicle. On the other hand, selective dimpling of certain strategic vehicle zones ought to reduce such reluctance.

Besides drag reduction, dimpling offers the additional benefit of enhanced heat transfer, as evidenced by both experiments (Abney et al. 2013; Choi et al. 2013) and CFD calculations using RSD (Rodriguez 2016, 2017). As noted in Fig. 3.13, judicious dimpling patterns exhibit significant heat transfer enhancement. This is

Fig. 3.12 Drag reduction in
a vehicle using dimpling

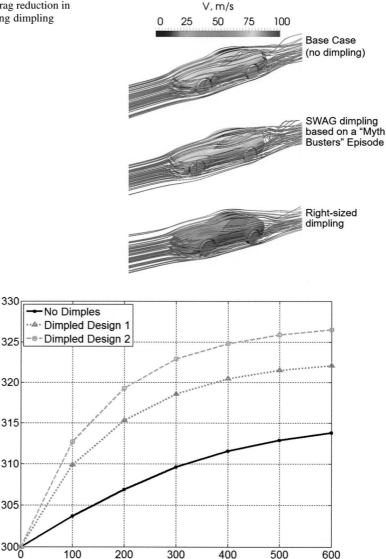

Fig. 3.13 Significant heat transfer increase as a result of right-size surface dimpling

not surprising, because dimpling can bring turbulence closer to the wall, and
therefore, more heat is transferred from the wall and carried away by the fluid.
Figure 3.13 shows the impact of dimpling on a solar energy collector.

But, can flat surfaces benefit from dimpling, especially when the Eiffel paradox
presumably applies to highly curved surfaces? Yes, recent experiments have shown

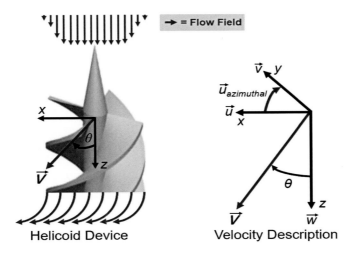

Fig. 3.14 Helicoid swirl device and associated velocity distribution

that dimpled *flat* surfaces exhibit notable drag and heat transfer improvements (Choi et al. 2013). The surface-engineering gains also apply to internal channels with circular and hexagonal dimples, ribs, and grooved surfaces (Bilen et al. 2009; Sripattanapipat et al. 2019).

In addition, researchers have recently focused on biomimetics, whereby surface features found in nature have been applied onto engineering designs, oftentimes resulting in more optimized systems. For example, shark skin features have been used for drag reduction and antifouling properties (Pu et al. 2016). Note that these surface features, as opposed to "dimples" manifested as surface cavities, have nano and micro features that reside on the *exterior* of the surface (and hence are sometimes called "pimples" instead of dimples, or more simply, exterior dimples) (Mustak et al. 2015). That is, pimples are extensions outside of the surface that are sufficiently large such that they can be sensed by human touch. These features help trap air pockets, thereby allowing sharks to more easily flow (slice) through water. Drag reduction can also be achieved in airfoils and wings by using zig-zag turbulators (Popelka et al. 2011) and vortex generators (Popelka et al. 2011; EDF 2019). In such case, the zig-zag turbulators increased the lift-to-drag ratio of airfoils by approximately 11%. Other variations to the theme include exterior dimples (Mustak et al. 2015), as well as noncircular dimples (e.g. hexagonal and rectangular) (Hong and Asai 2017; Sripattanapipat et al. 2019).

Another area with strong potential includes static (motionless) swirl-generating surfaces, as shown in Fig. 3.14. Swirling promotes azimuthal momentum at the expense of the axial momentum, as shown in Fig. 3.15. It is this 3D rotational motion that generates additional mixing of mass, heat, and momentum in *focused regions*, as shown in Fig. 3.16 for a swirling jet (Rodriguez 2011). The figure is based on LES calculations of a swirling jet and includes swirl angles from 30° to slightly past the formation of the central recirculation zone (CRZ), which occurs at a swirl angle of

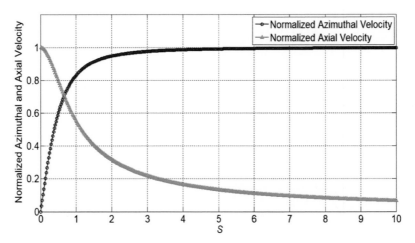

Fig. 3.15 Normalized azimuthal and axial velocities

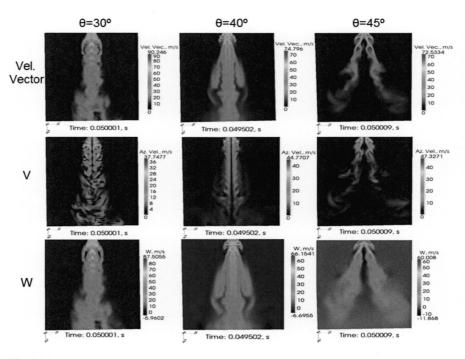

Fig. 3.16 Velocity vector (total velocity), azimuthal velocity, and axial velocity as functions of a jet flow with moderate swirl angles

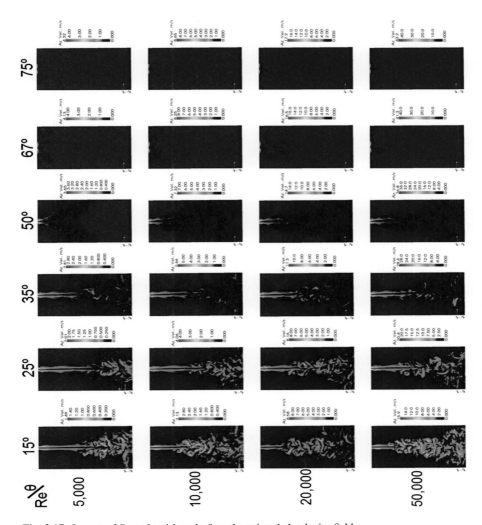

Fig. 3.17 Impact of *Re* and swirl angle θ on the azimuthal velocity field

about 42–45°. For higher angles thereafter, the azimuthal velocity diminishes rapidly (Loitsyanskiy 1953; Rodriguez et al. 2012); see Fig. 3.15. This critical transition can also be noticed in both Figs. 3.16 and 3.17. Figure 3.17 is an extension of Fig. 3.16 and shows LES calculations for the azimuthal velocity as a function of swirl angle θ ranging from 15 to 75° and *Re* from 5000 to 50,000. For readers interested in swirl BCs, Sect. 6.2.2 describes how to model circular boundary conditions.

The literature is filled with many swirl applications, including chemical species mixing and combustion (Nirmolo 2007; Yongqiang 2008; Valera-Medina et al. 2009; Stein 2009), heat transfer (King 2005; Rodriguez 2011, 2016; Illyas et al. 2019), aircraft design and safety (Batchelor 1964; Kavsaoglu and Schetz 1989;

Rusak and Lamb 1999; Pandya et al. 2003; Nelson 2004; Whitehead 2009), and nuclear reactors (Kim et al. 2007; Lavante and Laurien 2007; Nematollahi and Nazifi 2007; Johnson 2008; Laurien et al. 2010; Rodriguez 2011). Recent approaches evaluate the minimum pressure while identifying the most stable vortices (Nakayama et al. 2014); this approach has a strong potential to optimize swirler design. Furthermore, it is noted that whereas there are more than 20 distinct axisymmetric vortices (e.g., Chepura, Loitsyanskiy, Batchelor, Lamb-Oseen, Burgers, Gortler, Rankine, Sullivan, Newman, etc.), these vortices have various key behaviors in common (Rodriguez et al. 2012). One shared characteristic is that their azimuthal velocity is sine-like, so they can be expressed using a truncated Laurent series. Furthermore, their azimuthal velocities are bounded on one extreme by the forced-Rankine vortex and by the Lamb-Oseen vortex on the other (Rodriguez et al. 2012).

Finally, consider internal flow in a pipe, as the fluid is pumped with a variable-speed pump that undergoes a velocity increase. As the flow transitions from laminar to turbulent flow, the pressure drop will increase rapidly as the system transitions from linear to nonlinear dynamics. That is, as *Re* increases, the flow physics undergoes a drastic transformation: *whereas the pressure drop is a linear function for laminar flow in a pipe, the pressure drop increases as an exponential function of the velocity for turbulent flow, namely,*

$$\Delta P_{\text{lam}} \sim Re, \tag{3.56A}$$

while

$$\Delta P_{\text{turb}} \sim Re^{1.8}. \tag{3.56B}$$

Clearly, design advantages can be obtained by running a system in the turbulent regime, but doing so must be accomplished without unduly increasing the pressure drop, which can result in diminished performance in other areas. This is a cost/benefit issue that warrants optimization.

And this reverts us to Osborne Reynolds' terse, cryptic, and happy moment at his lab (Reynolds 1883):

The result was very happy.

Reynolds' epiphany came immediately after he substituted the pressure P with the pressure drop from Darcy's experiments, which closely followed Poiseuille's law for laminar flow. Using a relationship somewhat similar to Stokes' work 33 years earlier (Stokes 1850),

$$\frac{\mu u}{xP} \approx \text{constant}, \tag{3.57A}$$

Reynolds formulated his dimensionless number, Re,

$$\frac{\rho x u}{\mu} > 2000 \text{ (unstable; turbulent)}. \tag{3.57B}$$

In his "aha moment," Reynolds realized there was a relationship based on Re to an exponential power from 1.0 to 1.72 for the pressure drop, as the flow transitioned from laminar to turbulent flow, respectively. That is, the dynamics changed from linear to nonlinear during the transition (compare Eqs. 3.56A and 3.57C):

$$\Delta P_{\text{turb,Reynolds}} \sim Re^{1.72}. \tag{3.57C}$$

And happier even more so, the 1.72 relationship was obtained experimentally by Reynolds, whereas the 1.8 relationship (Eq. 3.56B) was obtained through analytical means.

3.8 Problems

3.1 A cylindrical pipe with $D = 0.5$ m has an average velocity of 12.5 m/s. The fluid consists of pressurized air at 600 K and 5.0×10^6 Pa. Use LIKE to find the integral, Taylor, and Kolmogorov eddy length, time, and velocity.

3.2 An equilateral triangular duct with side $S = 0.5$ m has an average velocity of 6.33 m/s. The fluid consists of liquid sodium at 1 atmosphere and 450 K. Use LIKE to find the integral, Taylor, and Kolmogorov eddy length, time, and velocity.

3.3 A flat plate 10 m long is exposed to liquid sodium flow at $U_\infty = 6.33$ m/s. The sodium is at 1 atmosphere and 450 K. Use LIKE to find the maximum integral, Taylor, and Kolmogorov eddy length, time, and velocity. Hint: because Re grows WRT plate length, find Re_x at $x = L$.

3.4 It is desired to design an ethanol distillation plant with a cylindrical condenser pipe that is shaped as a spiral. The mass flow rate is 0.15 kg/s, $\mu = 0.0011$ kg/m-s, and $\rho = 789$ kg/m^3. It is desired to run under turbulent flow so that the heat transfer is sufficiently large. Assume $Re = 3000$. Find D.

3.5 Consider SAE 40 oil flowing in a cylindrical motor manifold with $D = 0.07$ m, at an average velocity of 1.25 m/s, $\mu = 0.0113$ kg/m-s, and $\rho = 829.1$ kg/m^3. Suppose that you want to use a k-ω RANS turbulence model and want to compare the CFD output with your own, back-of-the-envelope values for k, ω, ν_t, and peak turbulence velocity. Calculate the four quantities.

3.6 An engineer would like to use the LES turbulence model to simulate the eddies up to the Taylor length scale. The system consists of a 3D rectangular duct with internal flow, with water flowing at an average velocity of 7.5 m/s. The duct has side $S = 0.11$ m. The water is at 350 K and 5.7×10^5 Pa. Find λ. Hint: use the LIKE algorithm to obtain the necessary input for the λ length scale.

3.7 A system analyst would like to apply a RANS-based turbulence model for flow parallel to a flat plate that measures 2.0 m long. The fluid is a heavy gas with $\rho = 10.3$ kg/m^3 and $\mu = 2.4 \times 10^{-7}$ kg/m-s. The flow has a free stream velocity of 4.5 m/s. If the first computational node is placed at $y^+ = 1$, how far away from the wall will it be?

3.8 A pipe with $D = 1.0$ m has water flowing at an average velocity of 5.0 m/s and is held steady for data acquisition. Then, the water velocity is increased in 5.0 m/s increments and held steady once again, and so forth, until reaching 50.0 m/s. The water is at 350 K and 5.7×10^5 Pa. Use LIKE and the formula for the length scale for Kolmogorov eddies to plot η for the entire velocity range.

3.9 Repeat Problem 3.8, but now plot the viscous sublayer thickness for the entire velocity range.

3.10 For those who had the fortune (or misfortune!) of solving Problems 3.8 and 3.9, overlay the viscous sublayer thickness curve with the Kolmogorov eddy length scale curve. Is the Kolmogorov scale larger than the viscous sublayer? If so, why might that be?

3.11 A pipe with an elliptical cross section has a semiminor axis = 0.2 m and a semimajor axis = 0.3 m. The fluid consists of liquid sodium at 1 atmosphere and 500 K. Find the viscous sublayer thickness, y at $y^+ = 5$, as well as the integral, Taylor, and Kolmogorov eddy length scales.

3.12 Show that Spalding's equation reduces to the laminar viscous sublayer for small y^+.

3.13 Show that Spalding's equation reduces to the log layer for large values of y^+.

3.14 Consider a turbulent flow in a rough pipe with $\lambda = 0.001$ m. The pipe has a mass flow rate of 0.045 kg/s, $\rho = 14.7$ kg/m^3, $\mu = 3.77 \times 10^{-5}$ kg/m-s, and $D = 0.1$ m. Find the peak turbulent velocity. Recall that the peak velocity is at the pipe centerline and use an appropriate log law.

3.15 A turbulent flow in a rough pipe with $\lambda = 0.002$ m and $D = 0.1$ m has a mass flow rate at 0.05 kg/s, $\rho = 15.0$ kg/m^3, and $\mu = 3.89 \times 10^{-5}$ kg/m-s. Plot the velocity distribution from the wall to the centerline. Make sure there are sufficient points to clearly show the various layers, and label the regions.

3.16 Assume an isotropic flow with $k \sim t^{-1.25}$. Find an approximation for the Taylor length scale λ as a function solely of the kinematic viscosity ν and the Taylor time scale, $t = \tau_\lambda$.

References

Abney, A., et al. (2013). *Hypersonic boundary-layer transition experiments in the boeing/AFOSR Mach-6 quiet tunnel*, 51st AIAA aerospace sciences meeting.

Andersson, B., et al. (2012). *Computational fluid dynamic for engineers*. Cambridge, UK: Cambridge University Press.

Babinsky, H., & Edwards, J. A. (1997). Large scale roughness influence on turbulent hypersonic boundary layers approaching compression corners. *Journal of Spacecraft and Rockets, 34*(1), 70.

Baek, S., & Kim, M. (2013). Flight trajectory of a golf ball for a realistic game. *International Journal of Innovation and Technology, 4*(3), 346.

Basse, N. T. (2017). Turbulence intensity and the friction factor for smooth- and rough-wall pipe flow. *Fluids, 2*, 30.

Batchelor, G. K. (1964). Axial flow in trailing line vortices. *Journal of Fluid Mechanics, 20, Part 4*, 645.

Bilen, K., et al. (2009). The investigation of groove geometry effect on heat transfer for internally grooved tubes. *Applied Thermal Engineering, 29*, 753–761.

Bird, R., Stewart, W., & Lightfoot, E. (2007). *Transport phenomena* (2nd ed.). New York: Wiley.

Bolaños, S. J., & Vernescu, B. (2017). Derivation of the Navier slip and slip length for viscous flows over a rough boundary. *Physics of Fluids, 29*, 057103.

Buschmann, M. H., & Gad-el-Hak, M. (2003). Debate concerning the mean-velocity profile of a turbulent boundary layer. *AIAA, 41*(4), 565.

CFD-Online. (2017a). Estimating the turbulence length scale, https://www.cfd-online.com/Wiki/Turbulence_length_scale. Accessed on Oct 31.

CFD-Online. (2017b). Turbulence intensity, https://www.cfd-online.com/Wiki/Turbulence_intensity. Accessed on Oct 31.

Chapman, D. R., & Kuhn, G. D. (1986). The limiting behavior of turbulence near a wall. *Journal of Fluid Mechanics, 70*, 265.

Choi, E. Y., Choi, Y. D., & Kwak, J. S. (2013). Effect of dimple configuration on heat transfer coefficient in a Rib-Dimpled channel. *Journal of Thermophysics and Heat Transfer, 27*(4), 653.

Corrsin, S., & Kistler, A. L. (1954). *Free-stream boundaries of turbulent flows*, The Johns Hopkins University, NACA, Report 1244.

EDF. (2019). 3M wind vortex generator fact sheet. EDF Renewable Services, www.edf-re.com.

Eiffel, G. (1913). Sur la Resistance des Spheres dans l'air en Mouvement. *Comtes Rendus, 156*.

Fage, A., & Townend, H. C. H. (1932). An examination of the turbulent flow with an ultramicroscope. *Proceedings of the Royal Society of London, 135*, 656.

Fluent Guidelines. (2012). "Turbulence", an introduction to ANSYS Fluent, Lecture 7.

George, W. K. (2013). *Lectures in turbulence for the 21st century*. London: Dept. of Aeronautics, Imperial College of London.

Hinze, J. (1987). *Turbulence*. New York: McGraw-Hill Classic Textbook Reissue Series.

Hoerner, S. F. (1992). *Fluid-dynamic drag*, Hoerner fluid dynamics, PO Box 21992, Bakersfield Ca, Ph. (661) 665–1065.

Hong, S., & Asai, T. (2017). Aerodynamic effects of dimples on soccer ball surfaces. Heliyon, Vol. 3, Art. e00432.

Illyas, S. M., Bapu, B. R. R., & Rao, V. V. S. (2019). Experimental analysis of heat transfer and multi objective optimization of swirling jet impingement on a flat surface. *Journal of Applied Fluid Mechanics, 12*(3), 803.

Johnson, G. A. (2008). *Power conversion system evaluation for the Next Generation Nuclear Plant (NGNP)*, Proc. international congress on advances in nuclear power plants (ICAPP 08), American Nuclear Society, Paper 8253, Anaheim, CA.

Kavsaoglu, M. S., & Schetz, J. A. (1989). Effects of swirl and high turbulence on a jet in a crossflow. *Journal of Aircraft, 26*(6), 539.

Kawahara, G. (2009). Theoretical interpretation of coherent structures in near-wall turbulence. *Fluid Dynamics Research, 41*, 064001.

Kim, J., Moin, P., & Moser, R. (1987). Turbulence statistics in fully developed channel flow at low Reynolds number. *Journal of Fluid Mechanics, 177*, 133.

Kim, M.-H., Lim, H.-S., & Lee, W.-J. (2007). *A CFD analysis of a preliminary cooled-vessel concept for a VHTR*. Korea Atomic Energy Research Institute.

King, A. (2005). *Heat transfer enhancement and fluid flow characteristics associated with jet impingement cooling*, PhD Diss., Curtin University of Technology, *c*.

Klebanoff, P. S. (1954). *Characteristics of turbulence in a boundary layer with zero pressure gradient*, NACA, Technical Note 3178.

Kolmogorov, A. N. (1942). Equations of turbulent motion of an incompressible fluid, Izv. Akad. Nauk SSSR, Seria fizicheska VI, No. 1–2. (Translated by D. B. Spalding.)

Kontis, K., & Lada, C. (2005). *Flow control effectiveness at high speed flows*, Proceedings of the fifth European symposium on aerothermodynamics for space vehicles.

Kontis, K., Lada, C., & Zare-Behtash, H. (2008). Effect of dimples on glancing shock wave turbulent boundary layer interactions. *Shock Waves, 17*, 323.

Laurien, E., Lavante, D. V., & Wang, H. (2010, Oct 18–20). *Hot-gas mixing in the annular channel below the core of high-power HTR's*, Proceedings of the 5th international topical meeting on high temperature reactor technology, HTR 2010–138, Prague, Czech Republic.

Lavante, D. V., & Laurien, E. (2007). 3-D simulation of hot gas mixing in the lower plenum of high-temperature reactors. *International Journal for Nuclear Power, 52*, 648.

Loitsyanskiy, L. G. (1953). The propagation of a twisted jet in an unbounded space filled with the same fluid. *Prikladnaya Matematika i Mekhanika, 17*(1), 3.

Mandal, B. C., & Mazumdar, H. P. (2015). The importance of the Law of the Wall. *International Journal of Applied Mechanics and Engineering, 20*(4), 857.

Mansour, N. N., Kim, J., & Moin, P. (1988). *Near-wall k-ε turbulence modeling*, NASA Technical Memorandum 89461.

McKeon, B. J., et al. (2004). Friction factors for smooth pipe flow. *Journal of Fluid Mechanics, 511*, 41.

Minin, I. V., & Minin, O. V. (Eds.). (2011). Computational fluid dynamics of turbulent flow; P. Majumdar, Chapter 9 author. In *Computational fluid dynamics technologies and applications*. Intech Web.

Mustak, R., et al. (2015). Effect of different shaped dimples on airfoils. *Proceedings of the International Conference on Mechanical Engineering and Renewable Energy*, Bangladesh, ICMERE2015-PI-230.

MythBusters. (2009). Clean car vs. dirty car, https://en.wikipedia.org/wiki/MythBusters_(2009_season)#Episode_127_–_"Clean_Car_vs._Dirty_Car", Episode No. 127.

Nakayama, K., et al. (2014). A unified definition of a vortex derived from vortical flow and the resulting pressure minimum. *Fluid Dynamics Research, 46*.

Nazarenko, S. (2000). Exact solutions for near-wall turbulence theory. *Physics Letters A, 264*, 444.

Nelson, R. C. (2004, Aug 16–19). *The trailing vortex wake hazard: Beyond the takeoff and landing corridors*, Atmospheric Flight Mechanics Conference and Exhibit, AIAA 2004–5171, Providence, Rhode Island.

Nematollahi, M. R., & Nazifi, M. (2007). *Enhancement of heat transfer in a typical pressurized water reactor by new mixing vanes on spacer grids*. Shiraz, Iran: ICENES.

Nirmolo, A. (2007). *Optimization of radial jets mixing in cross-flow of combustion chambers using computational fluid dynamics*, PhD Diss., Otto-von-Guericke U. of Magdeburg, Germany.

Oberlack, M. (2001). A unified approach for symmetries in plane parallel turbulent shear flows. *Journal of Fluid Mechanics, 427*, 299.

Osterlund, J. M. (1999). *Experimental studies of zero pressure-gradient turbulent boundary layer flow*, Royal Institute of Technology, PhD Diss.

Pandya, S. A., Murman, S. M., & Sankaran, V. (2003, June 23–26). *Unsteady computations of a jet in crossflow with ground effect*, 33rd AIAA fluid dynamics conference and exhibit, Orlando, Florida, AIAA 2003–3890.

Patankar, S. V., & Minkowycz, W. J. (2017). In Memoriam Professor D. Brian spalding (1923-2016). *International Journal of Heat and Mass Transfer, 109*, 657.

Popelka, L., et al. (2011). Boundary layer transition, separation and flow control on airfoils, wings and bodies in numerical, wind-tunnel and in-flight studies. *Technical Soaring, 35*(4).

Prandtl, L. (1945). "Uber ein neues Formelsystem fur die ausgebildete Turbulenz", ("On a New Formulation for the Fully Developed Turbulence" or "On a New Equation System for Developed Turbulence"), Nacr. Akad. Wiss. Gottingen, Math-Phys. Kl.

Prasath, M. S., & Angelin, S. I. (2017). Effect of dimples on aircraft wing. *Global Research and Development Journal for Engineering, 2*(5).

Pu, X., Li, G., & Huang, H. (2016). Preparation, anti-fouling and drag-reduction properties of a biomimetic shark skin surface. *Biology Open, 5*, 389.

Reynolds, O. (1883). An experimental investigation of the circumstances which determine whether the motion of water shall be direct or sinuous, and of the Law of Resistance in parallel channels. *Philosophical Transactions of the Royal Society of London, 174*. (Can also be obtained through JSTOR, http://www.jstor.org).

Richardson, L. F. (1922). *Weather Prediction by Numerical Processes*. Boston: Cambridge University Press.

Rodriguez, I., et al. (2015). On the flow past a circular cylinder from critical to super-critical Reynolds numbers: wake topology and vortex shedding. *International Journal of Heat and Fluid Flow, 55*.

Rodriguez, S. (2011, May). *Swirling Jets for the Mitigation of Hot Spots and Thermal Stratification in the VHTR Lower Plenum*, PhD Diss., University of New Mexico.

Rodriguez, S. (2016, July). *Next-generation heat transfer surfaces*. Sandia National Laboratories, Tracking No. 476209.

Rodriguez, S. (2017). *Right-sized dimple evaluator v. 2.0*. Sandia National Laboratories, Copyright SCR # 2243.0.

Rodriguez, S., et al. (2012). *Towards a unified Swirl Vortex model*, AIAA 42nd fluid dynamics conference, AIAA 2012–3354, New Orleans.

Rodriguez, S., Grieb, N., & Chen, A. (2017). *Dimpling aerodynamics for a ford mustang GT*. Sandia National Laboratories, SAND2017–6522.

Rusak, Z., & Lamb, D. (1999). Prediction of Vortex breakdown in leading-edge vortices above slender delta wings. *Journal of Aircraft, 36*(4), 659.

Russo, F., & Basse, N. T. (2016). Scaling of turbulence intensity for low-speed flow in smooth pipes. *Flow Measurement and Instrumentation, 52*, 101.

Sandborn, V. A. (1955). *Experimental evaluation of momentum terms in turbulent pipe flow*, National Advisory Committee for Aeronautics, NACA TN 3266, Technical Note.

Schlichting, H., & Gersten, K. (2000). *Boundary layer theory* (8th ed.). New York: Springer.

Schmitt, F. G. (2007). About Boussinesq's turbulent viscosity hypothesis: Historical remarks and a direct evaluation of its validity. *Comptes Rendus Mecanique, 335*, 617.

Sekaran, M., & Naik, B. A. (2011). *Dimples-an effective way to improve in performance*, the first international conference on interdisciplinary research and development, Thailand, May 31 through June 1.

Shesterikov, I., et al. (2012). Direct evidence of eddy breaking and tilting by edge sheared flows observed in the TEXTOR Tokomak. *Nuclear Fusion, 52*(4), 042004.

Sodja, J. (2007). *Turbulence models in CFD*. University of Ljubljana.

Shih, T. H., et al. (1999). A generalized wall function, NASA/TM, ICOMP-99-08.

Spalding, D. B. (1961). A single formula for the 'Law of the Wall'. *ASME, Journal of Applied Mechanics, 28*, 455.

Sripattanapipat, S., et al. (2019). Numerical simulation of turbulent flow and heat transfer in hexagonal dimpled channel. The 33rd Conference of Mechanical Engineering Network of Thailand, *HTE-020*, 1–6.

Stein, O. (2009, March). *Large eddy simulation of combustion in swirling and opposed jet flows*, PhD Diss., Imperial College London.

Stokes, G. G. (1850). On the effect of the internal friction of fluids on the motion of pendulums. *Transactions of the Cambridge Philosophical Society, 9*, 1–99.

Tabeling, P. (2009). *Introduction to microfluidics*. Oxford: Oxford University Press.

Taylor, G. (1935). Statistical theory of turbulence. *Royal Society of London, Series A, Mathematical and Physical Sciences*. Can also be downloaded from rspa.royalsocietypublishing.org. (Refer to pages 438 and 439).

Ting, L. L. (2003). *Effects of dimple size and depth on golf ball aerodynamic performance*, Proceedings of FEDSM '03, Fourth ASME-JSME joint fluids engineering conference.

Valera-Medina, A., Syred, N., & Griffiths, A. (2009, Jan 5–8). *Characterization of large coherent structures in a Swirl Burner under combustion conditions*, 47th AIAA aerospace sciences meeting including the New Horizons forum and aerospace exposition, AIAA 2009–646, Orlando, Florida.

van Nesselrooij, M., et al. (2016). Drag reduction by means of dimpled surfaces in turbulent boundary layers. *Experiments in Fluids, 57*(142), 1.

von Karman, T. (1939). The analogy between fluid friction and heat transfer. *Transactions of the ASME, 61*, 705.

von Karman, T. (1940). "Mechanical similitude and turbulence", Technical Memorandums, National Advisory Committed for Aeronautics, NACA-TM-611.

von Karman, T., & Howarth, L. (1938). On the statistical theory of isotropic turbulence. *Proceedings of the Royal Society of London. Series A, 164*, 192. (The title page has "de" Karman, which also means "von" Karman in German, which means "of").

White, F. (1991). *Viscous fluid flow* (2nd ed.). New York: McGraw-Hill.

Whitehead, E. J. (2009). *The stability of multiple wing-tip vortices*, The University of Adelaide, PhD Diss.

Wilcox, D. C., *Turbulence modeling for CFD*, 3rd Ed., printed on 2006 and 2010. (2006 k-ω model).

Yongqiang, F. (2008, Jan). *Aerodynamics and combustion of axial swirlers*, PhD Diss., University of Cincinnati.

Chapter 4
RANS Turbulence Modeling

An ideal model should introduce the minimum amount of complexity while capturing the essence of the relevant physics.

—David Wilcox, 2006

Abstract The turbulent kinetic energy equation is derived and explained in full detail. The motivation and development of RANS-based models is provided, with the aim of generating a deeper understanding of turbulence phenomena. This includes the detailed descriptions for key k-ε, k-ω, and SST hybrid models. Fundamental RANS terms are explained, such as turbulent kinematic viscosity, production, and decay. RANS models are evaluated and compared, and the best overall turbulence model is suggested. Model applicability, best performance regions, and deficiencies are discussed for zero-, one-, and two-equation RANS models. Compelling reasons for avoiding the standard k-ε are provided. Multiple insights regarding ties associated with the development of k-ε and k-ω models are presented, such as the Taylor scale and eddy dissipation.

Wilcox summarized it quite well: an "ideal" turbulence model ought to require a minimal number of equations and variables, but at the same time, it should be able to model features that are important for designers, researchers, engineers, physicists, mathematicians, and so forth. Unfortunately, the literature shows that turbulence models tend to become more complex as their ability to capture physics increases. A somewhat analogous view for turbulence models is that they ought to capture as many of the physical features displayed in turbulent flows, while having a minimal number of ad hoc fits, stitches, and patches. Furthermore, such model should be purely based on variables that reflect the correct physics. For example, if the turbulent kinetic energy k refers to the energy of the *large* integral eddies, would it be acceptable to couple k with the *small* Kolmogorov eddies? These issues, as well as how turbulence models were derived, and their applicability (or lack thereof), are covered in the sections that follow.

© Springer Nature Switzerland AG 2019
S. Rodriguez, *Applied Computational Fluid Dynamics and Turbulence Modeling*,
https://doi.org/10.1007/978-3-030-28691-0_4

4.1 The Mechanics Behind Reynolds Stress Transport Models

At this point, Chap. 4 extends the theory and modeling presented in Chap. 2 in Sects. 2.4 and 2.5. It is now a matter of moving forward toward closure, based on the desired Reynolds-averaged Navier-Stokes (RANS) modeling complexity; there are zero-, one-, and two-equation models, as well as the so-called second-order closure models. The latter are also referred as second-moment closure models, stress transport models, Reynolds stress models (RSM), and Reynolds stress transport (RST) models. Note that these models are not the same as shear stress transport (SST) models, which unfortunately have similar-sounding names.

The following sections extend turbulence modeling by zeroing in on the unanswered issues that remained from Chap. 2. This is particularly so for the Reynolds stress tensor R, to point out what is still needed to have a full set of n independent turbulence equations, thereby solving n unknowns. Then, it is shown that applying the Boussinesq approximation is not sufficient for closure, so the path forward is to perform a set of mathematical operations that lead to additional independent equations, with the aim of finally providing closure to the turbulence problem. In particular, the R tensor is derived mathematically in its exact form from the unoperated Navier-Stokes momentum equation by taking its first moment based on the fluctuating velocity, and then a time-average is performed. The resultant R tensor is used in RSM models and is therefore part of the motivation. But most importantly, the k PDE is derived from R by taking its trace, and it is this k PDE that is used in most zero-, one-, and two-equation RANS models (e.g., Prandtl, k-ε, k-ω, SST, etc.).

4.2 The Derivation of the R Tensor for Turbulence Modeling

Because the RANS closure process is rather long-winded, it is sketched from a "bird's-eye view" (roadmap) in Fig. 4.1a, with the hope that the overall approach is clear, especially once the full details of the derivation are covered! The figure shows how the Navier-Stokes momentum PDE is time-averaged to get the *exact* RANS momentum PDE, which is then used to derive the *exact* Reynolds stress tensor R PDE. In this context, "exact" refers to the PDE version that was derived mathematically using no approximations, while "solvable" refers to the PDE version that was derived from the "exact" PDE using engineering approximations. In summary, the overview shown in Fig. 4.1a is covered in details in Sects. 4.2 and 4.3.

To begin the process, recall that if the kinematic viscosity is constant and the gravitational term is negligible (or is lumped with the P term), the original (unoperated) Navier-Stokes equation is simply

a

(Navier - Stokes $\xrightarrow{\text{TIME-AVERAGE}}$ *Exact* RANS)

$$\frac{\partial u_i}{\partial t} + u_j \frac{\partial u_i}{\partial x_j} = \nu \frac{\partial^2 u_i}{\partial x_j \partial x_j} - \frac{1}{\rho} \frac{\partial P}{\partial x_i} \quad \xrightarrow{\text{TIME-AVERAGE}}$$

$$\frac{\partial \bar{u}_i}{\partial t} + \bar{u}_j \frac{\partial \bar{u}_i}{\partial x_j} = \nu \frac{\partial^2 \bar{u}_i}{\partial x_j \partial x_j} - \frac{1}{\rho} \frac{\partial \bar{P}}{\partial x_i} - \frac{\partial \overline{u_i' u_j'}}{\partial x_j} \quad (\textit{Exact } \text{RANS})$$

(*Exact* RANS $\xrightarrow{\text{MULT. BY } U',\ \text{TIME AVE.}}$ *Exact R* PDE)

$$\frac{\partial \bar{u}_i}{\partial t} + \bar{u}_j \frac{\partial \bar{u}_i}{\partial x_j} = \nu \frac{\partial^2 \bar{u}_i}{\partial x_j \partial x_j} - \frac{1}{\rho} \frac{\partial \bar{P}}{\partial x_i} - \frac{\partial \overline{u_i' u_j'}}{\partial x_j} \quad \xrightarrow{\text{MULT BY } U',\ \text{TIME AVE.}}$$

$$\frac{\partial \overline{u_i' u_j'}}{\partial t} + \bar{u}_k \frac{\partial \overline{u_i' u_j'}}{\partial x_k} = -\left(\overline{u_i' u_k'} \frac{\partial \bar{u}_j}{\partial x_k} + \overline{u_j' u_k'} \frac{\partial \bar{u}_i}{\partial x_k} \right) + 2\nu \overline{\frac{\partial u_i'}{\partial x_k} \frac{\partial u_j'}{\partial x_k}}$$

$$+ \frac{\partial}{\partial x_k}\left(\nu \frac{\partial \overline{u_i' u_j'}}{\partial x_k} \right) + \frac{1}{\rho}\left(\overline{u_i' \frac{\partial P'}{\partial x_j}} + \overline{u_j' \frac{\partial P'}{\partial x_i}} \right) + \frac{\partial}{\partial x_k}\left(\overline{u_i' u_j' u_k'} \right) \quad (\textit{Exact } R\ \text{PDE})$$

b

(Navier - Stokes $\xrightarrow{\text{TIME-AVERAGE}}$ *Exact* RANS)

$$\frac{\partial u_i}{\partial t} + u_j \frac{\partial u_i}{\partial x_j} = \nu \frac{\partial^2 u_i}{\partial x_j \partial x_j} - \frac{1}{\rho} \frac{\partial P}{\partial x_i} \quad \xrightarrow{\text{TIME-AVERAGE}}$$

$$\frac{\partial \bar{u}_i}{\partial t} + \bar{u}_j \frac{\partial \bar{u}_i}{\partial x_j} = \nu \frac{\partial^2 \bar{u}_i}{\partial x_j \partial x_j} - \frac{1}{\rho} \frac{\partial \bar{P}}{\partial x_i} - \frac{\partial \overline{u_i' u_j'}}{\partial x_j} \quad (\textit{Exact } \text{RANS})$$

(*Exact* RANS $\xrightarrow{\text{MULT. BY } U',\ \text{TIME AVE.}}$ *Exact R* PDE)

$$\frac{\partial \bar{u}_i}{\partial t} + \bar{u}_j \frac{\partial \bar{u}_i}{\partial x_j} = \nu \frac{\partial^2 \bar{u}_i}{\partial x_j \partial x_j} - \frac{1}{\rho} \frac{\partial \bar{P}}{\partial x_i} - \frac{\partial \overline{u_i' u_j'}}{\partial x_j} \quad \xrightarrow{\text{MULT BY } U',\ \text{TIME AVE.}}$$

$$\frac{\partial \overline{u_i' u_j'}}{\partial t} + \bar{u}_k \frac{\partial \overline{u_i' u_j'}}{\partial x_k} = -\left(\overline{u_i' u_k'} \frac{\partial \bar{u}_j}{\partial x_k} + \overline{u_j' u_k'} \frac{\partial \bar{u}_i}{\partial x_k} \right) + 2\nu \overline{\frac{\partial u_i'}{\partial x_k} \frac{\partial u_j'}{\partial x_k}}$$

$$+ \frac{\partial}{\partial x_k}\left(\nu \frac{\partial \overline{u_i' u_j'}}{\partial x_k} \right) + \frac{1}{\rho}\left(\overline{u_i' \frac{\partial P'}{\partial x_j}} + \overline{u_j' \frac{\partial P'}{\partial x_i}} \right) + \frac{\partial}{\partial x_k}\left(\overline{u_i' u_j' u_k'} \right) \quad (\textit{Exact } R\ \text{PDE})$$

Fig. 4.1 (**a**) Bird's eye view showing how the Navier-Stokes momentum PDE is used to derive the *exact* RANS PDE and the *exact R* PDE. (**b**) Bird's eye view showing how the *exact R* PDE is used to derive the *exact* and *solvable k* PDEs

$$\frac{\partial u_i}{\partial t} + u_j \frac{\partial u_i}{\partial x_j} = \nu \frac{\partial^2 u_i}{\partial x_j \partial x_j} - \frac{1}{\rho} \frac{\partial P}{\partial x_i}. \tag{4.1}$$

Recall that the RANS momentum equation was derived in Chap. 2 by replacing u with $\bar{u} + u'$ and simplifying. In its vector-tensor notation, the equation appears as

$$\frac{\partial \bar{u}_i}{\partial t} + \bar{u}_j \frac{\partial \bar{u}_i}{\partial x_j} = -\frac{\partial \bar{P}}{\partial x_i} + 2\frac{\partial \left(\nu \bar{S}_{ji}\right)}{\partial x_j} - \frac{\partial \left(\overline{u'_j u'_i}\right)}{\partial x_j} \tag{4.2A}$$

where the mean strain rate tensor for an incompressible fluid is

$$\bar{S}_{ij} = \frac{1}{2}\left(\frac{\partial \bar{u}_i}{\partial x_j} + \frac{\partial \bar{u}_j}{\partial x_i}\right). \tag{4.2B}$$

At this point, the mean strain rate tensor can be included in the momentum equation, and the vector-tensor equation can be expanded in the x-, y-, and z-directions of a Cartesian system. The PDE for the x-momentum is

$$\frac{\partial \bar{u}}{\partial t} + \left(\bar{u}\frac{\partial \bar{u}}{\partial x} + \bar{v}\frac{\partial \bar{u}}{\partial y} + \bar{w}\frac{\partial \bar{u}}{\partial z}\right) = -\frac{1}{\rho}\frac{\partial \bar{P}}{\partial x} + \nu\left(\frac{\partial^2 \bar{u}}{\partial x^2} + \frac{\partial^2 \bar{u}}{\partial y^2} + \frac{\partial^2 \bar{u}}{\partial z^2}\right)$$
$$- \left(\frac{\partial \overline{u'u'}}{\partial x} + \frac{\partial \overline{u'v'}}{\partial y} + \frac{\partial \overline{u'w'}}{\partial z}\right), \tag{4.3A}$$

and similarly for the y-momentum,

$$\frac{\partial \bar{v}}{\partial t} + \left(\bar{u}\frac{\partial \bar{v}}{\partial x} + \bar{v}\frac{\partial \bar{v}}{\partial y} + \bar{w}\frac{\partial \bar{v}}{\partial z}\right) = -\frac{1}{\rho}\frac{\partial \bar{P}}{\partial y} + \nu\left(\frac{\partial^2 \bar{v}}{\partial x^2} + \frac{\partial^2 \bar{v}}{\partial y^2} + \frac{\partial^2 \bar{v}}{\partial z^2}\right)$$
$$- \left(\frac{\partial \overline{v'u'}}{\partial x} + \frac{\partial \overline{v'v'}}{\partial y} + \frac{\partial \overline{v'w'}}{\partial z}\right), \tag{4.3B}$$

and for the z-momentum,

$$\frac{\partial \bar{w}}{\partial t} + \left(\bar{u}\frac{\partial \bar{w}}{\partial x} + \bar{v}\frac{\partial \bar{w}}{\partial y} + \bar{w}\frac{\partial \bar{w}}{\partial z}\right) = -\frac{1}{\rho}\frac{\partial \bar{P}}{\partial z} + \nu\left(\frac{\partial^2 \bar{w}}{\partial x^2} + \frac{\partial^2 \bar{w}}{\partial y^2} + \frac{\partial^2 \bar{w}}{\partial z^2}\right)$$
$$- \left(\frac{\partial \overline{w'u'}}{\partial x} + \frac{\partial \overline{w'v'}}{\partial y} + \frac{\partial \overline{w'w'}}{\partial z}\right). \tag{4.3C}$$

The primed terms are a direct consequence of the time-averaging that was performed earlier (see Chap. 2). An inspection of the last term in parenthesis on the RHS of each of the above x-, y-, and z-momentum equations shows how the R tensor is assembled. In particular, each of the three PDEs contributes three primed quantities, for a total of nine terms, namely,

$$\vec{\vec{R}} = - \begin{bmatrix} \overline{u'u'} & \overline{u'v'} & \overline{u'w'} \\ \overline{v'u'} & \overline{v'v'} & \overline{v'w'} \\ \overline{w'u'} & \overline{w'v'} & \overline{w'w'} \end{bmatrix} = -\overline{u'_i u'_j}. \qquad (4.4)$$

The three terms on the top column of the tensor correspond to the x-momentum Reynolds stresses, followed by the middle column with its three terms corresponding to the y-momentum, and likewise for the bottom column.

R has many useful properties, including positive definiteness. Fortunately, the tensor is symmetric as well, so only six of the nine Reynolds stress terms are unknown. For example, $R_{xy} = R_{yx}$.

This is where the linear Boussinesq approximation provides additional closure information, e.g., more relationships that can be substituted onto the RANS momentum equations. Recall the definition for the Boussinesq approximation, viz.,

$$R_{ij} \equiv 2\nu_t \overline{S}_{ij} - \frac{2}{3}kI_{ij} = \nu_t \left(\frac{\partial \overline{u}_i}{\partial x_j} + \frac{\partial \overline{u}_j}{\partial x_i} \right) - \frac{2}{3}k\delta_{ij}. \qquad (4.5A)$$

For convenience, the entire set of nine components are summarized here (though there are only six independent components). As shown in Example 2.16 in Chap. 2, the individual components are rather straightforward expressions of the Boussinesq approximation. The three x-momentum Reynolds stresses are

$$R_{xx} = -\overline{u'u'} = 2\nu_t \frac{\partial \overline{u}}{\partial x} - \frac{2}{3}k, \qquad (4.5B)$$

$$R_{xy} = -\overline{u'v'} = \nu_t \left(\frac{\partial \overline{u}}{\partial y} + \frac{\partial \overline{v}}{\partial x} \right), \qquad (4.5C)$$

and

$$R_{xz} = -\overline{u'w'} = \nu_t \left(\frac{\partial \overline{u}}{\partial z} + \frac{\partial \overline{w}}{\partial x} \right), \qquad (4.5D)$$

while the three y-momentum Reynolds stresses are

$$R_{yx} = -\overline{v'u'} = \nu_t \left(\frac{\partial \overline{v}}{\partial x} + \frac{\partial \overline{u}}{\partial y} \right), \qquad (4.5E)$$

$$R_{yy} = -\overline{v'v'} = 2\nu_t \frac{\partial \overline{v}}{\partial y} - \frac{2}{3}k, \qquad (4.5F)$$

and

$$R_{yz} = -\overline{v'w'} = \nu_t \left(\frac{\partial \overline{v}}{\partial z} + \frac{\partial \overline{w}}{\partial y} \right). \qquad (4.5G)$$

Finally, the three z-momentum Reynolds stresses are

$$R_{zx} = -\overline{w'u'} = \nu_t \left(\frac{\partial \overline{w}}{\partial x} + \frac{\partial \overline{u}}{\partial z} \right), \tag{4.5H}$$

$$R_{zy} = -\overline{w'v'} = \nu_t \left(\frac{\partial \overline{w}}{\partial y} + \frac{\partial \overline{v}}{\partial z} \right), \tag{4.5I}$$

and

$$R_{zz} = -\overline{w'w'} = 2\nu_t \frac{\partial \overline{w}}{\partial z} - \frac{2}{3}k. \tag{4.5J}$$

At this point, it is desirable to incorporate the R terms into the RANS PDE. For example, the x-direction Cartesian RANS equation transforms into a more manageable, computationally solvable expression:

$$
\begin{aligned}
\frac{\partial \overline{u}}{\partial t} &+ \left(\overline{u} \frac{\partial \overline{u}}{\partial x} + \overline{v} \frac{\partial \overline{u}}{\partial y} + \overline{w} \frac{\partial \overline{u}}{\partial z} \right) = -\frac{1}{\rho} \frac{\partial \overline{P}}{\partial x} + \nu \left(\frac{\partial^2 \overline{u}}{\partial x^2} + \frac{\partial^2 \overline{u}}{\partial y^2} + \frac{\partial^2 \overline{u}}{\partial z^2} \right) \\
&+ \frac{\partial \left[\nu_t \left(\frac{\partial \overline{u}}{\partial x} + \frac{\partial \overline{u}}{\partial x} \right) - \frac{2}{3}k \right]}{\partial x} + \frac{\partial \left[\nu_t \left(\frac{\partial \overline{u}}{\partial y} + \frac{\partial \overline{v}}{\partial x} \right) \right]}{\partial y} + \frac{\partial \left[\nu_t \left(\frac{\partial \overline{u}}{\partial z} + \frac{\partial \overline{w}}{\partial x} \right) \right]}{\partial z}.
\end{aligned} \tag{4.6A}
$$

The same can be applied to the y- and z-momentum equations, resulting in the following two PDEs, respectively:

$$
\begin{aligned}
\frac{\partial \overline{v}}{\partial t} &+ \left(\overline{u} \frac{\partial \overline{v}}{\partial x} + \overline{v} \frac{\partial \overline{v}}{\partial y} + \overline{w} \frac{\partial \overline{v}}{\partial z} \right) = -\frac{1}{\rho} \frac{\partial \overline{P}}{\partial y} + \nu \left(\frac{\partial^2 \overline{v}}{\partial x^2} + \frac{\partial^2 \overline{v}}{\partial y^2} + \frac{\partial^2 \overline{v}}{\partial z^2} \right) \\
&+ \frac{\partial \left[\nu_t \left(\frac{\partial \overline{v}}{\partial x} + \frac{\partial \overline{u}}{\partial y} \right) \right]}{\partial x} + \frac{\partial \left[\nu_t \left(\frac{\partial \overline{v}}{\partial y} + \frac{\partial \overline{v}}{\partial y} \right) - \frac{2}{3}k \right]}{\partial y} + \frac{\partial \left[\nu_t \left(\frac{\partial \overline{v}}{\partial z} + \frac{\partial \overline{w}}{\partial y} \right) \right]}{\partial z}
\end{aligned} \tag{4.6B}
$$

and

$$
\begin{aligned}
\frac{\partial \overline{w}}{\partial t} &+ \left(\overline{u} \frac{\partial \overline{w}}{\partial x} + \overline{v} \frac{\partial \overline{w}}{\partial y} + \overline{w} \frac{\partial \overline{w}}{\partial z} \right) = -\frac{1}{\rho} \frac{\partial \overline{P}}{\partial z} + \nu \left(\frac{\partial^2 \overline{w}}{\partial x^2} + \frac{\partial^2 \overline{w}}{\partial y^2} + \frac{\partial^2 \overline{w}}{\partial z^2} \right) \\
&+ \frac{\partial \left[\nu_t \left(\frac{\partial \overline{w}}{\partial x} + \frac{\partial \overline{u}}{\partial z} \right) \right]}{\partial x} + \frac{\partial \left[\nu_t \left(\frac{\partial \overline{w}}{\partial y} + \frac{\partial \overline{v}}{\partial z} \right) \right]}{\partial y} + \frac{\partial \left[\nu_t \left(\frac{\partial \overline{w}}{\partial z} + \frac{\partial \overline{w}}{\partial z} \right) - \frac{2}{3}k \right]}{\partial z}.
\end{aligned} \tag{4.6C}
$$

Notice that at this point, the six unknown, independent R components are resolved thanks to the Boussinesq approximation. But in doing so, the process introduced two new variables, k and ν_t! So, at this point, closure is closer, but not *fully achieved*. Fortunately, patience is a good trait of turbulence modelers!

The above three PDEs can *almost* be solved numerically for the time-averaged turbulence velocities, $\bar{u}$, $\bar{v}$, and $\bar{w}$. Two items are needed at this point for closure: (1) an expression for k must somehow be derived for the large, energy-bearing eddies and (2) a relationship for ν_t, which is usually derived using dimensional arguments as functions of k and some characteristic length or equivalent. Certainly, most modern expressions for ν_t are based on dimensional arguments that usually involve k, but not always—it all depends on the pedigree of the model. In any case, this describes how RANS turbulent transport models were developed—the number of unknown variables determines the number of transport and auxiliary equations that are required for mathematical closure. This means that if there are n unknown turbulence variables, there must be n independent equations to solve the flow; but if the transport equations add additional m unknowns, the model must also supply m additional auxiliary equations, and so forth, until closure is achieved using the same number of independent equations as unknowns.

At this point, the next step is to develop a rigorous mathematical expression for R, which will then be used to derive k.

4.3   Mathematical Derivation of R

The R tensor is derived mathematically in its exact form as follows. The first step is to take the unoperated "laminar" Navier-Stokes momentum equation:

$$\frac{\partial u_i}{\partial t} + u_k \frac{\partial u_i}{\partial x_k} = \nu \frac{\partial^2 u_i}{\partial x_k \partial x_k} - \frac{1}{\rho} \frac{\partial P}{\partial x_i} \qquad (4.7)$$

and apply a so-called first moment based on the fluctuating velocities u_i' and u_j'. Basically, each term in the Navier-Stokes equation is multiplied by the fluctuating velocities, and then the instantaneous u_i and P variables are replaced with the appropriate Reynolds decomposition expressions (e.g., $u_i = \bar{u}_i + u_i'$ and $P = \bar{P} + P'$). Following this, the terms are time-averaged, and the final step is to simplify the terms.

Due to the number of mathematical steps performed during this approach, it is easier to do the RANS equation term by term and then assemble the results. *Another advantage of solving each term individually is that it is easier to see and to appreciate how new turbulence physics terms arise* (e.g., *this is where the turbulence production, dissipation, and turbulence triple velocity correlation transport terms first appear*). The step-by-step approach is covered in what follows, with many details shown for the momentum accumulation term, as it serves as an ideal example for the technique that is used for all the terms.

4.3.1   The Momentum Accumulation Term

The overall procedure begins by taking the momentum term and multiplying it by the fluctuating velocities:

$$
\underbrace{\frac{\partial u_i}{\partial t}}_{\text{starting point}} \quad \rightarrow \quad \underbrace{u_i'\frac{\partial u_j}{\partial t} + u_j'\frac{\partial u_i}{\partial t}}_{\text{mult.momentum term with } u'} \quad \rightarrow \quad \underbrace{u_i'\left[\frac{\partial\left(\overline{u}_j + u_j'\right)}{\partial t}\right] + u_j'\left[\frac{\partial\left(\overline{u}_i + u_i'\right)}{\partial t}\right]}_{\text{Substitute Reynolds decomposition velocity}}
$$

$$
\rightarrow \quad \underbrace{\overline{u_i'\left[\frac{\partial\left(\overline{u}_j + u_j'\right)}{\partial t}\right]} + \overline{u_j'\left[\frac{\partial\left(\overline{u}_i + u_i'\right)}{\partial t}\right]}}_{\text{Apply time average}} \quad \rightarrow \quad \text{simplify.}
$$

$$(4.8\text{A})$$

The simplification expands the time-averaged partial differential by multiplying the terms and applying Eq. 2.34L $\left(\overline{a+b} = \overline{a} + \overline{b}\right)$,

$$
\overline{u_i'\left[\frac{\partial\left(\overline{u}_j + u_j'\right)}{\partial t}\right]} + \overline{u_j'\left[\frac{\partial\left(\overline{u}_i + u_i'\right)}{\partial t}\right]} = \overline{u_i'\left(\frac{\partial\overline{u}_j}{\partial t}\right)} + \overline{u_i'\left(\frac{\partial u_j'}{\partial t}\right)} + \overline{u_j'\left(\frac{\partial\overline{u}_i}{\partial t}\right)}
$$

$$
+ \overline{u_j'\left(\frac{\partial u_i'}{\partial t}\right)}. \qquad (4.8\text{B})
$$

Recall from Sect. 2.5 that time-averaging the product of a fluctuating velocity and a time-averaged velocity is zero,

$$
\overline{u'\overline{u}} = 0,
$$

and

$$
\overline{\frac{\partial u}{\partial t}} = \frac{\partial\overline{u}}{\partial t}.
$$

Thus, the first and third terms are zero,

$$
\overline{u_i'\left(\frac{\partial\overline{u}_j}{\partial t}\right)} + \overline{u_i'\left(\frac{\partial u_j'}{\partial t}\right)} + \overline{u_j'\left(\frac{\partial\overline{u}_i}{\partial t}\right)} + \overline{u_j'\left(\frac{\partial u_i'}{\partial t}\right)}. \qquad (4.8\text{C})
$$

The two remaining terms can be lumped into a single term by recognizing that they are formed from the derivative of a product, namely,

$$\overline{u_i'\left(\frac{\partial u_j'}{\partial t}\right)} + \overline{u_j'\left(\frac{\partial u_i'}{\partial t}\right)} = \overline{\left[\frac{\partial\left(u_i'u_j'\right)}{\partial t}\right]} \quad \text{(derivative product rule)} \tag{4.8D}$$

Finally, recalling the definition for the Reynolds stress R, the above expression reduces to

$$\overline{\left[\frac{\partial\left(u_i'u_j'\right)}{\partial t}\right]} = -\frac{\partial\widetilde{R}_{ij}}{\partial t}\left(\begin{array}{l}\text{Recall }\widetilde{R}_{ij} = -\overline{u_i'u_j'}.\\ \text{Substitute definition for Reynolds stress.}\end{array}\right). \tag{4.8E}$$

Thus, the key result for the transient term is the transformation of u into the stress tensor $\widetilde{R}_{ij}$:

$$\frac{\partial u_j}{\partial t} \rightarrow -\frac{\partial\widetilde{R}_{ij}}{\partial t}. \tag{4.8F}$$

Recall that the transformed terms will be assembled later.

4.3.2 The Convective Term

Just as was done to the transient (accumulation) term above, now take the convective term and perform the same procedure. In particular, take $u_k\frac{\partial u_i}{\partial x_k}$ and multiply it by the fluctuating velocities:

$$\overline{u_i'\left(u_k\frac{\partial u_j}{\partial x_k}\right) + u_j'\left(u_k\frac{\partial u_i}{\partial x_k}\right)}. \tag{4.9A}$$

Next, substitute the Reynolds decomposition velocity, and apply the time average operator:

$$\overline{u_i'(\overline{u}_k + u_k')\frac{\partial\left(\overline{u}_j + u_j'\right)}{\partial x_k}} + \overline{u_j'(\overline{u}_k + u_k')\left[\frac{\partial(\overline{u}_i + u_i')}{\partial x_k}\right]}. \tag{4.9B}$$

At this point, it is now a matter of simplifying the expression. In this case, use the FOIL multiplication acronym (front, outside, inside, last) to expand each of the above two terms:

$$\left[\overline{u_i'\overline{u}_k \left(\frac{\partial \overline{u}_j}{\partial x_k} \right)} + \overline{u_i'\overline{u}_k \left(\frac{\partial u_j'}{\partial x_k} \right)} + \overline{u_i'u_k' \left(\frac{\partial \overline{u}_j}{\partial x_k} \right)} + \overline{u_i'u_k' \left(\frac{\partial u_j'}{\partial x_k} \right)} \right]$$

$$+ \left[\overline{u_j'\overline{u}_k \left(\frac{\partial \overline{u}_i}{\partial x_k} \right)} + \overline{u_j'\overline{u}_k \left(\frac{\partial u_i'}{\partial x_k} \right)} + \overline{u_j'u_k' \left(\frac{\partial \overline{u}_i}{\partial x_k} \right)} + \overline{u_j'u_k' \left(\frac{\partial u_i'}{\partial x_k} \right)} \right] . \tag{4.9C}$$

Recall that $\overline{\overline{u}u'} = 0$ and $\overline{u'\overline{u}} = 0$, and rearrange a little,

$$\left\{ \left[\overline{u_i'\overline{u}_k \left(\frac{\partial u_j'}{\partial x_k} \right)} + \overline{u_j'\overline{u}_k \left(\frac{\partial u_i'}{\partial x_k} \right)} \right] + \left[\overline{u_i'u_k' \left(\frac{\partial u_j'}{\partial x_k} \right)} + \overline{u_j'u_k' \left(\frac{\partial u_i'}{\partial x_k} \right)} \right] \right\}$$

$$+ \left\{ \overline{u_j'u_k' \left(\frac{\partial \overline{u}_i}{\partial x_k} \right)} + \overline{u_i'u_k' \left(\frac{\partial \overline{u}_j}{\partial x_k} \right)} + \overline{u_j'\overline{u}_k \left(\frac{\partial \overline{u}_j}{\partial x_k} \right)} + \overline{u_j'\overline{u}_k \left(\frac{\partial \overline{u}_i}{\partial x_k} \right)} \right\} . \tag{4.9D}$$

Now combine the derivatives in anticipation of obtaining $\widetilde{R}_{ij}$:

$$\left\{ \overline{\overline{u}_k \left[\frac{\partial \left(u_i'u_j' \right)}{\partial x_k} \right]} + \overline{u_j'u_k' \left(\frac{\partial \overline{u}_i}{\partial x_k} \right)} + \overline{u_i'u_k' \left(\frac{\partial \overline{u}_j}{\partial x_k} \right)} + \overline{u_k' \left[\frac{\partial \left(u_i'u_j' \right)}{\partial x_k} \right]} \right\}$$

$$= -\overline{u}_k \frac{\partial \widetilde{R}_{ij}}{\partial x_k} - \widetilde{R}_{ik} \frac{\partial \overline{u}_j}{\partial x_k} - \widetilde{R}_{jk} \frac{\partial \overline{u}_i}{\partial x_k} + \frac{\partial \left(\overline{u_i'u_j'u_k'} \right)}{\partial x_k} \quad \left(\text{Recall } \widetilde{R}_{ij} = -\overline{u_i'u_j'}. \right) \tag{4.9E}$$

The key results for the convective term transformation are

$$u_k \frac{\partial u_i}{\partial x_k} \rightarrow \underbrace{-\overline{u}_k \frac{\partial \widetilde{R}_{ij}}{\partial x_k}}_{\substack{\text{Stress} \\ \text{convection}}} \quad \underbrace{-\widetilde{R}_{ik} \frac{\partial \overline{u}_j}{\partial x_k} - \widetilde{R}_{jk} \frac{\partial \overline{u}_i}{\partial x_k}}_{\text{Production term: ``eddy factory''}} + \underbrace{\frac{\partial \left(\overline{u_i'u_j'u_k'} \right)}{\partial x_k}}_{\substack{\text{triple velocity} \\ \text{correlation}}} . \tag{4.9F}$$

The second term on the RHS accounts for the production of eddies. It should come as no surprise that operators applied onto the momentum convection term unveil a "hidden" eddy-production term, which accounts for eddies that arise from flow instabilities in the mean flow. *That is, eddy production is a consequence of the nonlinear convective term.* It is also very insightful that production is the product of the stress and velocity gradient of the mean flow. Velocity gradients are largest near the wall, and this is precisely where the production term is the largest, in the approximate range of $7 < y^+ < 25$, and thereafter drops exponentially as y^+ increases. As a rule of thumb, this y^+ behavior is applicable for Re in the range of 3000–40,000

(Mansour et al. 1988; Wilcox 2006). Note as well that the triple velocity correlation term originated from the convective term.

4.3.3 The Viscous Term

Now apply the same procedure onto the viscous term, $\nu \frac{\partial^2 u_i}{\partial x_k \partial x_k}$, such that

$$\nu \overline{\left[u_i' \left(\frac{\partial^2 u_j}{\partial x_k \partial x_k} \right) + u_j' \left(\frac{\partial^2 u_i}{\partial x_k \partial x_k} \right) \right]} = \nu u_i' \frac{\overline{\partial^2 \left(\bar{u}_j + u_j' \right)}}{\partial x_k \partial x_k} + \nu u_j' \frac{\overline{\partial^2 \left(\bar{u}_i + u_i' \right)}}{\partial x_k \partial x_k}. \quad (4.10\text{A})$$

Recalling that both $\overline{u'\bar{u}}$ and $\overline{u' \frac{\partial \bar{u}}{\partial x}}$ are equal to zero, and expanding the terms, reduces to

$$\nu \left(\overline{u_i' \frac{\partial^2 \bar{u}_j}{\partial x_k \partial x_k}} + \overline{u_i' \frac{\partial^2 u_j'}{\partial x_k \partial x_k}} + \overline{u_j' \frac{\partial^2 \bar{u}_i}{\partial x_k \partial x_k}} + \overline{u_j' \frac{\partial^2 u_i'}{\partial x_k \partial x_k}} \right). \quad (4.10\text{B})$$

Applying the mathematical trick that

$$\overline{u_i' \frac{\partial^2 u_j'}{\partial x_k \partial x_k}} = \frac{\overline{\partial \left(u_i' \frac{\partial u_j'}{\partial x_k} \right)}}{\partial x_k} + \overline{\frac{\partial u_i'}{\partial x_k} \frac{\partial u_j'}{\partial x_k}} \quad (4.10\text{C})$$

and in an analogous manner,

$$\overline{u_j' \frac{\partial^2 u_i'}{\partial x_k \partial x_k}} = \frac{\overline{\partial \left(u_j' \frac{\partial u_i'}{\partial x_k} \right)}}{\partial x_k} + \overline{\frac{\partial u_j'}{\partial x_k} \frac{\partial u_i'}{\partial x_k}}. \quad (4.10\text{D})$$

Note also that

$$\overline{\frac{\partial u_i'}{\partial x_k} \frac{\partial u_j'}{\partial x_k}} = \overline{\frac{\partial u_j'}{\partial x_k} \frac{\partial u_i'}{\partial x_k}}. \quad (4.10\text{E})$$

Substituting the above mathematical equivalents onto the time-averaged, Reynolds-decomposed, first-moment viscous term results in the following expression:

$$\nu \left\{ \left[\frac{\partial \left(\overline{u_i' \frac{\partial u_j'}{\partial x_k}} \right)}{\partial x_k} + \frac{\partial \left(\overline{u_j' \frac{\partial u_i'}{\partial x_k}} \right)}{\partial x_k} \right] - 2 \overline{\frac{\partial u_i'}{\partial x_k} \frac{\partial u_j'}{\partial x_k}} \right\}. \qquad (4.10\text{F})$$

The first two terms in the above expression can be lumped, so Eq. 4.10F reduces to

$$\nu \left\{ \left[\frac{\partial^2 \left(\overline{u_i' u_j'} \right)}{\partial x_k \partial x_k} \right] - 2 \overline{\frac{\partial u_i'}{\partial x_k} \frac{\partial u_j'}{\partial x_k}} \right\}, \qquad (4.10\text{G})$$

and in anticipation of swapping the product of the two fluctuating velocities with the Reynolds stress,

$$\nu \left\{ \left[\frac{\partial}{\partial x_k} \left(\frac{\partial}{\partial x_k} \left(\overline{u_i' u_j'} \right) \right) \right] - 2 \overline{\frac{\partial u_i'}{\partial x_k} \frac{\partial u_j'}{\partial x_k}} \right\} = -\nu \frac{\partial^2 \widetilde{R}_{ij}}{\partial x_k \partial x_k} - 2\nu \overline{\frac{\partial u_i'}{\partial x_k} \frac{\partial u_j'}{\partial x_k}}. \qquad (4.10\text{H})$$

The key results for the viscous term transformation are therefore

$$\nu \frac{\partial^2 u_i}{\partial x_k \partial x_k} \quad \rightarrow \quad - \underbrace{\nu \frac{\partial^2 \widetilde{R}_{ij}}{\partial x_k \partial x_k}}_{\text{Viscous stress}} \quad - \underbrace{2\nu \overline{\frac{\partial u_i'}{\partial x_k} \frac{\partial u_j'}{\partial x_k}}}_{\text{Eddy dissipation}}. \qquad (4.10\text{I})$$

$$\text{diffusion}$$

The second term on the RHS accounts for eddy dissipation (decay). It comes as no surprise that operators applied onto the momentum viscous term unveil a "hidden" eddy-damping term. This "new" term fosters the decay of eddies, until all their turbulent energy reverts to the mean flow as heat, thereby causing the eddy's swirling motion to cease. That is, eddy decay is a consequence of the viscous damping term, which applies the brakes onto eddies.

4.3.4   The Pressure Term

Finally, apply the same procedure to the pressure term, $\frac{1}{\rho} \frac{\partial P}{\partial x_i}$.

Here, there are no instantaneous velocities, but there is an instantaneous pressure that can be decomposed using Reynolds decomposition, $P = \overline{P} + P'$.

Other than this change, the procedure is the same as was used for the other three momentum terms. Namely,

$$\overline{u_i' \frac{\partial P}{\partial x_j}} + \overline{u_j' \frac{\partial P}{\partial x_i}} = \overline{u_i' \frac{\partial (\overline{P} + P')}{\partial x_j}} + \overline{u_j' \frac{\partial (\overline{P} + P')}{\partial x_i}}. \tag{4.11A}$$

Now take the derivatives, and recall that because $\overline{u'\overline{P}}$ is zero, then $\overline{u' \frac{\partial \overline{P}}{\partial x}}$ must also be zero, thereby yielding

$$\overline{u_i' \frac{\partial \overline{P}}{\partial x_j}} + \overline{u_i' \frac{\partial P'}{\partial x_j}} + \overline{u_j' \frac{\partial \overline{P}}{\partial x_i}} + \overline{u_j' \frac{\partial P'}{\partial x_i}}. \tag{4.11B}$$

Using the property that $\overline{a + b} = \overline{a} + \overline{b}$ leads to the simplification of the pressure term as

$$\overline{u_i' \frac{\partial P'}{\partial x_j}} + \overline{u_j' \frac{\partial P'}{\partial x_i}}. \tag{4.11C}$$

The key result for the pressure term transformation is that

$$\frac{1}{\rho} \frac{\partial P}{\partial x_i} \quad \longrightarrow \quad \underbrace{\frac{1}{\rho} \left(\overline{u_i' \frac{\partial P'}{\partial x_j}} + \overline{u_j' \frac{\partial P'}{\partial x_i}} \right)}_{\substack{\text{Stress transport due to} \\ \text{pressure-velocity fluctuations}}}. \tag{4.11D}$$

4.3.5 Assembly of the R Reynolds Stress PDE

At last, the above key results that were derived for each transformed momentum term are assembled, and dividing by -1 results in the sought-after expression for the Reynolds stress $\widetilde{R}_{ij}$ PDE:

$$\underbrace{\frac{\partial \widetilde{R}_{ij}}{\partial t}}_{1} + \underbrace{\overline{u}_k \frac{\partial \widetilde{R}_{ij}}{\partial x_k}}_{2} = \underbrace{-\left(\widetilde{R}_{ik} \frac{\partial \overline{u}_j}{\partial x_k} + \widetilde{R}_{jk} \frac{\partial \overline{u}_i}{\partial x_k} \right)}_{3} + \underbrace{2\nu \overline{\frac{\partial u_i'}{\partial x_k} \frac{\partial u_j'}{\partial x_k}}}_{4} + \underbrace{\frac{\partial}{\partial x_k} \left(\nu \frac{\partial \widetilde{R}_{ij}}{\partial x_k} \right)}_{5}$$

$$+ \underbrace{\frac{1}{\rho} \left(\overline{u_i' \frac{\partial P'}{\partial x_j}} + \overline{u_j' \frac{\partial P'}{\partial x_i}} \right)}_{6} + \underbrace{\frac{\partial}{\partial x_k} \left(\overline{u_i' u_j' u_k'} \right)}_{7}. \tag{4.12}$$

A description of the Reynolds stress PDE terms is as follows:

Term 1 = transient (accumulation) stress rate of change

Term 2 = stress rate of change associated with convection

Term 3 = production terms that arise from the product of the stress $\widetilde{R}$ (calculated via the Boussinesq approximation) and the mean velocity gradients. The production term quantifies the rate at which the mean flow imparts energy onto eddies

Term 4 = stress dissipation rate. This represents the rate at which the stress-generated eddies revert their turbulent energy back into the flow's internal energy. The dissipation stress tensor is defined as

$$\varepsilon_{ij} \equiv 2\nu \overline{\frac{\partial u_i'}{\partial x_k} \frac{\partial u_j'}{\partial x_k}}, \tag{4.13}$$

Term 5 = rate of viscous (molecular) stress diffusion

Term 6 = turbulent stress transport rate associated with the eddy pressure and velocity fluctuations

Term 7 = diffusive stress transport rate resulting from the triple-correlation eddy velocity fluctuations

Note that Terms 1, 2, and 5 are described exactly in a mathematical and physical sense and thus have no need for the dreaded and infamous "drastic surgery" described by Wilcox. Term 3 is reasonably modeled (most often) using the Boussinesq approximation. However, if more complex flows need to be modeled, then Boussinesq can be replaced with a more appropriate (elaborate) expression, such as those that include rotational and higher-order terms.

On the other hand, the remaining three terms, the dissipation, the pressure-velocity fluctuation correlation, and the eddy fluctuations associated with the triple velocity correlation, are not known and thus require "drastic surgery." As so fondly described by Wilcox (2006),"In essence, Reynolds averaging is a brutal simplification that loses much of the information contained in the Navier-Stokes equation."

And this certainly applies to the R PDE. Note that Wilcox applied the term "drastic surgery" to both the k and ε PDEs. However, the transformation of the k PDE requires some drastic engineering measures, but certainly not to the degree of the ε PDE, as will be shown later.

Recall that RSM models only use six transport PDEs because R is symmetric, so the nine PDEs reduce to six. Further, RSM involves no k PDE because the stresses are calculated individually. On the other hand, RANS assumes that k is the sum of the diagonal stresses:

$$\frac{1}{2}\overline{u_i'u_i'} = \frac{1}{2}\left(\overline{u'^2} + \overline{v'^2} + \overline{w'^2}\right). \tag{4.14}$$

As far as the historical records show, Chou was the first to develop the R PDE, as expressed in more modern terms (Chou 1940 (refer to his report, Eqs. 3.1 and 3.4)). He elaborated other turbulence terms in later papers (Chou 1945; Chou and Chou 1995).

But what does k have to do with R? Everything, as will be shown next; k is a simplified case of R!

4.4   Development of the k PDE and a More Formal Definition for k

Again, because the RANS closure process is rather long-winded, Fig. 4.1b shows a "bird's eye view" to demonstrate how R is used to derive the k PDE, which is used in most RANS-based turbulence models. Therefore, Fig. 4.1b shows an overview of the entire process, while Sect. 4.4 provides the details.

As formidable as the Reynolds stress tensor PDE appears, it can be vastly simplified by taking the mathematical trace of all its terms, thereby generating a new expression for k. This is achieved as follows. First, consider the trace for the definition of tensor $\widetilde{R}_{ij}$:

$$\text{tr}\left(\widetilde{R}_{ij}\right) = \widetilde{R}_{ii} \equiv -\overline{u_i' u_i'}. \tag{4.15}$$

Recalling the definition for k first proposed by Prandtl in 1945,

$$k = \frac{1}{2}\overline{u_i' u_i'} \tag{4.16A}$$

or

$$\overline{u_i' u_i'} = 2k. \tag{4.16B}$$

Therefore,

$$\widetilde{R}_{ii} = -\overline{u_i' u_i'} = -2k. \tag{4.17}$$

Next, take the trace of R (Eq. 4.12), thereby yielding

$$\underbrace{\frac{\partial \widetilde{R}_{ii}}{\partial t}}_{1} + \underbrace{\overline{u}_k \frac{\partial \widetilde{R}_{ii}}{\partial x_k}}_{2} = \underbrace{2\widetilde{R}_{ij}\frac{\partial \overline{u}_i}{\partial x_j}}_{3} + \underbrace{\varepsilon_{ii}}_{4} + \underbrace{\frac{\partial}{\partial x_k}\left(\nu \frac{\partial \widetilde{R}_{ii}}{\partial x_k}\right)}_{5} + \underbrace{\frac{2}{\rho}\left(\overline{u_i'\frac{\partial P'}{\partial x_i}}\right)}_{6}$$

$$+ \underbrace{\frac{\partial}{\partial x_k}\left(\overline{u_i' u_i' u_k'}\right)}_{7}. \tag{4.18}$$

Note that the two production terms are combined into a single term (Term 3). At this point, $\widetilde{R}_{ii}$ in Eq. 4.18 can be replaced with its equivalent (i.e., $-2k$ from Eq. 4.17). Hence, the *exact* k PDE is obtained at last:

$$\underbrace{\frac{\partial k}{\partial t}}_{1} + \underbrace{\bar{u}_i \frac{\partial k}{\partial x_i}}_{2} = \underbrace{\widetilde{R}_{ij} \frac{\partial \bar{u}_i}{\partial x_j}}_{3} - \underbrace{\varepsilon_{ii}}_{4} + \underbrace{\frac{\partial \left(\nu \frac{\partial k}{\partial x_i}\right)}{\partial x_i}}_{5} - \underbrace{\frac{1}{\rho} \frac{\partial \left(\overline{P'u_i'}\right)}{\partial x_i}}_{6} - \underbrace{\frac{1}{2} \frac{\partial}{\partial x_i} \left(\overline{u_j'u_j'u_i'}\right)}_{7}. \quad (4.19)$$

It is pointed out that the *solvable k* PDE originates from the *exact k* PDE; that is, once the *exact* form undergoes some "drastic surgery" (engineering approximations), it becomes solvable and is used in most RANS turbulence models. The transformation will be discussed shortly.

In the meantime, the *exact k* PDE includes the following terms:

Term 1 = transient change (accumulation)

Term 2 = convection

Term 3 = production term that arises from the product of the Boussinesq stress $\widetilde{R}_{ij}$ and the mean velocity gradients. This is a metric for the rate at which the mean flow imparts energy onto eddies

Term 4 = dissipation rate. This represents the rate at which k is converted back into internal energy

Term 5 = viscous (molecular) diffusion of turbulence energy

Term 6 = turbulent transport associated with the eddy pressure and velocity fluctuation

Term 7 = diffusive turbulent transport resulting from the eddy triple-correlation velocity fluctuations

For the k PDE, Terms 1, 2, and 5 are described exactly in a mathematical and physical sense, with no need for "drastic surgery" conversions. The production term is reasonably modeled using the Boussinesq approximation or some form of nonlocal, nonequilibrium approach (Speziale and Eringen 1981; Hamba 2005; Schmitt 2007; Hamlington and Dahm 2009; Wilcox 2006; Spalart 2015). On the other hand, the remaining three terms, dissipation, eddy fluctuation due to the triple velocity correlation, and the pressure-velocity fluctuation correlation, are not known and thus require the dreaded "drastic surgery" described by Wilcox. *On the brighter side, this situation applies to three terms in the k PDE, whereas the ε PDE requires drastic surgery for six terms, perhaps making the k PDE 50% more palatable!*

The expression for Term 4, $\varepsilon_{ii} = \varepsilon$, depends on which model is used. For example, Prandtl's one-equation model assumes that

$$\varepsilon = \frac{C_D k^{3/2}}{\ell}. \quad (4.20)$$

On the other hand, the standard k-ε (SKE) model uses the moment approach to derive an ε PDE, analogous to the development of the k PDE.

For Terms 6 and 7, dimensional arguments loosely based on gradient transport are employed (refer to Chap. 2). Thus,

$$-\overline{u_i'\phi'} = \nu_t \frac{\partial \overline{\phi}}{\partial x_i}. \tag{4.21}$$

Then, the two terms are lumped, with the following bold (and desperate!) assertion that

$$-\frac{1}{\rho}\overline{P'u_i'} - \frac{1}{2}\overline{u_j'u_j'u_i'} \approx \frac{\nu_t}{\sigma_k}\frac{\partial k}{\partial x_i}. \tag{4.22}$$

Of course, neither $\overline{P'u_i'}$ nor $\overline{u_j'u_j'u_i'}$ have much to do with either ν_t or the spatial gradient of k (but doesn't the expression seem respectable?). To say the least, the analytical expression of the two terms is currently unknown, so they are reconfigured as a wishful and fanciful expression with no physical or mathematical basis other than having the appropriate dimensional arguments! Presumably, the constant-valued "correction factor" σ_k helps the user make adjustments if experimental data should be available. In any case, the "drastic surgery" Terms 6 and 7 are lumped as follows:

$$\underbrace{-\frac{1}{\rho}\frac{\partial\left(\overline{P'u_i'}\right)}{\partial x_i}}_{6} \underbrace{-\frac{1}{2}\frac{\partial}{\partial x_i}\left(\overline{u_j'u_j'u_i'}\right)}_{7} = -\frac{\partial\left(\frac{1}{\rho}\overline{P'u_i'} + \frac{1}{2}\overline{u_j'u_j'u_i'}\right)}{\partial x_i} \approx \frac{\partial\left(\frac{\nu_t}{\sigma_k}\frac{\partial k}{\partial x_i}\right)}{\partial x_i}. \tag{4.23}$$

So yes, the lumped term has a dubious origin. On the other hand, does someone have a better idea? What else can turbulence researchers do when presented with such a dilemma? It is difficult to obtain accurate measurements for experimental pressure fluctuations and triple-correlation terms. Fortunately, the above approximation may be fairly harmless under simple turbulent flows. For example, DNS at $Re = 3200$ shows that Term 7 is of the same magnitude as the production term for $y^+ < 7$ (the viscous sublayer), while production was much larger for $10 < y^+ < 100$. For $y^+ > 100$, the two approached each other asymptotically (Mansour et al. 1988). On the other hand, the pressure gradient term (Term 6) was relatively smaller than the production term for $y^+ < 50$ and was of comparable magnitude for $y^+ > 50$. It is noted that these differences are expected to be magnified as Re increases. In summary, the lumping of Terms 6 and 7 to form the product of the turbulent kinematic viscosity and the spatial gradient of the turbulent kinetic energy can result in significant issues in complex flows involving high Re and two-phase flows (Mansour et al. 1988; Sawko 2012).

Nevertheless, it is unfortunate that the triple velocity correlation is "wished away" with a term that does not reflect its behavior. This is especially so, because the triple velocity correlation originates from the convective term, which is also the source for the crucial production term. Therefore, a more rigorous methodology to estimate the term's turbulent behavior is highly desirable.

Despite any derivation issues, the k PDE is at last in a form that is solvable as a result of the aforementioned transformations, approximations, and an expression for the turbulent kinematic viscosity, ν_t,

$$\frac{\partial k}{\partial t} + \bar{u}_i \frac{\partial k}{\partial x_i} = \tilde{R}_{ij} \frac{\partial \bar{u}_i}{\partial x_j} - \varepsilon + \frac{\partial\left(\nu \frac{\partial k}{\partial x_i}\right)}{\partial x_i} + \frac{1}{\sigma_k}\frac{\partial\left(\nu_t \frac{\partial k}{\partial x_i}\right)}{\partial x_i}. \tag{4.24}$$

Notice that if the RANS zero-, one-, and two-equation models use the above k PDE, then they assume ν_t is *isotropic*. Once an expression for ν_t is known (e.g., Prandtl-Kolmogorov for the k-ε model, k/ω for the k-ω model, etc.), then the entire set of equations required for closure is complete, and computational turbulence modeling may proceed at last!

Generally (with some exceptions), the lower the number of transport equations used, the faster the model will run for simple turbulent flows; by contrast, more complex behaviors can be analyzed as a larger number of transport equations are used but at the expense of longer computational time.

In any case, now that the k PDE has been derived, it is noted that its full expression, albeit an approximation, is rather complex. A full expansion of the k PDE shows just how complex the k behavior can be:

$$\frac{\partial k}{\partial t} + \bar{u}\frac{\partial k}{\partial x} + \bar{v}\frac{\partial k}{\partial y} + \bar{w}\frac{\partial k}{\partial z} =$$

$$\left\{\left[\nu_t\left(\frac{\partial \bar{u}}{\partial x} + \frac{\partial \bar{u}}{\partial x}\right) - \frac{2}{3}k\right]\frac{\partial \bar{u}}{\partial x} + \nu_t\left(\frac{\partial \bar{u}}{\partial y} + \frac{\partial \bar{v}}{\partial x}\right)\frac{\partial \bar{u}}{\partial y} + \nu_t\left(\frac{\partial \bar{u}}{\partial z} + \frac{\partial \bar{w}}{\partial x}\right)\frac{\partial \bar{u}}{\partial z}\right\}$$

$$+\left\{\nu_t\left(\frac{\partial \bar{v}}{\partial x} + \frac{\partial \bar{u}}{\partial y}\right)\frac{\partial \bar{v}}{\partial x} + \left[\nu_t\left(\frac{\partial \bar{v}}{\partial y} + \frac{\partial \bar{v}}{\partial y}\right) - \frac{2}{3}k\right]\frac{\partial \bar{v}}{\partial y} + \nu_t\left(\frac{\partial \bar{v}}{\partial z} + \frac{\partial \bar{w}}{\partial y}\right)\frac{\partial \bar{v}}{\partial z}\right\}$$

$$+\left\{\nu_t\left(\frac{\partial \bar{w}}{\partial x} + \frac{\partial \bar{u}}{\partial z}\right)\frac{\partial \bar{w}}{\partial x} + \nu_t\left(\frac{\partial \bar{w}}{\partial y} + \frac{\partial \bar{v}}{\partial z}\right)\frac{\partial \bar{w}}{\partial y} + \left[\nu_t\left(\frac{\partial \bar{w}}{\partial z} + \frac{\partial \bar{w}}{\partial z}\right) - \frac{2}{3}k\right]\frac{\partial \bar{w}}{\partial z}\right\}$$

$$-\varepsilon + \frac{\partial}{\partial x}\left[\left(\nu + \frac{\nu_t}{\sigma_k}\right)\frac{\partial k}{\partial x}\right] + \frac{\partial}{\partial y}\left[\left(\nu + \frac{\nu_t}{\sigma_k}\right)\frac{\partial k}{\partial y}\right] + \frac{\partial}{\partial z}\left[\left(\nu + \frac{\nu_t}{\sigma_k}\right)\frac{\partial k}{\partial z}\right].$$

$$\tag{4.25}$$

- *Chou's k PDE*

The development of the k PDE has a colorful history, with many insights. The earliest version is attributed here to Chou because he derived the R PDE (see Sect. 4.6.3.5) and then proceeded to "*contract* the indices...to find the equation of '*energy transport*'" (Chou 1940). *Note that "tensor contraction" is a generalized form of the trace.* Chou's original 1940 k PDE is as follows:

$$-R_{kj}\varepsilon_{kj} + \frac{1}{2}\frac{\partial\left[\overline{\left(\bar{u}_j + u'_j\right)q^2}\right]}{\partial x_j} = -\frac{1}{\rho}\frac{\partial\left(\overline{u'_k P'}\right)}{\partial x_k} + \frac{1}{2}v\nabla^2 q^2 - vg^{mn}\overline{\frac{\partial u'_k}{\partial x_m}}$$

$$\times \frac{\partial u'_k}{\partial x_n}, \qquad (4.26A)$$

where, in Chou's notation,

$$R_{kk} \equiv q^2 = \overline{u'_i u'_i}, \qquad (4.26B)$$

$$\varepsilon_{kj} \equiv S_{kj} \text{ (Chou chose } \varepsilon \text{ as the } deformation \text{ tensor, not } dissipation), \qquad (4.26C)$$

and

$$g^{mn} = 1. \qquad (4.26D)$$

Upon rearrangement, multiplication by 2, and some substitution, Eq. 4.26A becomes

$$-2R_{kj}S_{kj} + \frac{\partial\left[\overline{\left(\bar{u}_j + u'_j\right)q^2}\right]}{\partial x_j} = -\frac{2}{\rho}\frac{\partial\left(\overline{u'_k P'}\right)}{\partial x_k} + v\nabla^2 k - 2v\overline{\frac{\partial u'_k}{\partial x_m}\frac{\partial u'_k}{\partial x_n}}, \qquad (4.27A)$$

where the second term on the LHS becomes

$$\frac{\partial\left[\overline{\left(\bar{u}_j + u'_j\right)q^2}\right]}{\partial x_j} = \frac{\partial\left(\overline{\bar{u}_j q^2}\right)}{\partial x_j} + \frac{\partial\left(\overline{u'_j q^2}\right)}{\partial x_j} = \frac{\partial\left(\overline{\bar{u}_j k}\right)}{\partial x_j} + \frac{\partial\left(\overline{u'_j u'_i u'_i}\right)}{\partial x_j}$$

$$= \bar{u}_j\frac{\partial k}{\partial x_j} + k\frac{\partial\bar{u}_j}{\partial x_j} + \frac{\partial\left(\overline{u'_j u'_i u'_i}\right)}{\partial x_j}. \qquad (4.27B)$$

Therefore, Chou's PDE is

$$-2R_{kj}S_{kj} + \bar{u}_j\frac{\partial k}{\partial x_j} + k\frac{\partial\bar{u}_j}{\partial x_j} + \frac{\partial\left(\overline{u'_j u'_i u'_i}\right)}{\partial x_j} = -\frac{2}{\rho}\frac{\partial\left(\overline{u'_k P'}\right)}{\partial x_k} + v\nabla^2 k - 2v\overline{\frac{\partial u'_k}{\partial x_m}\frac{\partial u'_k}{\partial x_n}}. \qquad (4.28)$$

At this point, it is convenient to recall the *modern* definition for the *dissipation* symbol ε, namely,

$$\varepsilon_{\text{diss}} \equiv 2\nu \overline{\frac{\partial u_k'}{\partial x_m} \frac{\partial u_k'}{\partial x_n}}, \tag{4.29}$$

which is actually the last term in Chou's PDE equation.

Therefore, Chou's k PDE, in more *modern* terms, and somewhat more reorganized, is as follows at this point:

$$\overline{u}_j \frac{\partial k}{\partial x_j} + k \frac{\partial \overline{u}_j}{\partial x_j} = 2R_{kj}S_{kj} - \varepsilon + \frac{\partial \left(\nu \frac{\partial k}{\partial x_i} \right)}{\partial x_i} - \frac{2}{\rho} \frac{\partial \left(\overline{u_k' P'} \right)}{\partial x_k} - \frac{\partial \left(\overline{u_j' u_i' u_i'} \right)}{\partial x_j}. \tag{4.30}$$

Of course, now that turbulence modelers have more information, the last two terms in Chou's PDE formulation can be lumped using dimensional arguments based on gradient transport (as discussed earlier in this section), whereby

$$-\frac{1}{\rho} \overline{u_i' P'} - \frac{1}{2} \overline{u_j' u_i' u_i'} \approx \frac{\nu_t}{\sigma_k} \frac{\partial k}{\partial x_i}. \tag{4.31}$$

Therefore, Chou's 1940 k PDE has the following equivalent modern expression:

$$\overline{u}_j \frac{\partial k}{\partial x_j} + k \frac{\partial \overline{u}_j}{\partial x_j} = 2R_{ij} \frac{\partial \overline{u}_i}{\partial x_j} - \varepsilon + \frac{\partial \left(\nu \frac{\partial k}{\partial x_i} \right)}{\partial x_i} + \frac{2}{\sigma_k} \frac{\partial \left(\nu_t \frac{\partial k}{\partial x_i} \right)}{\partial x_i}. \tag{4.32}$$

Consequently, except for the missing transient term and an extra term, Chou's k PDE is the same as the modern k PDE. Nevertheless, it is pointed out that the transient term is missing in Chou's formulation because he attempted to apply average values "with respect to time over a period τ." *Intriguingly, Chou also included a k transport term on the LHS $\left(k \frac{\partial \overline{u}_j}{\partial x_j} \right)$, which is not present in the modern k PDE.*

- *Toward a More Formal Definition of k*

To better understand k, consider it as a fundamental metric of the energy contained within the swirling, dynamic, 3D, high-vorticity flow sheets that form eddies. It is this energy that shapes the fluid so that it rolls into curls, eddies, and all sorts of coherent structures—this is Reynolds' *eddying motion*. The eddy velocity is based on clusters of fluid that move in a coherent fashion, with fluctuating velocity u'; it is the square of the fluctuating velocities that provides a measure of the eddy's turbulent kinetic energy. If a mathematical definition of a root-mean-square (RMS) for a generalized n-space involving n velocities is considered, then

$$u' \equiv \sqrt{\frac{1}{n} \left(\overline{u_1'^2} + \cdots + \overline{u_n'^2} \right)}. \tag{4.33A}$$

For example, u' can be expressed in its generalized, anisotropic form in a Cartesian, 3D space as a fluctuating velocity RMS (Kolmogorov 1942; Spiegel 1999; SimScale 2018):

$$u' = \sqrt{\frac{1}{3}\left(\overline{u'^2} + \overline{v'^2} + \overline{w'^2}\right)}. \tag{4.33B}$$

Because the square of the three velocities is summed and then divided by three, the quantity inside the square root represents an *average* velocity in this situation.

The velocity fluctuations for an isotropic flow can be expressed as a function of the turbulence kinetic energy, indicating that higher velocity fluctuations occur for the more energetic eddies:

$$u' = \sqrt{\frac{2}{3}k}. \tag{4.33C}$$

If the above two equations are set equal and solved for k, then

$$k = \frac{1}{2}\left(\overline{u'^2} + \overline{v'^2} + \overline{w'^2}\right), \tag{4.34}$$

which is the exact expression for the turbulent kinetic energy that was first used by Prandtl in the development of his one-equation turbulence transport model (Prandtl 1945). Note that the above expression uses 1/2 instead of 1/3 (which is what Kolmogorov used in his 1942 turbulence model). Thus, the Prandtl formulation does not strictly follow the mathematical definition of an RMS, though it is quite often referred as such in the literature. That is, the magnitude of the constant that relates the fluctuating velocities and k is 1/2, even though 1/3 is the expected RMS quantity for the RHS of Eq. 4.34.

In any case, k can be expressed as a simplified function of the fluctuating velocities for isotropic eddies, whereby

$$\overline{u'^2} = \overline{v'^2} = \overline{w'^2}, \tag{4.35}$$

and therefore, in a rather circuitous logic, the expression for Eq. 4.33C is obtained:

$$k = \frac{3}{2}\overline{u'^2}. \tag{4.36}$$

However, the above expression for k arises more rigorously from the integral of the energy spectral density across all the eddy wave numbers and is represented by the Greek letter kappa (κ, not to be confused with "k"). The function is shown in Fig. 3.1 in Chap. 3, and its integration yields the total kinetic energy k held by the entire eddy spectrum, from the largest to the smallest eddies:

$$k = \int_{0}^{\infty} E(\kappa)\, d\kappa = \frac{3}{2}\overline{u'^2}. \tag{4.37}$$

In this context, the wavenumber (eddy size) κ is

$$\kappa = \frac{2\pi}{\widetilde{\lambda}} \text{ or } \frac{1}{\kappa} = \frac{\widetilde{\lambda}}{2\pi}, \tag{4.38}$$

where the 2π factor comes into play because κ is in terms of *radians* per unit length.

Thus, the integration across all wavelengths yields the same total kinetic energy as the isotropic flow approximation. Stated differently, it is imperative that $k = \frac{1}{2}\left(\overline{u'^2} + \overline{v'^2} + \overline{w'^2}\right)$ is used if $k = \frac{3}{2}\overline{u'^2}$ is to be satisfied; by the same token, using $k = \frac{1}{3}\left(\overline{u'^2} + \overline{v'^2} + \overline{w'^2}\right)$ will result in an inconsistency (i.e., $k = \overline{u'^2}$ in this case). The modern literature uses the Prandtl notation fairly exclusively, and this is perhaps because it is consistent with the total kinetic energy for isotropic flow. For this reason, the Prandtl k is the preferred version in this book. For convenience, Table 4.1 shows some expressions and approximations for k.

4.5   The Choice of Transport Variables

So, why are there various choices for transport variables and which might be best? When viewed solely from a pure dimensionless perspective (as many turbulence researchers have often done), the situation reduces to the following issue: How can a mathematical expression for ν_t be derived such that it is formed solely from relevant transport variables? This reasoning, of course, is diametrically opposite to the approach whereby only turbulence principles are considered. Nevertheless, the dimensionless approach has resulted in plenty of great numerical simulations, and it is difficult to argue with success! Thus, as discussed previously, k is used because its square root yields a reasonable transport eddy scale—a turbulence velocity. So, researchers reason that if analytical expressions for ν_t and k are not available, then which dimensionally correct combination of k and some other transport variable can form ν_t, thereby reaching mathematical closure? In other words, closure is sought such that

$$\nu_t = \nu_t(k; \omega, \varepsilon, \ell, t \ldots). \tag{4.39A}$$

In SI units, ν_t is in units of m²/s. Thus, the goal is to find a second transport variable that is (1) compatible with k, (2) has a plausible chance of describing the eddy behavior as transport variable "x," and (3) conveniently combines with k to form an expression that has units of m²/s, and thus,

Table 4.1 Expressions and approximations for k

Origin	k formula or similar expression	Isotropic k (or similar)	Notes
Kolmogorov (1942)	$k = \frac{1}{3}\left(\overline{u'^2} + \overline{v'^2} + \overline{w'^2}\right)$	$k_{\mathrm{iso}} = \overline{u'^2}$	Based on a formal RMS representation. Kolmogorov used "b" as his symbol for "k"
Prandtl (1945)	$k = \frac{1}{2}\left(\overline{u'^2} + \overline{v'^2} + \overline{w'^2}\right)$	$k_{\mathrm{iso}} = \frac{3}{2}\overline{u'^2}$	Used predominantly in the modern literature; this is the preferred version
Chou (1945)	$q^2 = \overline{u_i' u_i'}$ Per Chou, this is based on "the root-mean-square of the velocity fluctuation" (Chou 1945). Therefore, $q^2 = \overline{u_i' u_i'} = \frac{1}{3}\left(\overline{u'^2} + \overline{v'^2} + \overline{w'^2}\right)$, which is consistent with Kolmogorov	$q^2_{\mathrm{iso}} = \frac{1}{3}q^2 = \frac{1}{3}\left(\overline{u'^2} + \overline{v'^2} + \overline{w'^2}\right) = \overline{u'^2}$	q^2 is still used in the literature
RMS formula	For an n-dimension system, $u' \equiv \sqrt{\frac{1}{n}\left(\overline{u_1'^2} + \cdots + \overline{u_n'^2}\right)}$ For Cartesian systems, $u' = \sqrt{\frac{1}{3}\left(\overline{u'^2} + \overline{v'^2} + \overline{w'^2}\right)}$	$q^2_{\mathrm{iso}} = \overline{u'^2}$	Based on the mathematical expression for RMS

$$\nu_t = \nu_t(k, x). \tag{4.39B}$$

This is very convenient, as the introduction of the second transport variable is required to form part of the expression for ν_t; by doing so, no more new variables are introduced, so closure is easier to achieve.

Some examples of this dimensionless approach are shown in Table 4.2, which lists reasonable pairs of variables in Column 1. Prandtl used the following expressions for his one-transport model (Prandtl 1945):

$$\nu_t = C_D \frac{k^2}{\varepsilon} \tag{4.40}$$

and

$$\varepsilon = C_D \frac{k^{3/2}}{\ell}. \tag{4.41}$$

Certainly, this approach is not confined to one or two transport variables, and models with three or more variables can be found in the literature (Wilcox 2006; Bna et al. 2012).

In general, the above models can be summarized with a single equation, such that

$$\nu_t = \alpha k^\beta x^\gamma \tag{4.42}$$

where

k = kinetic energy (first transport variable),
x = second transport variable, and
α, β, and γ are constants chosen such that ν_t is in units of length squared per unit time.

Values for the constants are summarized in Table 4.3 for several well-known turbulence models.

Table 4.2 Transport variable development for turbulence kinematic viscosity relationships

Transport variable pair	Variable formulation	Turbulence kinematic viscosity expression
$\{k, \omega\}$	$\omega \approx \frac{k^{1/2}}{\ell}$	$\nu_t \approx \frac{k}{\omega}$ (Kolmogorov 1942; Wilcox 2006)
$\{k, \varepsilon\}$	$\varepsilon \approx \frac{k^{3/2}}{\ell}$	$\nu_t \approx \frac{k^2}{\varepsilon}$ (Chou 1945; Taylor suggested $\varepsilon \approx \frac{k^{3/2}}{\ell}$ in 1935)
$\{k, \ell\}$	$\ell \approx \frac{k^{3/2}}{\varepsilon}$	$\nu_t \approx k^{1/2}\ell$ (Rotta 1951, 1962; Taylor suggested $\varepsilon \approx \frac{k^{3/2}}{\ell}$ in 1935)
$\{k, t\}$	$t \approx \frac{\ell}{k^{1/2}}$	$\nu_t = C_\mu kt$ (t = turbulence dissipation time) (Zeierman and Wolfshtein 1986)

Table 4.3 Transport variables for turbulence equations based on $\nu_t = \alpha k^\beta x^\gamma$

ν_t	x description	α	β	γ
$\nu_t = \frac{k}{\omega}$	$x = \omega$, eddy frequency (Kolmogorov k-ω, 2006 k-ω)	1	1	-1
$\nu_t = k^{1/2}\ell$	$x = \ell$, eddy length scale Prandtl one-equation model with constitutive relationship for ℓ; Rotta two-equation model (Rotta 1962) $\ell = \ell(y) = \kappa y = 0.41y$	1	1/2	1
$\nu_t = C_\mu \frac{k^{3/2}}{\ell}$	$x = \ell$, eddy length scale, two-equation model	$\alpha = C_\mu = 0.09$	3/2	-1
$\nu_t = C_\mu \frac{k^2}{\varepsilon}$ (Prandtl-Kolmogorov relationship)	$x = \varepsilon$, dissipation rate (Taylor, Chou, SKE, RNG, RKE)	$\alpha = C_\mu = 0.09$; $C_\mu = 0.09f(Re)$	2	-1
$\nu_t = \alpha\ell$	$x = \ell$, eddy length scale	α is a constant in units of "average" eddy velocity	0	1
$\nu_t = C_\mu kt$	$x = t$ (or τ), eddy time scale (Zeierman and Wolfshtein 1986; Speziale et al. 1992)	$\alpha = C_\mu = 0.09$	1	1
$\nu_t = C\frac{k}{\omega}$	$x = \omega$. The square of the parameter is solved using a PDE for ω^2, which is the mean vorticity scale for the energetic (large) eddies (Saffman 1970)	$\alpha = C$, where C is a "universal" constant with magnitude of 1.0	1	-1

Other formulations for these relationships can be found in the literature (Rodi 1993; Andersson et al. 2012), such as

$$\phi = k^\alpha \ell^\beta. \tag{4.43}$$

4.6   Top RANS Turbulence Models

4.6.1   Zero-Equation Models

The zero-equation models (also known as algebraic turbulence models) require no additional transport closure equations and hence the naming convention, "zero."

In general, zero-equation models tend to have the following advantages:

- Fastest computationally.
- Easiest to code and tend to be very robust numerically.
- Great for developing analytical solutions and theoretical insights from PDEs (e.g., the Prandtl mixing-length model can be readily substituted into a RANS PDE).
- Form the simplest turbulence models, while providing some physical insights.
- Because of the way the models were developed and calibrated ("fine-tuned" for those who embrace these models and "fudged" for those who do not prefer them),

these models reasonably predict certain complex turbulent flows (e.g., Baldwin-Lomax is reasonable for turbomachinery, aerospace, and applications that have attached, thin boundary layers).

Some general disadvantages of the zero-equation models are as follows:

- Very limited applicability. They may be great for thin boundary layers but fall flat outside of their intended, limited space!
- Provide no transport of turbulent scales (no velocity, length, or some other appropriate variable).
- Successful mostly for very simple flows. (Wilcox found such models as a "pleasant surprise despite its theoretical shortcomings.")

Despite the complexity of turbulent flows, algebraic models serve a unique and respectable role in turbulence modeling, and there are dozens of such models. Provided the caveats and limitations are understood, three well-known and reasonable models include Prandtl mixing length, Baldwin-Lomax, and Cebeci-Smith.

4.6.2 One-Equation Models

One equation models are also known as "k-algebraic" models. Generally based on the k PDE, these models are therefore predominantly built on one turbulence scale, namely, the velocity. Various popular models are still used in the literature, including Prandtl (see Sect. 4.6.2.1), Baldwin-Barth, and Spalart-Allmaras. The Spalart-Allmaras model is good for supercritical fluids, turbomachinery, and aerospace applications (refer to Sect. 6.3). Despite a lack of buoyancy term, Spalart-Allmaras shows surprisingly great results for heat transfer in supercritical fluids (Otero et al. 2018), and good comparisons with experimental data for certain turbulent flows (e.g., far wakes and mixing layers), but is not as good for jets (radial, round, and plane) (Wilcox 2006).

4.6.2.1 Prandtl's One-Equation Model

For one-equation models, a turbulence transport variable is used; such variable is predominantly the turbulence kinetic energy, k, in conjunction with auxiliary closure expressions. Ludwig Prandtl developed the first *one-equation* full-closure turbulence transport model, where k was his transport variable of choice (Prandtl 1945).

Note that the first *two-equation* model was developed by Kolmogorov 3 years earlier, in 1942 (refer to Sect. 4.6.3.1). Kolmogorov used a k transport PDE (using b for his naming convention) but did not include a full set of closure parameters. Also, note that the first modern k PDE was derived independently by Chou in 1940, 5 years before Prandtl's one-equation turbulence model (Chou 1940); refer to

Sect. 4.4. In any case, Prandtl's k transport formulation continues to serve as the modern standard for the k PDE and is expressed as follows,

$$\frac{\partial k}{\partial t} + \bar{u}_j \frac{\partial k}{\partial x_j} = R_{ij} \frac{\partial \bar{u}_i}{\partial x_j} - \varepsilon + \frac{\partial}{\partial x_j} \left[\left(\nu + \frac{\nu_t}{\sigma_k} \right) \frac{\partial k}{\partial x_j} \right], \qquad (4.44\text{A})$$

where

$$R_{ij} = \nu_t \left(\frac{\partial \bar{u}_i}{\partial x_j} + \frac{\partial \bar{u}_j}{\partial x_i} \right) - \frac{2}{3} k \delta_{ij}. \qquad (4.44\text{B})$$

Prandtl provided closure by defining dissipation, the turbulent kinematic viscosity, and the eddy length scale, respectively,

$$\varepsilon = \frac{C_D k^{3/2}}{\ell}, \qquad (4.44\text{C})$$

$$\nu_t = \ell k^{1/2}, \qquad (4.44\text{D})$$

and

$$\ell = \kappa y. \qquad (4.44\text{E})$$

The closure coefficients are

$$\sigma_k = 1.0, \qquad (4.44\text{F})$$

$$\kappa = 0.41 = \text{Kolmogorov's constant,} \qquad (4.44\text{G})$$

$$C_D = 0.3 \text{ (for shear flows),} \qquad (4.44\text{H})$$

and

$$C_D \approx C_\mu = 0.085 \text{ (in general).} \qquad (4.44\text{I})$$

Not surprisingly, Prandtl's model is very useful for simplified geometries, such as flow across flat surfaces, pipe flow, and Couette flow. Best of all, it is quite easy to program. For example, it is possible to use straightforward solution approaches, such as forward-time, central space (FTCS) finite differences because of the relatively non-rigid nature of the k PDE. For the interested reader, Chap. 7 includes FORTRAN coding for the Prandtl turbulence model.

On the other hand, the ad hoc nature of the C_D coefficient limits its applicability. The linear expression for ℓ also limits its applicability, though more physics-intensive relationships exist that can add modeling accuracy, such as van Driest formulations.

4.6.3   Two-Equation Models

As a result of the ubiquitous advent of computers in the workplace, the two-equation turbulence models gained popularity and formed the basis for much of the turbulence simulations over the past 40 years. As their name implicates, these models are primarily based on finding two key variables that are self-consistent and suitable for forming an expression having the same units as ν_t; these variables must also uniquely describe the turbulence scale (eddy) behavior. For example, turbulent flow has eddies with length scales (e.g., ℓ, λ, η) that move about chaotically at some velocity u'. These eddies move about with a characteristic velocity v, move about with acceleration a, decay within certain time t or frequency ω, dissipate with some magnitude ε, and so forth.

For the two-equation model closure, k is usually chosen, as well as one additional transport variable. Typically, the second closure equation involves one of the following variables: ω, ε, τ (time), ζ (enstrophy), a (acceleration), and so forth. The list can be as extensive as the researcher's imagination, so long as dimensionless analysis allows the combination of the two transport variables to form a mathematical expression that has units of length squared per unit time—the units for ν_t. Note that since the early 1980s, the accuracy of the k PDE has compared successfully with many experiments and DNS simulations, but the same cannot be said for many choices for the second transport variable. Furthermore, the introduction of a second transport variable PDE can cause numerical difficulties and yield anomalous results if not coded and coupled correctly.

Practically all two-equation turbulence models employ the Boussinesq approximation and use the k transport variable. Though they are generally suitable for many applications, two-equation models are notoriously inaccurate for nonequilibrium, nonisotropic flows, and hence inclusion of nonlocal; nonequilibrium approaches are required, instead of using the traditional Boussinesq approximation (Speziale and Eringen 1981; Hamba 2005; Schmitt 2007; Hamlington and Dahm 2009; Wilcox 2006; Spalart 2015).

4.6.3.1   Kolmogorov 1942 k-ω

This is the first two-equation transport model ever developed and is now approaching the century mark (Kolmogorov 1942). Kolmogorov's original model is discussed here for various reasons. First, it was the precursor of the modern k-ω turbulence models. Secondly, it forms the basis for other two-equation RANS models, and finally, it sheds light on the applicability of the k-ε models. Though developed too early to take advantage of modern computational power, Kolmogorov's model contributed to modern turbulence analysis by showing that turbulence can be modeled with momentum-like transport variables that reflect key scales such as velocity, length, and so forth.

Kolmogorov began his model by noting that turbulence flows consist of "turbulent pulsations" (eddies) that range from large to small scales that are superimposed onto the mean flow. The large scales are the integral eddies, which correspond to the length scale ℓ found in modern turbulence research (Kolmogorov originally used "L"). He also used "λ" to refer to the smallest eddy scale at which the fluid viscosity dampens eddies; note that in modern notation, λ is used for Taylor eddies. In Kolmogorov's honor, the smallest eddies are called "Kolmogorov eddies," with modern notation using the symbol "η."

In any case, Kolmogorov proceeded with his two-equation model development by noting that the large eddies constantly extract energy from the mean flow, and in turn, their energy is transferred onto smaller eddies. Finally, only the smallest eddies (the Kolmogorov eddies) participate in a decay process whereby they transform from tiny chaotic structures into tiny laminar regions as the dissipation of their energy is brought forth by the viscous force of the fluid. Kolmogorov described the transport of the turbulent kinetic energy as a "stream of energy, flowing constantly." That is, he envisioned that a continuous transfer of energy existed in the fluid continuum. The process is now called "cascading." *These phenomena insights are crucial, as dissipation primarily applies to Kolmogorov eddies, but energy transfer involves all three scales—integral, Taylor, and Kolmogorov.* In fact, Kolmogorov referred to his second transport variable (ω) as applicable to all (three) scales, as evidenced by his remark (Kolmogorov 1942), "A fundamental characteristic of the turbulent motion *at all scales is the quantity ω* which stands for the rate of dissipation in unit volume and time."

Thus, if equilibrium is assumed, then the decay rate equals the production rate, and ω rightfully describes the energy dissipation rate under equilibrium conditions as experienced throughout the cascade—from the integral to the Kolmogorov scales. This physical description of turbulence energy flow strongly suggests which variables are appropriate and, most importantly, *which form self-consistent pairs of variables for the development of two-equation turbulence models.*

Kolmogorov continued his model development by postulating that ω is a "fundamental" quantity that is applicable to the entire length spectrum, because the eddy energy transfer is assumed constant—in equilibrium. Certainly, this need not always be the case in all turbulent flows but is a reasonable approximation under many situations—it forms a good starting point. Kolmogorov envisioned the fundamental quantity ω as representing the dissipation rate in a flow volume per unit time or as a "mean frequency." Thus, ω is construed from an equilibrium point of view, applicable to the entire eddy spectrum, whereby turbulent energy transfer occurs as a "continuous stream of energy" from the mean flow, and that cascades throughout all eddy scales. This approach is consistent with Hinze, who stated (Hinze 1987), "ε is on the one hand practically equal to the dissipation in the highest-wavenumber range and on the other hand practically equal to the work done by the energy-containing eddies, which is the energy supplied to the smaller eddies."

In addition, Saffman considered ω^2 as the mean vorticity scale for the energetic (large) eddies (Saffman 1970).

Therefore, k and ω are self-consistent turbulence transport properties, where k represents the large eddies that typically hold about 80% of the total turbulent kinetic energy, while ω considers the energy dissipation of the large eddies (as well as the smaller eddies, because Kolmogorov considered the *equilibrium* cascade energy transfer as affecting all three scales). Said briefly, both k and ω are associated with the large eddy scale size and are thus a self-consistent pair of transport variables.

Kolmogorov defined his turbulence model using three partial differential equations, as follows. He started with the time-averaged velocity for the RANS-based momentum equation:

$$\frac{D\bar{u}_i}{Dt} = F_i - \frac{\partial}{\partial x_i}\left(\frac{\bar{p}}{\rho} + b\right) + A\sum_j \frac{\partial}{\partial x_j}\left[\frac{b}{\omega}\left(\frac{\partial \bar{u}_i}{\partial x_j} + \frac{\partial \bar{u}_j}{\partial x_i}\right)\right]. \qquad (4.45\text{A})$$

Then, the turbulent mean frequency ω is modeled with the following transport equation:

$$\frac{D\omega}{Dt} = -\frac{7}{11}\omega^2 + A'\sum_j \frac{\partial}{\partial x_j}\left[\frac{b}{\omega}\frac{\partial \omega}{\partial x_j}\right], \qquad (4.45\text{B})$$

while the averaged root-mean-square of the fluctuating velocities, b, is

$$\frac{Db}{Dt} = -b\omega + \frac{1}{3}A\frac{b}{\omega}\varepsilon + A''\sum \frac{\partial}{\partial x_j}\left[\frac{b}{\omega}\frac{\partial b}{\partial x_j}\right]. \qquad (4.45\text{C})$$

Note that Kolmogorov used the following notation:

$$\frac{D}{Dt} = \text{substantial derivative,}$$

$$F_i = \text{external force,}$$

$$\bar{u}_i = \text{time-averaged velocity,}$$

and

$$\bar{P} = \text{time-averaged pressure.}$$

In Kolmogorov's context, b is akin to k. That is, he defined the RMS fluctuation velocity:

$$b \equiv \frac{1}{3} \sum_j \overline{u_j'^2}. \tag{4.45D}$$

Thus, Kolmogorov's turbulent kinetic energy is

$$b = \frac{1}{3} \left(\overline{u'^2} + \overline{v'^2} + \overline{w'^2} \right) \approx k, \tag{4.45D$'$}$$

which is an RMS quantity. As noted in Sect. 4.4 (under "Toward a more formal definition of k"), Prandtl used a slightly different definition (Prandtl 1945; Spalding 1991), namely, $k = \frac{1}{2} \left(\overline{u'^2} + \overline{v'^2} + \overline{w'^2} \right)$.

Finally, using Kolmogorov's nomenclature,

$$\varepsilon = \sum_{i,j} \left(\frac{\partial \overline{u}_i}{\partial x_j} + \frac{\partial \overline{u}_j}{\partial x_i} \right)^a. \tag{4.45E}$$

In this context, a, A, A', and A'' are closure constants that have yet to be determined experimentally. Note that Kolmogorov's ε symbol in the third partial differential equation (the b transport PDE, Eq. 4.45C) is defined explicitly in Eq. 4.45E; it is not to be confused with the modern symbol for dissipation. Basically, the first k-ε model was developed 3 years *after* Kolmogorov's k-ω turbulence model, and hence the naming confusion did not arise from Kolmogorov. In any case, letting $a = 1$ shows that in his context, ε is actually twice the modern mean strain rate tensor:

$$\varepsilon = \sum_{i,j} \left(\frac{\partial \overline{u}_i}{\partial x_j} + \frac{\partial \overline{u}_j}{\partial x_i} \right) \equiv 2 \overrightarrow{\overrightarrow{S}}. \tag{4.45E$'$}$$

For the sake of clarity, Kolmogorov's ε (i.e., his mean strain rate) will *not* be used outside of this section. Kolmogorov's turbulent mean frequency is

$$\omega \equiv c \frac{b^{1/2}}{L} \approx c \frac{k^{1/2}}{\ell} \tag{4.45F}$$

where

ℓ = characteristic eddy length.

As noted by various researchers, Kolmogorov's ω transport equation does not include a production term (Spalding 1991; Bulicek and Malek 2000; Wilcox 2006; Mielke and Naumann 2015). The literature indicates that this issue can be addressed to diverse degrees of success by including an ω source boundary at the wall (Spalding 1991) or a fixed nonhomogeneous Dirichlet boundary (Bulicek and Malek 2000), as well as periodic boundaries (Mielke and Naumann 2015).

Wilcox was the first to introduce an ω production term onto the ω transport PDE, as noted in his 1988 k-ω model, where the production term P_ω is

$$P_\omega = \alpha \frac{\omega}{k} R_{ij} \frac{\partial \overline{u}_i}{\partial x_j}.$$ (4.46A)

However, it is noted that Kolmogorov did include a b production term that is a function of ω, namely,

$$P_k = \frac{1}{3} A \frac{b}{\omega} \varepsilon = \frac{2}{3} A \frac{b}{\omega} \overline{S}$$ (4.46B)

(recall Kolmogorov's non-standard definition for ε, namely, $\varepsilon = 2\overline{S}$).

Because the transport equations are coupled and P_k is a function of ω, then at least there is a connection between turbulent kinetic energy production and ω in the Kolmogorov turbulence model.

Note that ω is in units of inverse time and is therefore a fundamental turbulence frequency associated with the time it takes for eddies to transfer their energy. Based on dimensionless arguments, the frequency is a function of the turbulent kinetic energy k and the dissipation ε,

$$\omega \approx \frac{\varepsilon}{k}.$$ (4.47)

Wilcox expressed Kolmogorov's ω transport model in more modern terms (Wilcox 2006) and upgraded the model by adding an ω production term (the first term on the RHS of the arrow in Eq. 4.48A), viz.,

$$\underbrace{-\frac{7}{11}\omega^2 + A'\sum_j \frac{\partial}{\partial x_j}\left[\frac{b}{\omega}\frac{\partial \omega}{\partial x_j}\right]}_{\text{Kolmogorov 1942}} \rightarrow \underbrace{\gamma \frac{\omega}{k} R_{ij} \frac{\partial \overline{u}_i}{\partial x_j} - \beta \omega^2 + \frac{\partial}{\partial x_j}\left[(\nu + \sigma \nu_t)\frac{\partial \omega}{\partial x_j}\right]}_{\text{Wilcox 2006}}$$

(4.48A)

whereby

$$\beta = \frac{7}{11},$$ (4.48B)

$$A' = \sigma,$$ (4.48C)

and

$$\nu_t = \frac{b}{\omega}.$$ (4.48D)

4.6.3.2   Wilcox 1988 *k-ω*

The 1988 model is included here to show the evolutionary progress of the *k-ω* models. Though the 1988 *k-ω* can be sensitive to free stream boundary conditions (Menter 1992; Menter et al. 2003; Wilcox 2006), *it cannot be overemphasized that this issue was **fixed** in the 2006 k-ω version. To be clear, the 2006 k-ω model supersedes the 1988 and 1998 k-ω versions.*

Using his prodigious turbulence modeling skills, Wilcox added his own upgrades, and like an accomplished craft master, he carefully blended Prandtl's *k* transport turbulence PDE (Prandtl 1945), Kolmogorov's *ω* transport PDE (Kolmogorov 1942), and upgrades from Saffman (1970), to produce the 1988 *k-ω* version.

Note that Saffman computed ω^2 transport, which is the mean vorticity scale for the energetic (large) eddies; this once again confirms the notion that if *k* is computed (which is clearly based on the larger eddies), then the associated second transport parameter ought to be based on a larger scale as well, assuming consistency matters! And clearly, *ω* satisfies this constraint. In the literature, $\zeta = \frac{1}{2}\omega^2 = $ enstrophy and is used in *k-ζ* models. Saffman's turbulent kinematic viscosity is $\nu_t = \gamma \frac{k}{\omega}$, where *γ* is a "universal" constant valued at 1.0.

But rather than computing ω^2, Wilcox, just like Kolmogorov, considered the transport of the eddy frequency (or specific dissipation rate), *ω*; the 1988 Wilcox *k-ω* turbulence model is discussed next (Wilcox 1988a, 1993a, b).

The *k* PDE is

$$\frac{\partial k}{\partial t} + \bar{u}_i \frac{\partial k}{\partial x_i} = R_{ij} \frac{\partial \bar{u}_i}{\partial x_j} - \beta^* k\omega + \frac{\partial}{\partial x_i}\left[(\nu + \sigma^* \nu_t)\frac{\partial k}{\partial x_i}\right], \qquad (4.49A)$$

while the turbulent frequency *ω* (specific dissipation rate) is

$$\frac{\partial \omega}{\partial t} + \bar{u}_i \frac{\partial \omega}{\partial x_i} = \gamma \frac{\omega}{k} R_{ij} \frac{\partial \bar{u}_i}{\partial x_j} - \beta \omega^2 + \frac{\partial}{\partial x_i}\left[(\nu + \sigma \nu_t)\frac{\partial \omega}{\partial x_i}\right]. \qquad (4.49B)$$

The model has the following five closure coefficients (Wilcox 1988a, 1988b, 1993a, b; Bredberg 2000):

$$\beta = \frac{3}{40}, \qquad (4.49C)$$

$$\beta^* = \frac{9}{100}, \qquad (4.49D)$$

$$\gamma = \frac{5}{9}, \qquad (4.49E)$$

$$\sigma = \frac{1}{2}, \qquad (4.49F)$$

and

$$\sigma^* = \frac{1}{2}.$$

(4.49G)

The closure relationship is

$$\nu_t = \frac{k}{\omega}.$$

(4.49H)

A similar turbulence model, but with no alignment between the Reynolds stress tensor and the mean strain rate, was developed the same year (Wilcox 1988b). Wilcox also developed a 1988 k-ω variant that is suitable for laminar to turbulent transition (Wilcox 1994). Not surprisingly, the literature indicates that there are many other k-ω hybrid models (Wilcox 1993a, b, 2006; Bredberg 2000; Gorji et al. 2014; NASA1 2018), so the specific version should always be noted by the analyst.

The 1988 model is elegant because it captures the major elements of transport for both k and ω, while having neither limiters nor blending functions, and requires only a minimal number of closure coefficients. The model is very useful for low Re and near-wall boundary layers. But again, given the choice, the 2006 k-ω is vastly preferable over the 1988 k-ω.

4.6.3.3   Wilcox 1998 k-ω

A decade later, Wilcox overhauled his 1988 k-ω model to generate the 1998 k-ω model (Wilcox 2004). On the one hand, the 1998 version resulted in significant advances for k-ω models, as Wilcox experimented with various recent turbulence modeling advances. For example, this was the first time whereby he added a cross-diffusion term, albeit it had a strong ω function (to the third power). By comparison, note that the 2006 model has a functionality to the first power (see Sect. 4.6.3.4); it is very likely that the lower power dependency significantly decreased k-ω numerical stiffness for the 2006 model. Other 1998 model advancements included a vorticity tensor, which is certainly ideal for modeling rotating coherent structures.

To say the least, these ideas fostered Wilcox's attempt to improve the 1988 k-ω, and many of these concepts would later prove very beneficial in his future k-ω models, especially the 2006 k-ω. But, on the other hand, the 1998 k-ω model resulted in undesirable de-evolution in k-ω development, mainly because the 1998 model included a disconcertingly large number of closure coefficients and blending functions. And yes, this direction went against Wilcox's fundamental belief that "an ideal model should introduce the minimum amount of complexity while capturing the essence of the relevant physics." For these reasons, and because the 2006 k-ω model replaces the 1998 model, a description of the 1998 version is not included here.

4.6.3.4   Wilcox 2006 k-ω

Fortunately, a little less than a decade after his 1998 model, in a stroke of genius, Wilcox not only reduced the number of blending functions in half but also improved its modeling characteristics in a significant manner (Wilcox 2006, 2007a, b). For example, Wilcox added a revised cross-diffusion term to vastly reduce boundary-condition sensitivity to the free stream. In addition, the new cross-diffusion model had a weaker ω dependency (first- vs. third-power dependency compared to the 1998 version). Then, Wilcox went further by adding a stress limiter to better simulate flow detachment, incompressible flows, as well as transonic flows; neither the 1988 nor 1998 versions include the limiter. Finally, the 2006 model only requires six closure coefficients (not including the limiter).

The 2006 transport k and ω models are, respectively,

$$\frac{\partial k}{\partial t} + \overline{u}_i \frac{\partial k}{\partial x_i} = R_{ij} \frac{\partial \overline{u}_i}{\partial x_j} - \beta^* k\omega + \frac{\partial}{\partial x_i} \left[\left(\nu + \sigma^* \frac{k}{\omega} \right) \frac{\partial k}{\partial x_i} \right] \tag{4.50A}$$

and

$$\frac{\partial \omega}{\partial t} + \overline{u}_i \frac{\partial \omega}{\partial x_i} = \alpha \frac{\omega}{k} R_{ij} \frac{\partial \overline{u}_i}{\partial x_j} - \beta \omega^2 + \frac{\partial}{\partial x_i} \left[\left(\nu + \sigma \frac{k}{\omega} \right) \frac{\partial \omega}{\partial x_i} \right] + \sigma_d \frac{1}{\omega} \frac{\partial k}{\partial x_i} \frac{\partial \omega}{\partial x_i}. \tag{4.50B}$$

The Reynolds stresses (R) are obtained from the Boussinesq approximation.

The 2006 model has the following closure relationships. Starting with the turbulent kinematic viscosity,

$$\nu_t = \frac{k}{\widetilde{\omega}}. \tag{4.50C}$$

The specific dissipation rate (i.e., the frequency) has the following stress limiter for improving separated flows and incompressible flows up to transonic Ma:

$$\widetilde{\omega} = \max \left(\omega, C_{\lim} \sqrt{\frac{2S_{ij}S_{ij}}{\beta^*}} \right). \tag{4.50D}$$

The stress limiter only becomes active for large mean strain rate (S_{ij}) magnitudes, such as occur for strongly separated flows and high Ma flows.

The blending function f_β is defined as

$$f_\beta = \frac{1 + 85\chi_\omega}{1 + 100\chi_\omega}. \tag{4.50E}$$

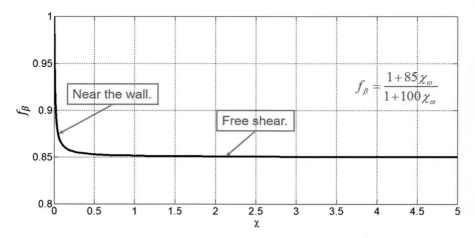

Fig. 4.2 The 2006 k-ω blending function

The blending function ranges from a minimum of 0.85 as χ approaches large values for free shear up to a peak value of 1.0 as χ approaches 0 near the wall, as shown in Fig. 4.2.

In this context, χ is the so-called non-dimensional vortex stretching and is defined as

$$\chi_\omega \equiv \left| \frac{\Omega_{ij}\Omega_{jk}S_{ki}}{(\beta^*\omega)^3} \right|. \tag{4.50F}$$

Notice the introduction of the vorticity tensor Ω, which is an ideal function for quantifying rotational fluid mechanics. Because turbulent coherent structures involve 3D sheets that fold with spiral-like curvature, involving the mean rotational tensor is highly intuitive. The mean rotational and mean strain rate tensors, respectively, are defined as

$$\Omega_{ij} = \frac{1}{2}\left(\frac{\partial \overline{u}_i}{\partial x_j} - \frac{\partial \overline{u}_j}{\partial x_i} \right) \tag{4.50G}$$

and

$$S_{ij} = \frac{1}{2}\left(\frac{\partial \overline{u}_i}{\partial x_j} + \frac{\partial \overline{u}_j}{\partial x_i} \right). \tag{4.50H}$$

The inclusion of χ_ω is an attempt to quantify vortex stretching as the energy cascade proceeds from the larger to the smaller eddies; rightly so, $\chi_\omega = 0$ for 2D flows. However, the vortex stretching phenomenon is much more complex than anticipated. For example, recent investigations for homogeneous isotropic flows show that the larger eddies tend to be perpendicular to the fastest-stretching eddies, while the orientation of the larger eddies, compared with the stretching direction associated with the smaller vortices, is about 45° (Hirota et al. 2017).

The last term of the ω PDE model is the cross-diffusion term, which was included to minimize sensitivity to boundary conditions. In particular, the near-wall and free shear effects are conveniently addressed via the product of the k and ω spatial gradients. More specifically, the leading coefficient is zero near the wall, so cross-diffusion is suppressed. The cross-diffusion term is of higher consequence for free shear flows, so ω production is increased (i.e., the term is nonzero). The cross-diffusion term is defined as follows:

$$\sigma_d = \begin{cases} 0, & \text{if } \dfrac{\partial k}{\partial x_i}\dfrac{\partial \omega}{\partial x_i} \le 0 \text{ (near the wall)} \\[3mm] \sigma_{do}, & \text{if } \dfrac{\partial k}{\partial x_i}\dfrac{\partial \omega}{\partial x_i} > 0 \text{ (free shear)} \end{cases} . \tag{4.50I}$$

Finally, the model has the following closure coefficients:

$$\alpha = \frac{13}{25}, \tag{4.50J}$$

$$\beta = \beta_0 f_\beta, \tag{4.50K}$$

$$\beta_0 = 0.0708, \tag{4.50L}$$

$$\beta^* = \frac{9}{100} = C_\mu, \tag{4.50M}$$

$$\sigma = \frac{1}{2}, \tag{4.50N}$$

$$\sigma_{do} = \frac{1}{8}, \tag{4.50O}$$

and

$$\sigma^* = \frac{3}{5}. \tag{4.50P}$$

The stress limiter coefficient is

$$C_{\lim} = \frac{7}{8}. \tag{4.50Q}$$

Based on the above definitions and dimensional arguments, the following quantities can be obtained, if so desired:

$$\ell = \frac{k^{1/2}}{\omega},$$ (4.50R)

$$t = \frac{1}{\omega},$$ (4.50S)

and

$$\varepsilon = \beta^* \omega k.$$ (4.50T)

Wilcox proudly tested his 2006 model on approximately 100 relevant test cases, including:

- Attached boundary layers
- Free shear flows (far wake, mixing layer; plane, round, and radial jets)
- Backward-facing steps
- Shock-separated flows
- Low to high Re
- Ma ranging from incompressible to hypersonic flows
- Fluids undergoing heat transfer

Therefore, the model is highly recommended for free shear flows, including far wakes, mixing layers, strongly separated flows, as well as plane, round, and radial jets (Wilcox 2006). It is also suitable for flows with a high degree of swirl (Rodriguez 2011). Comparisons with experimental data are typically very good, and 1988 k-ω issues involving free shear flows, strongly separated flows, and high Ma flows have been substantially resolved. Therefore, the 2006 k-ω is a good, all-around model for both near-wall and free stream turbulence and is therefore a great contender vs. Menter's SST model. But in the final balance, it must be pointed out that the 2006 k-ω model does not have the SKE issues discussed in Sects. 4.7 through 4.7.3 that rightfully apply onto the SKE portion of the SST. Therefore, this thrusts the 2006 k-ω as the best two-equation RANS model. For more guidelines regarding this assertion, refer to Eq. 2.48 and the associated paragraph immediately below.

Finally, note that the 2006 k-ω clearly supersedes the 1988 and 1998 k-ω models because it eliminates issues regarding sensitivity to boundary conditions and versatility (Wilcox 2006). Indeed, unlike its 1988 and 1998 predecessors, the 2006 k-ω is suitable for both free shear and near-wall flows and subsonic, supersonic, and hypersonic flows. Therefore, modern analysts and researchers should not use the earlier versions if the 2006 k-ω is available. To do otherwise is done at the expense of finding "issues" that were corrected long ago by Wilcox. This is pointed out here explicitly, as there are many papers in the 2006–2019 time frame that still use the 1988 k-ω model and whose work is therefore not cited here as a courtesy.

4.6.3.5   Standard k-ε (SKE) Model

The two-equation transport model, now known as the standard k-ε (SKE) turbulence model, traces its origins to Pei Yuan Chou's eddy decay research, whose ideas were later adapted and extended by many researchers. *Chou developed the notion that k can be associated with vorticity (ω) decay.* Note that what is referred in more modern times as the dissipation ε term based on the derivatives of the fluctuating velocities was derived independently by Chou. Furthermore, Chou *associated ε decay with the Taylor length scale, λ* (Chou 1940, 1945). Ironically, Chou proceeded to develop a vorticity transport dissipation model based on ω for Taylor eddies (not Kolmogorov eddies); the present author considers this as highly ironic because the origin of the k-ε model is based on the k-ω turbulence model! In Chou's words (Chou 1945), "Since Taylor's scale of micro-turbulence λ plays a very important role in the decay of turbulence, it is necessary to find the equation which governs the behaviour of this fundamental length. This equation is provided by the decay of the vorticity."

Note that Chou referred "dissipation" as "decay." His original ω PDE transport model is found in Eq. 1.4 of his original paper (Chou 1945) and is reproduced here for convenience, to trace how Chou's original concept for ω transport evolved (or de-evolved) into the ε transport model. Note that Chou uses w_i and U_i to denote the velocity fluctuations and mean velocities, respectively, and that τ represents his Reynolds stress. For convenience, τ is replaced with R. Chou's ω transport is

$$\frac{\partial \omega_{ik}}{\partial t} + U_j \frac{\partial \omega_{ik}}{\partial x_j} + \frac{\partial U_j}{\partial x_k}\frac{\partial w_i}{\partial x_j} - \frac{\partial U_j}{\partial x_i}\frac{\partial w_k}{\partial x_j} + w_j \frac{\partial \omega_{ik}}{\partial x_j} + \frac{\partial w_j}{\partial x_k}\frac{\partial w_i}{\partial x_j} - \frac{\partial w_j}{\partial x_i}\frac{\partial w_k}{\partial x_j} + w_j \frac{\partial \Omega_{ik}}{\partial x_j}$$

$$+ \frac{\partial w_j}{\partial x_k}\frac{\partial U_i}{\partial x_j} - \frac{\partial w_j}{\partial x_i}\frac{\partial U_k}{\partial x_j} = -\frac{1}{\rho}\left(\frac{\partial^2 R_{ji}}{\partial x_j \partial x_k} - \frac{\partial^2 R_{jk}}{\partial x_j \partial x_i} \right) + \nu \nabla^2 \omega_{ik}.$$

$$(4.51\text{A})$$

In this context, Ω is the mean vorticity expressed as

$$\Omega_{ik} = \frac{\partial U_i}{\partial x_k} - \frac{\partial U_k}{\partial x_i},$$

$$(4.51\text{B})$$

while the ω vorticity fluctuation is defined as

$$\omega_{ik} = \frac{\partial w_i}{\partial x_k} - \frac{\partial w_k}{\partial x_i}.$$

$$(4.51\text{C})$$

Chou also developed the first transport version of the Reynolds stress (R PDE) in 1940, whereby he introduced eddy decay (dissipation) (Chou 1940 (refer to Eqs. 3.1 and 3.4 in the cited reference)):

$$\frac{\partial R_{ik}}{\partial t} = w_j'\left(\frac{\partial U_i}{\partial x_j}w_k' + \frac{\partial U_k}{\partial x_j}w_i'\right) + U_j\frac{\partial R_{ik}}{\partial x_j} + w_j'\frac{\partial R_{ik}}{\partial x_j} = -\frac{1}{\rho}\left(\frac{\partial p'}{\partial x_i}w_k' + \frac{\partial p'}{\partial x_k}w_i'\right)$$

$$-\frac{1}{\rho}\left(\frac{\partial R_{ij}}{\partial x_j}w_k' + \frac{\partial R_{kj}}{\partial x_j}w_i'\right) + \nu\nabla^2 R_{ik} - \nu g_{mn}\left(\overline{\frac{\partial w_i'}{\partial x_m}\frac{\partial w_k'}{\partial x_n}} + \overline{\frac{\partial w_i'}{\partial x_n}\frac{\partial w_k'}{\partial x_m}}\right).$$

$$(4.52)$$

Years later, he elaborated and expanded various terms in the R PDE (Chou 1945; Chou and Chou 1995). With a desire to focus on the issue of dissipation, Chou's decay term is

$$\frac{\partial R_{ik}}{\partial t} \approx -\nu g_{mn}\left(\overline{\frac{\partial w_i'}{\partial x_m}\frac{\partial w_k'}{\partial x_n}} + \overline{\frac{\partial w_i'}{\partial x_n}\frac{\partial w_k'}{\partial x_m}}\right) = 2\nu g_{mn}\overline{\frac{\partial w_i'}{\partial x_m}\frac{\partial w_k'}{\partial x_n}} \equiv \varepsilon_{ik}. \qquad (4.53)$$

It is noted that this is the modern dissipation ε_{ik} term, with the metric tensor $g_{mn} = 1$. Chou proceeded by deriving a *relationship for eddy decay associated with the Taylor length scale, λ,*

$$\varepsilon_{ik} \equiv 2\nu g_{mn}\overline{\frac{\partial w_i'}{\partial x_m}\frac{\partial w_k'}{\partial x_n}} = -\frac{2\nu}{3\lambda^2}(k-5)q^2 g_{ik} + \frac{2\nu k}{\lambda^2}\overline{w_i'w_k'}, \qquad (4.54A)$$

where

$$q^2 = \overline{w_j'w_j'}. \qquad (4.54B)$$

Note that Chou developed transport PDEs for k (refer to Sect. 4.4 under "Chou's k PDE"), ω (Eqs. 4.51A through 4.51C), τ (i.e., R; Eq. 4.52), and an expression for ε = ε (k,λ,$\overline{w_i'w_k'}$) (Eqs. 4.54A and 4.54B). However, he did not develop a transport PDE for ε (Chou 1940). This was most likely because he clearly understood that ε was associated with λ, and eddy decay with ω, and thus there was no need for an ε PDE.

Continuing on, researchers developed triple velocity correlations in the hope that those equations could be used to achieve closure. Instead, the correlations generated fourth-order velocity correlations, and so forth, each time generating terms at the next higher order, in a process that can go on indefinitely (Chou 1940). As stated by Chou, ". . .and hence leads to nowhere." At this point, Chou chose to end the endless cycle by asserting that

$$\overline{u_i'u_j'u_k'u_l'} \approx \frac{1}{2}\left(\overline{u_i'u_j'}\,\overline{u_k'u_l'} + \overline{u_i'u_k'}\,\overline{u_l'u_j'} + \overline{u_i'u_l'}\,\overline{u_j'u_k'}\right). \qquad (4.55)$$

His justification is based on either assuming that the fluctuating velocities can be represented as sinusoidal functions of time or as "an independent hypothesis."

In 1941, Millionshtchikov developed a more generalized and rigorous expansion for the fourth-order velocity correlation based on homogeneous isotropic theory (Millionshtchikov 1941). Remarkably, its expression is nearly the same as Chou's approximation:

$$\overline{u'_i u'_j u'_k u'_l} \approx \overline{u'_i u'_j u'_k u'_l} + \overline{u'_i u'_k u'_j u'_l} + \overline{u'_j u'_k u'_i u'_l} = R_{ij}R_{kl} + R_{ik}R_{jl} + R_{jk}R_{il} \qquad (4.56)$$

The next progression in the k-ε model surfaced when the first ε transport PDE was finally completed by B. I. Davydov in 1961 (Davydov 1961):

$$\frac{\partial \varepsilon}{\partial t} + \bar{u}_i \frac{\partial \varepsilon}{\partial x_i} + 2\nu \frac{\partial \bar{u}_i}{\partial x_k} \left(\overline{\frac{\partial u'_i}{\partial x_l} \frac{\partial u'_k}{\partial x_l}} + \overline{\frac{\partial u'_l}{\partial x_i} \frac{\partial u'_l}{\partial x_k}} \right) + 2\nu \frac{\partial^2 \bar{u}_i}{\partial x_k \partial x_l} \overline{u'_k \frac{\partial u'_i}{\partial x_l}} + \nu \frac{\partial \left[\overline{u'_j \left(\frac{\partial u'_i}{\partial x_k} \right)^2} \right]}{\partial x_j}$$

$$+ 2\nu \overline{\frac{\partial u'_i}{\partial x_k} \frac{\partial u'_i}{\partial x_l} \frac{\partial u'_k}{\partial x_l}} + 2\nu \frac{\partial}{\partial x_k} \left(\overline{\frac{\partial p'}{\partial x_l} \frac{\partial u'_k}{\partial x_l}} \right) + 2\nu^2 \overline{\left(\frac{\partial^2 u'_i}{\partial x_k \partial x_l} \right)^2} = \nu \frac{\partial^2 \varepsilon}{\partial x_k^2}.$$

$$(4.57)$$

Note that a systematic, term-by-term comparison shows that Davydov's 1961 exact ε PDE has the same nine terms as the Hanjalic 1970 exact ε PDE (Hanjalic 1970). (For convenience, Hanjalic's exact ε PDE is discussed later in this section.) In any case, after some surgery on his ε PDE, Davydov reduced the exact PDE onto the following simplified expression:

$$\frac{\partial \varepsilon}{\partial t} + \bar{u}_i \frac{\partial \varepsilon}{\partial x_i} + \frac{\partial \left(u'_i \varepsilon \right)}{\partial x_i} + \alpha \frac{\varepsilon}{k} R_{ik} \frac{\partial \bar{u}_i}{\partial x_k} + \gamma \frac{\varepsilon^2}{k} = \nu \frac{\partial^2 \varepsilon}{\partial x_k^2}, \qquad (4.58A)$$

where some of Davydov's notations are expressed here in more modern terminology:

$$\varepsilon_{\text{Davydov}} \equiv Q \equiv \nu \overline{\left(\frac{\partial \bar{u}_i}{\partial x_k} \right)^2}, \qquad (4.58B)$$

$$C_k \equiv \nu \overline{u'_i \left(\frac{\partial u'_i}{\partial x_k} \right)^2} = \overline{u'_k \varepsilon} \text{ (dissipation flow)}, \qquad (4.58C)$$

$$R_{ij} = \overline{u'_i u'_j}, \qquad (4.58D)$$

$$R \equiv R_{ii} \equiv k \sim \frac{1}{t}, \qquad (4.58E)$$

and

$$\gamma = 4. \tag{4.58F}$$

Davydov chooses N to represent the turbulent kinematic viscosity:

$$N \equiv \nu_t \equiv -R_{ik} \frac{\partial \bar{u}_i}{\partial x_k} \left(\frac{\partial \bar{u}_l}{\partial x_m} \right)^{-2}. \tag{4.58G}$$

Again, a systematic, term-by-term comparison shows that except for one term, Davydov's 1961 simplified, modified ("drastic surgery") ε PDE is nearly the same expression as the 1970 Hanjalic simplified ε PDE (Hanjalic 1970). The only difference between the Davydov and Hanjalic PDEs is that Davydov considered the transport of dissipation through velocity fluctuations (gradient transport):

$$\frac{\partial \left(u_j' \varepsilon \right)}{\partial x_j}, \tag{4.59}$$

whereas Hanjalic considered diffusion associated with turbulent dissipation:

$$\frac{\partial}{\partial x_j} \left(\frac{\nu_t}{\sigma_\varepsilon} \frac{\partial \varepsilon}{\partial x_j} \right). \tag{4.60}$$

Of course, by using dimensional arguments based on gradient transport (refer to Chap. 2), Davydov's term becomes the same as Hanjalic's, except for the σ_ε coefficient. Namely, Davydov's dissipation term becomes

$$-\frac{\partial}{\partial x_j} \left(\overline{u_j' \phi'} \right)_{\text{Davydov}} = -\frac{\partial}{\partial x_j} \left(\overline{u_j' \varepsilon} \right)_{\text{Davydov}} = \frac{\partial}{\partial x_j} \left(\nu_t \frac{\partial \varepsilon}{\partial x_j} \right)_{\text{Davydov}}. \tag{4.61}$$

In this context, the comparison of both simplified ε PDEs shows that

$$\alpha_{\text{Davydov}} \rightarrow C_{I,\text{Hanjalic}} \tag{4.62}$$

and

$$\lambda_{\text{Davydov}} \rightarrow C_{2,\text{Hanjalic}}. \tag{4.63}$$

Nevertheless, unlike Hanjalic, Davydov solved $\frac{\partial \left(u_i' \varepsilon \right)}{\partial x_i}$ via an additional transport equation.

Hanjalic noted the similarities between Davydov's exact and simplified ε PDEs and those included in his research (Hanjalic 1970), but not to the extent discussed here, as will be shown later.

Just as with Kolmogorov's k-ω model, little progress was made toward Davydov's k-ε modeling approach until nearly a decade after its initial development (Harlow and Nakayama 1967, 1968); this was primarily due to the lack of computational power at the time. These researchers undertook a somewhat distinct approach from Davydov that was more analogous to Chou's vorticity decay transport. The formulation of this alternative ε transport PDE is as follows (Harlow and Nakayama1967, 1968):

$$
\begin{aligned}
\frac{\partial \varepsilon}{\partial t}+\bar{u}_k \frac{\partial \varepsilon}{\partial x_k} = {} & \frac{a\Delta \nu_t}{s^2}\frac{\partial \bar{u}_j}{\partial x_k}\frac{\partial \bar{u}_j}{\partial x_k}-\frac{2\nu \Delta'\varepsilon}{s^2}+\frac{a_2\Delta}{s^2}\frac{\partial}{\partial x_k}\left(\nu_t \frac{\partial k}{\partial x_k}\right) \\
& +\frac{\partial}{\partial x_k}\left[(\nu+a_3\nu_t)\frac{\partial \varepsilon}{\partial x_k}\right]+a_4\frac{\partial}{\partial x_k}\left(\frac{\Delta \nu_t}{2}\frac{\partial \varphi}{\partial x_k}\right),
\end{aligned}
\tag{4.64A}
$$

where ε is in s^{-2} (vs. m^2/s^3). The auxiliary constants and expressions are

$$
a \approx 2,
\tag{4.64B}
$$

$$
a_2 \approx a_3 \approx a_4 \approx 1,
\tag{4.64C}
$$

$$
\beta = 5,
\tag{4.64D}
$$

$$
\beta' = 2\beta,
\tag{4.64E}
$$

$$
\delta = 0.01,
\tag{4.64F}
$$

$$
\Delta = \beta(1+\delta\xi),
\tag{4.64G}
$$

$$
\Delta' = \beta'(1+\delta\xi),
\tag{4.64H}
$$

$$
\xi = \frac{\nu_t}{\nu},
\tag{4.64I}
$$

$$
s = \lambda\sqrt{1+\delta\xi},
\tag{4.64J}
$$

$$
\varphi = \frac{p'}{\rho},
\tag{4.64K}
$$

and $\lambda =$ Taylor length scale.

Finally, as shown next, Hanjalic extended the modified Davydov ε PDE by incorporating:

- A new approach for calculating dimensional arguments involving velocity and length scales for ν_t
- A term to compute turbulent diffusion of dissipation based on dimensional arguments in gradient transport (as discussed earlier in this section)
- A polynomial function based on the distance from the wall (Hanjalic 1970)

Similar to Chou's efforts (Chou 1940; Chou and Chou 1995), Hanjalic obtained a fluctuating velocity PDE whereby, as stated by him, "the time averaged parts are subtracted from the time dependent general Navier-Stokes equation (2.2)" (Hanjalic 1970). For convenience, the fluctuating velocity PDE is included here:

$$\frac{\partial u_i'}{\partial t} + \bar{u}_k \frac{\partial u_i'}{\partial x_k} = -u_k' \frac{\partial \left(\bar{u}_i + u_i' \right)}{\partial x_k} + \frac{\partial \left(\overline{u_i' u_k'} \right)}{\partial x_k} - \frac{\partial \left(\nu \frac{\partial u_i'}{\partial x_k} \right)}{\partial x_k} - \frac{1}{\rho} \frac{\partial \bar{P}}{\partial x_i}. \qquad (4.65)$$

Hanjalic then made the assertion that if the fluctuating velocity PDE is (Hanjalic 1970)

"differentiated with respect to x_ℓ, then multiplied by $2\nu \frac{\partial u_i'}{\partial x_\ell}$ and time averaged, it transforms into the exact transport equation for the variable $\nu \overline{\frac{\partial u_i'}{\partial x_\ell} \frac{\partial u_i'}{\partial x_\ell}} \equiv \varepsilon \dots$ which represents the dissipation of turbulent kinetic energy."

Like the earlier work presented by Davydov (1961), Hanjalic noted that ε could be modeled with a turbulence transport PDE. Hanjalic also postulated that the length scale for ε could be approximated with dissipating eddies *smaller* than those considered by Rotta (1951), to achieve a relationship whereby

$$\nu_t = k^{1/2} \ell = \nu_t(k, \varepsilon). \qquad (4.66)$$

Namely, Hanjalic stated (Hanjalic 1970), "One could, perhaps, argue that the integral scale, employed in Rotta's equation, is weighted much with the lowest wave numbers and is therefore not representative of the scale of the energy containing eddies which are mostly responsible for both the energy dissipation and diffusion processes."

Herein is the issue with most post-1961 k-ε versions: Hanjalic proposed that ε could be associated with a smaller length scale than the Taylor scale originally proposed for dissipation by Chou (1940, 1945). In particular, Hanjalic associated ε with a small-scale L_ε, whereby

$$\varepsilon \equiv \nu \overline{\frac{\partial u_i'}{\partial x_l} \frac{\partial u_i'}{\partial x_l}} = c_\varepsilon \frac{k^{3/2}}{L_\varepsilon}. \qquad (4.67)$$

Solving for the length scale,

$$L_\varepsilon = c_\varepsilon \frac{k^{3/2}}{\varepsilon}, \qquad (4.68)$$

and thus,

$$\nu_t = k^{1/2} \ell = k^{1/2} \left(c_\varepsilon \frac{k^{3/2}}{\varepsilon} \right) = c_\varepsilon \frac{k^2}{\varepsilon} = \nu_t(k, \varepsilon). \qquad (4.69)$$

In this context, c_ε is the modern C_μ. Therefore, the above relationship collapses onto the Prandtl-Kolmogorov relationship, which applies to the *larger* eddies, and most certainly not onto the *smaller* eddies.

Next, Hanjalic developed a third-order polynomial for calculating a new, geometry-dependent length scale, L_a. In Hanjalic's vernacular, this length scale was intended to "represent both L and L_ε" (Hanjalic 1970). Moreover, L_a was built around a polynomial function expressed as the distance from the wall. *To be clear, L represents the length of the energy carrying eddies in Hanjalic's model, whereas experimental data confirms that $L_\varepsilon \approx \eta$ (i.e., most of the dissipation occurs near the smallest scales), and hence it is not appropriate to have a single scale represent both L and L_ε.* Of course, eddy size depends heavily on fluid velocity, physical properties, and system size. Nevertheless, LIKE algorithm calculations for air and water systems for *Re* in the range of 3000 to 700,000 show that $\eta \sim 7 \times 10^{-5}$ to 1×10^{-4} m, $\lambda \sim 1 \times 10^{-3}$ to 1×10^{-2} m, and $\ell \sim 2 \times 10^{-3}$ to 5×10^{-2} m. Thus, Hanjalic's assertion that $L \approx L_\varepsilon$ is clearly incorrect, as it is more appropriate to say $\eta \ll \lambda \ll \ell$ or, alternatively, that $L_\varepsilon \ll L$.

At this point, the derivation of the modern "exact" ε PDE would be warranted, so it may shed additional light on the k-ε model. However, its derivation takes several dozen pages, assuming that one is careful and lucky enough not to make mathematical errors! Therefore, based on Hanjalic's version, and modernized somewhat, its entire expression is (Hanjalic 1970)

$$\underbrace{\frac{\partial \varepsilon}{\partial t}}_{1} + \underbrace{\overline{u}_j \frac{\partial \varepsilon}{\partial x_j}}_{2} = \underbrace{-2\nu \frac{\partial \overline{u}_i}{\partial x_j}\left(\overline{\frac{\partial u'_i}{\partial x_k}\frac{\partial u'_j}{\partial x_k}} + \overline{\frac{\partial u'_k}{\partial x_i}\frac{\partial u'_k}{\partial x_j}}\right)}_{3} - \underbrace{2\nu \frac{\partial^2 \overline{u}_i}{\partial x_k \partial x_j}\overline{u'_k \frac{\partial u'_i}{\partial x_j}}}_{4} - \underbrace{2\nu \overline{\frac{\partial u'_i}{\partial x_k}\frac{\partial u'_i}{\partial x_m}\frac{\partial u'_k}{\partial x_m}}}_{5}$$

$$\underbrace{-2\nu^2 \overline{\frac{\partial^2 u'_i}{\partial x_k \partial x_m}\frac{\partial^2 u'_i}{\partial x_k \partial x_m}}}_{6} + \underbrace{\nu \frac{\partial\left(\frac{\partial \varepsilon}{\partial x_j}\right)}{\partial x_j}}_{7} - \underbrace{\nu \frac{\partial\left(\overline{u'_j \frac{\partial u'_i}{\partial x_m}\frac{\partial u'_i}{\partial x_m}}\right)}{\partial x_j}}_{8} - \underbrace{2\frac{\nu}{\rho}\frac{\partial\left(\overline{\frac{\partial p'}{\partial x_m}\frac{\partial u'_j}{\partial x_m}}\right)}{\partial x_j}}_{9}$$

$$(4.70)$$

where

Term 1 = transient change of dissipation (accumulation term).

Term 2 = *mean* flow convection dissipation.

Term 3 = dissipation production term that arises from the product of the gradients of the fluctuating and mean velocities.

Term 4 = dissipation production term that generates additional dissipation based on the fluctuating and mean velocities.

Term 5 = the so-called "destruction" rate for dissipation, which is associated with eddy velocity fluctuation gradients.

Term 6 = another "destruction" term for dissipation, which arises from eddy velocity fluctuation diffusion.

Term 7 = term to account for the viscous diffusion associated with dissipation. Note that this is the only positive term on the RHS.

Term 8 = diffusive turbulent transport resulting from the eddy velocity fluctuations.

Term 9 = dissipation of turbulent transport arising from eddy pressure and fluctuation velocity gradients.

The above comprise the nine exact dissipation terms associated with the ε transport PDE. However, besides the two terms that form the substantial derivative (Terms 1 and 2) and the viscous diffusion of dissipation (Term 7), the remaining six terms are completely unknown, to a point whereby the PDE requires Wilcox's infamous "drastic surgery" to transform it into a *vastly different* equation that can be modeled (solved) numerically (Wilcox 2006). Some employ the phrase "surgically modified beyond recognition" (Schobeiri and Abdelfattah 2013). While others would agree with such description, they would perhaps use slightly less colorful assertions (Myong and Kasagi 1990; Rodi 1993; Andersson et al. 2012). In any case, because these terms are unknown, "drastic surgery" is exactly what Hanjalic was forced to perform on the exact ε PDE when he used analytical substitutions for the six PDE terms. Some of these surgery terms are still in use today.

In the aftermath of his model completion, Hanjalic applied it to a plane smooth channel, the boundary layer on a flat plate, a plane wall jet, a plane free jet, and a plane with a mixing layer. Except for the plane wall jet, whose predicted value was within 20 to 30% of the experimental value, the other four cases showed results that compared quite well with experimental data, usually to within 10% or less. However, only one predicted value was compared with the experimental data for each of the five cases (refer to Hanjalic's Table 4.4, page 189, Hanjalic 1970); when more detailed comparisons were made later in the Hanjalic reference, the plane mixing layer simulation vs. experimental data was not as good.

Moving forward a few years later, the k-ε model was revamped onto its more modern format in the early 1970s. This was a time period when various theoretical and computational efforts by Jones, Launder, and Sharma showed excellent potential vs. experimental data (Jones and Launder 1972, 1973; Launder et al. 1973; Launder and Sharma 1974). During this crucial 3-year period, the k-ε model evolved toward the following formulation, which is now commonly called the standard k-ε (SKE) model. The k PDE is written as

$$\frac{\partial k}{\partial t} + \bar{u}_j \frac{\partial k}{\partial x_j} = R_{ij} \frac{\partial \bar{u}_i}{\partial x_j} - \varepsilon + \frac{\partial}{\partial x_j}\left[\left(\nu + \frac{\nu_t}{\sigma_k}\right)\frac{\partial k}{\partial x_j}\right], \qquad (4.71A)$$

while the ε PDE is transformed through the magic of "drastic surgery" onto the following:

$$\underbrace{\frac{\partial \varepsilon}{\partial t}}_{\substack{1 \\ \text{transient}}} + \underbrace{\bar{u}_j \frac{\partial \varepsilon}{\partial x_j}}_{\substack{2 \\ \text{convection}}} = \underbrace{C_1 \frac{\varepsilon}{k} R_{ij} \frac{\partial \bar{u}_i}{\partial x_j}}_{\substack{3 \text{ and } 4 \\ \text{production}}} - \underbrace{C_2 \frac{\varepsilon^2}{k}}_{\substack{5 \text{ and } 6 \\ \text{decay}}}$$

$$+ \underbrace{\frac{\partial}{\partial x_j}\left[\left(\nu + \frac{\nu_t}{\sigma_\varepsilon}\right)\frac{\partial \varepsilon}{\partial x_j}\right]}_{\substack{7-9 \\ \text{diffusion}}}. \qquad (4.71\text{B})$$

As written in their original 1974 paper, the model includes an expanded Prandtl-Kolmogorov closure formulation,

$$\nu_t = C_\mu \frac{k^2}{\varepsilon}, \qquad (4.71\text{C})$$

that incorporates a damping function fit with Re_T as a turbulent Reynolds number (Launder and Sharma 1974):

$$C_\mu \equiv 0.09 e^{-\left[3.4/(1+Re_\mathrm{T}/50)^2\right]}. \qquad (4.71\text{D})$$

The turbulent Reynolds number is defined as

$$Re_\mathrm{T} = \frac{\rho k^2}{\mu \varepsilon}. \qquad (4.71\text{E})$$

Note that Re_T is based on the large eddies and that Launder and Sharma did not include C_μ in their 1974 Re_T expression. Therefore, the implied Re_T-based SKE large eddy length scale is $\ell_{SKE} = \frac{k^{3/2}}{\varepsilon}$, while the SKE Prandtl-Kolmogorov closure relationship implies $\ell_{\text{Prandtl-Kolmogorov}} = C_\mu \frac{k^{3/2}}{\varepsilon}$; this inconsistency is addressed in Sect. 4.7.2.

The final closure coefficients underwent "reoptimization" for high Re applications (Launder et al. 1973; Launder and Sharma 1974; Wilcox 2006), such that

$$C_1 = C_{\varepsilon 1} = 1.44, \qquad (4.71\text{F})$$

$$C_2 = C_{\varepsilon 2} = 1.92\left(1.0 - 0.3 e^{-Re_\mathrm{T}^2}\right), \qquad (4.71\text{G})$$

$$\sigma_\varepsilon = 1.3, \qquad (4.71\text{H})$$

and

$$\sigma_k = 1.0. \tag{4.71I}$$

The above formulation for the ε PDE shows that the nine terms for the "exact" PDE were collapsed into just five terms. Note that in the "transformed" ε PDE, only Terms 1, 2, and 7 survived in an intact manner—the remaining terms bear little, if any, resemblance to the exact equation! In other words, a comparison of the previous two ε PDE equations shows an utter lack of similitude once the transformative "drastic surgery" occurs, namely,

$$\underbrace{-2\nu\left(\frac{\partial u_i'}{\partial x_k}\frac{\partial u_j'}{\partial x_k}+\frac{\partial u_k'}{\partial x_i}\frac{\partial u_k'}{\partial x_j}\right)\frac{\partial \bar{u}_i}{\partial x_j}}_{3} \underbrace{-2\nu\frac{\partial^2 \bar{u}_i}{\partial x_k \partial x_j}u_k'\frac{\partial u_i'}{\partial x_j}}_{4} \rightarrow \underbrace{C_{\varepsilon 1}\frac{\varepsilon}{k}R_{ij}\frac{\partial \bar{u}_i}{\partial x_j}}_{\substack{3 \text{ and } 4 \\ \text{production}}}. \tag{4.72}$$

Ditto,

$$\underbrace{-2\nu\overline{\frac{\partial u_i'}{\partial x_k}\frac{\partial u_i'}{\partial x_m}\frac{\partial u_k'}{\partial x_m}}}_{5} \underbrace{-2\nu^2\overline{\frac{\partial^2 u_i'}{\partial x_k \partial x_m}\frac{\partial^2 u_i'}{\partial x_k \partial x_m}}}_{6} \rightarrow \underbrace{-C_{\varepsilon 2}\frac{\varepsilon^2}{k}}_{\substack{5 \text{ and } 6 \\ \text{decay}}}. \tag{4.73}$$

And ditto,

$$\underbrace{\nu\frac{\partial\left(\frac{\partial \varepsilon}{\partial x_j}\right)}{\partial x_j}}_{7} \underbrace{-\nu\frac{\partial\left(\overline{u_j'\frac{\partial u_i'}{\partial x_m}\frac{\partial u_i'}{\partial x_m}}\right)}{\partial x_j}}_{8} \underbrace{-2\frac{\nu}{\rho}\frac{\partial\left(\overline{\frac{\partial p'}{\partial x_m}\frac{\partial u_j'}{\partial x_m}}\right)}{\partial x_j}}_{9} \rightarrow \underbrace{\frac{\partial}{\partial x_j}\left[\left(\nu+\frac{\nu_t}{\sigma_\varepsilon}\right)\frac{\partial \varepsilon}{\partial x_j}\right]}_{\substack{7-9 \\ \text{diffusion}}}. \tag{4.74}$$

Certainly, some of this monkey business also goes on with k-ω models as well, but not to the degree whereby only three terms out of nine are modeled directly, and the remaining terms involve astonishing transformations that bear *absolutely no resemblance* to the "exact" terms.

Note that despite the claim that the dissipation eddies are small, and the less-than-rigorous transformation onto a form that can be modeled, the k-ε model remains popular and still enjoys substantial usage in the turbulence community. To be fair, the model has enjoyed many spectacular successes since its modern upgrades (Hanjalic 1970; Jones and Launder 1972, 1973; Launder et al. 1973; Launder and Sharma 1974; Rodi 1993; Gorji et al. 2014; Diaz and Hinz 2015). For example, the SKE has successfully modeled the following:

- High *Re* pipe flow (Jones and Launder 1972, 1973)
- Pipe flow with *Pr* in the range of 0.5–2000 and *Re* ~ 10,000 (Jones and Launder 1972, 1973)
- High *Re* spinning disk under heat transfer (*Re* ~ 1×10^5 to 1.5×10^6) (Launder and Sharma 1974)
- Free shear layer flow (e.g., high *Re*, isotropic, far away from walls) (Rodi 1993)
- Flows with nonexistent to small pressure gradients (Wilcox 2006)
- Simplified case of boundary layer flow over a flat plate (constant pressure, far away from the wall, no adverse pressure gradient, high *Re*) (Hanjalic 1970)
- Channel flow for *Re* in the range of 6000 to 150,000 (and higher, of course, because the SKE fares best at high *Re*) (Hanjalic 1970; Jones and Launder 1972, 1973; Gorgi et al. 2014)
- Plane wall and free jet (Hanjalic 1970)

However, the SKE is *not* suitable for:

- Adverse pressure gradients (Wilcox 1988a, 1993a, b, 2006; Fluent 2012; Argyropoulos and Markatos 2015)
- Separated flow (Wilcox 2006; Argyropoulos and Markatos 2015)
- High curvature (Argyropoulos and Markatos 2015; Fraczek and Wroblewski 2016)
- Large rotation/swirl flows (Fluent 2012; Andersson et al. 2012; Diaz and Hinz 2015)
- Round jet (Pope 1978; Wilcox 2006; Fluent 2012)
- Shock mixing (Dong et al. 2010)
- Asymmetric diffusers (El-Behery and Hamed 2009)
- Near-wall treatment: viscous sublayer; variable-pressure boundary wall ($30 \leq y^+ \leq 700$) (Wilcox 2006, 2012)
- Fluids at supercritical pressure (He et al. 2008; Bae et al. 2017)
- Large strain rate (regions with large velocity gradients, stagnation points) (Fluent 2012; Argyropoulos and Markatos 2015)
- Laminar to turbulent transition for heated flow (Abdollahzadeh et al. 2017)
- Situations where overdamping adversely impacts parameters that rely on wall-based quantities (e.g., wall heat transfer, wall friction) (Zhao et al. 2017)

Note that when a turbulence code under predicts ν_t, it said to be "overdamping," and this is certainly an issued associated with the SKE, which has a tendency to overdamp near the wall (Zhao et al. 2017).

The SKE's usage trend has seen decreased levels since the 1990s, when the *k-ω* models increased in maturity (Wilcox 2006) and other methods, such as large eddy simulation (LES) and direct numerical simulation (DNS), came into more usage. Another reason for the SKE's reduced usage may stem from a lackluster performance vs. the 2006 *k-ω* and SST models, when undergoing simulation of more complex turbulence systems (Menter 1992; Wilcox 1993a, b, 2006; Menter et al. 2003). In addition, the interested reader is encouraged to review Sections 4.7 through 4.7.3 for serious concerns regarding the SKE.

Certainly, dozens of upgrades have been proposed to improve the SKE. Notable improvements include the realizable k-ε model and the renormalization group-theory (RNG) k-ε models, with many successes, albeit with a nagging continuation of issues, all likely due to fundamental reasons that stem from the attempt to couple small (ε) and large (k) eddy scales into a single model. Hence, many k-ε improvements tend to gravitate toward functions that can blend the diametrically opposed scale behaviors. For example, instead of letting Re_T be based on a damping function as Hanjalic, Launder, and Sharma proposed, various authors now recommend a y^+ damping function and have noted better agreement with experimental data for heat transfer experiments involving supercritical fluids (Rodi and Mansour 1993; He et al. 2008; Bae et al. 2017).

Nevertheless, a fundamental issue for k-ε modeling remains: its association of the large turbulent kinetic energy scales with the small dissipative scales. It is very likely that, so long as the model relies on the two different scales (despite numerous blending attempts), issues will continue to crop up.

More recently, attempts to modify the k-ε model to include Taylor eddy behavior, as Chou originally proposed decades earlier, appear to bear much promise (Bae 2016; Bae et al. 2016, 2017). For these reasons, the Myong-Kasagi model is explored next.

4.6.3.6   Myong-Kasagi k-ε Model

The Myong-Kasagi (MK) k-ε model was developed from the SKE but with some fundamental paradigm changes associated with the ε PDE (Myong and Kasagi 1988; Myong et al. 1989; Myong and Kasagi 1990). The k PDE is the same, except for a single change in the diffusive term:

$$\sigma_k = 1.4 \quad (\text{vs.}1.0 \text{ in the SKE}). \tag{4.75A}$$

On the other hand, the MK ε PDE has key upgrades in the production and decay terms, as follows:

$$\underbrace{\frac{\partial \varepsilon}{\partial t}}_{\text{transient}} + \underbrace{\bar{u}_j \frac{\partial \varepsilon}{\partial x_j}}_{\text{convection}} = \underbrace{C_{\varepsilon 1} f_1 \frac{\varepsilon}{k} R_{ij} \frac{\partial \bar{u}_i}{\partial x_j}}_{\text{production}} - \underbrace{C_{\varepsilon 2} f_2 \frac{\varepsilon^2}{k}}_{\text{decay}} + \underbrace{\frac{\partial}{\partial x_j}\left[\left(\nu + \frac{\nu_t}{\sigma_\varepsilon}\right)\frac{\partial \varepsilon}{\partial x_j}\right]}_{\text{diffusion}}. \tag{4.75B}$$

The PDE includes the following blending function modifications:

$$C_{\varepsilon 2} = 1.8 \text{ (MK) vs. } 1.92\left(1.0 - 0.3e^{-Re_T^2}\right) \text{ (SKE)} \tag{4.75C}$$

and

$$f_\mu = \left(1 + \frac{3.45}{Re_T^{1/2}}\right)$$
$$\times \left(1 - e^{-y^+/70}\right) \text{ (MK)} \quad \text{vs.} \quad 0.09 e^{-[3.4/(1+Re_T/50)^2]} \text{ (SKE)}. \qquad (4.75\text{D})$$

Other researchers proposed a slightly modified blending function that more rapidly approaches the asymptotic value (Speziale et al. 1992):

$$f_\mu = \left(1 + \frac{3.45}{Re_T^{1/2}}\right) \tanh\left(\frac{y^+}{70}\right). \qquad (4.75\text{E})$$

The MK f_μ function was derived by adding the implied Prandtl-Kolmogorov length scale for the *large and Taylor eddies*,

$$L_D = L_\ell + L_\lambda = C_D \frac{k^{3/2}}{\varepsilon} + C_1 \sqrt{\frac{\nu k}{\varepsilon}}, \qquad (4.75\text{F})$$

and thereafter multiplying the lengths with a Van-Driest-like damping function. The Taylor length was obtained using dimensional arguments, whereby

$$\varepsilon \approx \frac{\nu k}{\lambda^2}. \qquad (4.75\text{G})$$

Note that Taylor's theoretical research indicates that $C_1 = \sqrt{10}$ for isotropic flows. The MK model includes the following new parameters:

$$f_1 = 1.0 \qquad (4.75\text{H})$$

and

$$f_2 = \left[1 - \frac{2}{9} e^{(Re_T/6)^2}\right] \left(1 - e^{-y^+/5}\right)^2. \qquad (4.75\text{I})$$

The $C_{\varepsilon 1}$ term was modified slightly:

$$C_{\varepsilon 1} = 1.4 \text{ (MK)} \quad \text{vs.} \quad 1.44 \text{ (SKE)}. \qquad (4.75\text{J})$$

Finally, the Taylor length scale is

$$\lambda \approx \left(\frac{\nu k}{\varepsilon}\right)^{1/2} = C_1 \left(\frac{\nu k}{\varepsilon}\right)^{1/2} = \left(10 \frac{\nu k}{\varepsilon}\right)^{1/2}. \qquad (4.76)$$

Ironically, this effort returns the k-ε back to Chou's earlier k-ω research premise, namely, k-ε ought to include the larger Taylor eddy length, thereby incorporating the behavior of eddies larger than Kolmogorov.

Even though it was developed in 1990, the inclusion of the Taylor eddy length scale continues to make this k-ε model worthy of recent investigation (Speziale et al. 1992; Bae 2016; Bae et al. 2016, 2017). Not surprisingly, modeling of the Taylor eddies increases turbulent production near the viscous sublayer and the buffer layer (Bae et al. 2017). Recent comparisons with experimental data show much promise (Speziale et al. 1992; Hrenya et al. 1995; Bae 2016; Bae et al. 2016, 2017), though there continue to be concerns due to overdamping (Zhao et al. 2017). The MK model was found as having the best overall performance of 10 k-ε RANS models (including SKE) during fully developed pipe flow in the range of $7000 \leq Re \leq 500{,}000$, despite an ε maldistribution (Hrenya et al. 1995). The MK velocity distribution vs. SST and Spalart-Allmaras was not as good for sCO2 modeling under heat transfer, though MK beat the two models for wall temperature distribution vs. distance (Otero et al. 2018).

4.6.3.7   Menter 2003 SST Model

With the wide usage of the SKE and the development of the 1988 k-ω version, it was soon evident in the late 1980s and early 1990s that each model has its pros and cons and that these are mutually exclusive in the sense of domain applicability. At the time, there was a need for a reliable model suitable for turbulence aeronautics flows that was specifically able to handle flow separation involving large, adverse pressure gradients. Menter shrewdly realized that the SKE resulted in reasonable flow simulations at high Re and away from the wall (free stream) but had shortcomings near the wall. On the other hand, the 1988 k-ω was very useful for low Re and near the wall boundary layer but was more sensitive than the SKE for free stream boundary conditions. Further, the 1988 k-ω model did not work well with pressure-induced separation (Menter 1992, 1993; Menter et al. 2003). Thus, in a very fortunate situation, the shortcomings of one model were already compensated by the other and vice versa. And as an extra bonus, there would be no need for additional damping functions near the wall.

So, Menter wondered, why not combine both models in such a way that each is used in the region where it excels and, in between, use some sort of weighted average or blending? Hence the 1992 shear stress transport (SST) model was born (Menter 1992, 1993, 1994). The model is also known as the SST k-ω and as Menter's SST model but is not to be confused with the similarly named "stress transport" models.

More specifically, each of the two RANS models is applied onto a region where it excels over the other, as follows:

- The boundary layer is computed by the 1988 k-ω.
- SKE is used for the free stream.

- A blending function F_1 computes the asymptotic turbulent behavior between the two distinct regions.
- The blending and stress limiter functions are based on the hyperbolic tangent (tanh) for rapid transition.
- To further improve the model, a function was used as a *stress limiter* to consider the impact of the mean strain rate on turbulent kinematic viscosity (analogous to the 2006 k-ω stress limiter; in fact, this was the 2006 k-ω stress limiter precursor).

The model dynamics are shown conceptually in Figs. 4.3 and 4.4.

Certainly, there are various SST hybrids. The model listed here is based on Menter's 2003 modifications, which debuted a little over a decade after the 1992 SST model (Menter et al. 2003). The primary differences between the original 1992 Menter model and the 2003 formulation are the replacement of vorticity with S and the replacement of 20 with 10 in the production limiter.

In the 2003 version, Menter used the following k and ω transport equations, respectively:

$$\frac{\partial k}{\partial t} + \bar{u}_j \frac{\partial k}{\partial x_j} = \widetilde{P}_k - \beta^* k\omega + \frac{\partial}{\partial x_j}\left[(\nu + \sigma_k \nu_t)\frac{\partial k}{\partial x_j}\right], \qquad (4.77\text{A})$$

where the production term is

$$P_k = \left[\nu_t\left(2S_{ij} - \frac{2}{3}\frac{\partial \bar{u}_k}{\partial x_k}\delta_{ij}\right) - \frac{2}{3}k\delta_{ij}\right]\frac{\partial \bar{u}_i}{\partial x_j} \qquad (4.77\text{B})$$

with

$$S_{ij} = \frac{1}{2}\left(\frac{\partial \bar{u}_i}{\partial x_j} + \frac{\partial \bar{u}_j}{\partial x_i}\right). \qquad (4.77\text{C})$$

The model is subject to a k turbulence production limiter that rightly so suppresses turbulence in regions where stagnated flow occurs, namely,

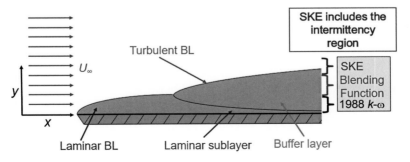

Fig. 4.3 Application domains for the SST model

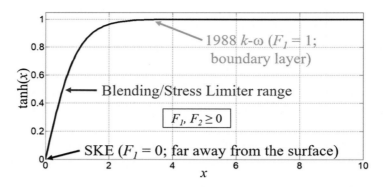

Fig. 4.4 Blending function application range for SST

$$\widetilde{P}_k = \min\left(P_k, 10\beta^* k\omega\right). \tag{4.77D}$$

The ω PDE is as follows:

$$\frac{\partial \omega}{\partial t} + \overline{u}_j \frac{\partial \omega}{\partial x_j} = \alpha \frac{\widetilde{P}_\omega}{\nu_t} - \beta\omega^2 + \frac{\partial}{\partial x_j}\left[(\nu + \sigma_\omega\nu_t)\frac{\partial \omega}{\partial x_j}\right] \\ + 2\sigma_{\omega 2}(1 - F_1)\frac{1}{\omega}\frac{\partial k}{\partial x_j}\frac{\partial \omega}{\partial x_j}. \tag{4.77E}$$

Note: this ω PDE version includes a correction made by NASA (NASA2 2018; Goldberg and Batten 2015); thus, the ω production term is not

$$P_\omega = \alpha S^2 \text{ (incorrect)} \tag{4.77F}$$

but is instead

$$P_\omega = R_{ij}\frac{\partial \overline{u}_i}{\partial x_j} \text{ (correct)}. \tag{4.77F'}$$

Production in the ω transport equation is also subject to a turbulence production limiter that suppresses turbulence in regions where stagnated flow occurs, namely,

$$\widetilde{P}_\omega = \min\left(P_\omega, 10\beta^* k\omega\right). \tag{4.77G}$$

The turbulent kinematic viscosity is

$$\nu_t = \frac{k}{\widetilde{\omega}} \tag{4.77H}$$

where

$$\widetilde{\omega} = \max\left(\omega, \frac{F_2 S}{a_1}\right). \tag{4.77I}$$

Therefore,

$$\nu_t = \frac{a_1 k}{\max\left(a_1 \omega, F_2 S\right)}, \tag{4.77J}$$

whereby

$$S = \sqrt{2 S_{ij} S_{ij}}. \tag{4.77K}$$

The blending function F_1 is defined as

$$F_1 = \tanh\left(\left\{\min\left[\max\left(\frac{\sqrt{k}}{\beta^* \omega y}, \frac{500\nu}{\omega y^2}\right), \frac{4\rho\sigma_{\omega 2} k}{CD_{k\omega} y^2}\right]\right\}^4\right), \tag{4.77L}$$

where y is the distance to the closest surface. The scalar $CD_{k\omega}$ is defined as

$$CD_{k\omega} = \max\left(2\rho\sigma_{\omega 2}\frac{1}{\omega}\frac{\partial k}{\partial x_j}\frac{\partial \omega}{\partial x_j}, 10^{-10}\right). \tag{4.77M}$$

The stress limiter function F_2 is

$$F_2 = \tanh\left\{\left[\max\left(\frac{2\sqrt{k}}{\beta^* \omega y}, \frac{500\nu}{\omega y^2}\right)\right]^2\right\}. \tag{4.77N}$$

Finally, due to blending, the closure coefficients for the SST are computed using the following weighted function:

$$\phi = \phi_1 F_1 + \phi_2(1 - F_1), \tag{4.77O}$$

where ϕ represents any of the Menter closure coefficients that have subscripts 1 and 2 (e.g., α_1, α_2, β_1, β_2, σ_{k1}, and σ_{k2}). In this context, "1" represents the 1988 k-ω model and "2" represents the SKE model.

The 2003 SST model closure coefficients are as follows:

$$a_1 = 0.31, \tag{4.77P}$$

$$\alpha_1 = \frac{5}{9}, \tag{4.77Q}$$

$$\alpha_2 = 0.44, \tag{4.77R}$$

$$\beta^* = 0.09, \tag{4.77S}$$

$$\beta_1 = \frac{3}{40}, \tag{4.77T}$$

$$\beta_2 = 0.0828, \tag{4.77U}$$

$$\sigma_{k1} = 0.85, \tag{4.77V}$$

$$\sigma_{k2} = 1.0, \tag{4.77W}$$

$$\sigma_\omega = 0.5, \tag{4.77X}$$

and

$$\sigma_{\omega 2} = 0.856. \tag{4.77Y}$$

Though originally conceived for turbulent aeronautics flows, the SST is commonly employed in many industries, with various modifications that extend the usefulness of the model, including rough surfaces (Hellsten and Laine 1997; Knopp et al. 2009), rotation (Hellsten 1997), and a version with no wall dependence (Goldberg and Batten 2015).

The 2003 SST model is ideal for:

- Adverse pressure gradients
- Separated flows
- Turbulent heat transfer
- Mixed low and high Re problems
- Of course, aerospace applications

Moving forward, past the 1992 and 2003 models, it is noted the 2006 k-ω model *does not* have the free stream boundary condition sensitivity of its 1988 predecessor, which was a key reason, if not the major reason, as to why the SST was developed. Thus, to be blunt, had the 2006 k-ω been developed two decades earlier, the need for the SST's blending of the SKE and the 1988 k-ω would not have existed. Nevertheless, the turbulence community is fortunate to have both the 2003 SST and the 2006 k-ω models. To say the least, both models are very good, all-around models suitable for both near-wall and free stream turbulence.

But, if a choice exists, the 2006 k-ω is recommended over the 2003 k-ω SST model, especially regarding higher Ma flows (supersonic through hypersonic flows) (Wilcox 2006). An exception to this recommendation is that the 1992 SST (which is very similar to the 2003 SST), generally outperforms the 2006 k-ω in the *transonic* regime (Wilcox 2006); as noted by Wilcox, the SST was fine-tuned for *transonic Ma*. Furthermore, the SST relies on the k-ε, which has two transport variables that are not exactly self-consistent in terms of their length scales (refer to Sections 4.7 through 4.7.3). Finally, there is the issue of the much higher number of closure formulations and constants used in the SST vs. the 2006 k-ω.

Note that the improvements in the 2006 k-ω model are attributable to the cross-diffusion term and the incorporation of blending. In this sense, the 2006 k-ω is very much SST-like. As will be shown in Sect. 4.8, the 2006 k-ω and the 2003 Menter SST are uncannily similar, after all!

Example 4.1 Consider a rectangular duct with $H = 0.005$ m, $W = 0.025$ m, average water velocity $= 0.247$ m/s, and $Re = 11,000$. It is desired to use various RANS models, so k, ε, ω, and ν_t must be pre-calculated for usage as input values. Apply the LIKE algorithm to estimate the values of the four variables.

Solution

$$D_h = \frac{4FA}{WP} = \frac{4HW}{2H + 2W} = \frac{2HW}{H + W} = \frac{2 * 0.005 * 0.025}{0.005 + 0.025} = 0.00833 \text{ m}$$

$$\ell = 0.07 D_h = 0.07 * 0.00833 = 0.000583 \text{ m}$$

$$I = 0.16 \, Re_h^{-1/8} = 0.16(11,000)^{-1/8} = 0.05$$

Now it is straightforward to calculate the input values:

$$k = \frac{3}{2}(\overline{u}I)^2 = \frac{3}{2}(0.247 * 0.05)^2 = 2.29 \times 10^{-4} \text{ m}^2/\text{s}^2$$

Recall $C_\mu = 0.09$. Therefore,

$$\varepsilon = C_\mu \frac{k^{3/2}}{\ell} = 0.09 \frac{\left(2.29 \times 10^{-4}\right)^{3/2}}{0.000583} = 5.35 \times 10^{-4} \text{ m}^2/\text{s}^3$$

$$\omega = \frac{k^{1/2}}{\ell} = \frac{\left(2.29 \times 10^{-4}\right)^{1/2}}{0.000583} = 26.0 \, 1/\text{s}$$

$$\nu_t \equiv \frac{\mu_t}{\rho} = \frac{C_\mu k^2}{\varepsilon} \text{ (Prandtl-Kolmogorov)} = \frac{0.09 \left(2.29 \times 10^{-4}\right)^2}{5.35 \times 10^{-4}}$$

$$= 8.82 \times 10^{-6} \text{ m}^2/\text{s}.$$

As a "sanity check,"

$$\nu \equiv \frac{\mu}{\rho} = \frac{2.32 x 10^{-4}}{942} = 2.46 \times 10^{-7} \text{ m}^2/\text{s},$$

while

$$\frac{\nu_t}{\nu} = \frac{8.82 \times 10^{-6}}{2.46 \times 10^{-7}} = 35.9.$$

Example 4.2 Find σ_k for the SST model.

Solution

$$\phi_1 = \sigma_{k1} = 0.85$$

$$\phi_2 = \sigma_{k2} = 1.0$$

From $\phi = \phi_1 F_1 + \phi_2(1 - F_1)$,

$$\sigma_k = \sigma_{k1} F_1 + \sigma_{k2}(1 - F_1) = 0.85 F_1 + 1.0(1 - F_1) = 1.0 - 0.15 F_1$$

If $F_1 = 1 \Rightarrow \sigma_k = 0.85$ (SKE).
If $F_1 = 0 \Rightarrow \sigma_k = 1.0$ (1988 k-ω).

4.7 Avoid the SKE Model?

That some researchers have noted the superiority of k-ω models over k-ε is not new (Wilcox 2006; Fluent 2012). A quote from the Fluent code developers points out that k-ω models have (Fluent 2012) "much better performance than k-ε models for boundary layer flows. For separation, transition, low Re effects, and impingement, k-ω models are more accurate than k-ε models."

As a pragmatic matter, the choice of turbulence model is oftentimes analogous to the choice of political party or religion. The author recognizes the psychological human factor associated with turbulence model selection and respects the user's desire to make choices based on numerous factors and circumstances.

However, as engineers, physicists, mathematicians, and technology enthusiasts, it is always important to reflect why *any* given transport variable is selected in turbulence models, such as k, ℓ, ω, ε, etc. Why should there be any "sacred cow" variables or combinations thereof? Throughout turbulence history, the easiest and most defensible choice is k for isotropic flows because it quantifies the eddy turbulent energy, so its square root provides a great metric for the fluctuating velocity. However, the choice for the second transport variable is neither as straightforward nor as easy to justify.

As noted in Sect. 4.6.3.5, the SKE has performed very well for certain flows but has had spectacular failures in others, especially for the more complex flows. At this point, three fundamental reasons for avoiding the usage of the SKE are elaborated in the sections that follow.

4.7.1 *Inconsistent SKE Transport and Eddy Scales*

To proceed, note that the total kinetic energy is held by the sum of all the eddies within the turbulent flow, which forms a continuous length spectrum for the integral, Taylor, and Kolmogorov eddies:

$$k_{\text{tot}} = \sum_{\bar{i}=1}^{n} k_{\ell,\bar{i}} + \sum_{\bar{j}=1}^{m} k_{\lambda,\bar{j}} + \sum_{\bar{k}=1}^{o} k_{\eta,\bar{k}}. \tag{4.78}$$

The Loitsianskii eddies (Hinze 1987) could be included as a separate class in the turbulent kinetic energy tally but are instead lumped as part of the large eddy group. Recall that the vast majority of the turbulent kinetic energy is held by the integral eddies, which is about 80%. The Taylor eddies take the larger fraction of the remaining turbulent kinetic energy, with the Kolmogorov eddies having a progressively smaller fraction, as shown by the curve in Fig. 3.1 of Chap. 3. Therefore, as a group ensemble,

$$k_\ell \gg k_\lambda \gg k_\eta. \tag{4.79}$$

Thus, k is most certainly a function of an eddy characteristic length, with that length being essentially a function of ℓ, with practically nothing having to do with η. It is therefore safe to express this mathematically and more succinctly as

$$k = k(\ell, \lambda) \tag{4.80}$$

and, conversely,

$$k \neq k(\eta). \tag{4.81}$$

So, attempting to hone in on a crucial point, but at the risk of being redundant, note that the transport variable k is based on the *larger* eddy scales, and not on the *smaller* scales. And of course, this makes physical sense; the larger eddies have the highest velocities and hence the highest kinetic energy.

On the *opposite* extreme of the highly energetic eddies is decay, whereby eddies are so small that they dissipate completely out of existence. This is where the eddy approach the tiny Kolmogorov scale, thereby surrendering their meager energy back to the main flow as the viscous force dampens them back to laminarity. As noted in Chap. 2, it is this dissipation ε (the change of turbulent kinetic energy per unit time) that defines the decay of the *small* scales:

$$\varepsilon \equiv \nu \overline{\frac{\partial u'_i}{\partial x_l} \frac{\partial u'_i}{\partial x_l}} = -\frac{\partial k}{\partial t}. \tag{4.82}$$

Recall that Kolmogorov's definition for the smallest eddies (Sect. 3.3) shows that eddy size is purely a function of ν and ε:

$$\eta = \left(\frac{\nu^3}{\varepsilon}\right)^{1/4}. \tag{4.83}$$

Stated differently, Kolmogorov's formulation indicates that

$$\varepsilon = \varepsilon(\eta, \nu), \tag{4.84}$$

thereby confirming that dissipation is purely a function of the smallest eddy scales and the kinematic viscosity.

Thus, it is clear that for SKE, k is based on ℓ, while ε is based on η, and failure to use consistent length scales will result in flawed turbulence models. In particular, the two transported SKE variables for the k-ε turbulence model are k, which applies strictly to the larger eddies, and ε, which applies to the smallest eddies. That is, eddy dissipation occurs at the smallest scales, as the flow approaches more isotropic conditions. Thus, dissipation usually occurs under conditions such as high Re, in free shear flows, the core, and away from the wall (Sondak 1992). By contrast, larger eddies tend to form in the buffer layer because of its large stresses and large velocity gradients (despite the lower Re in that region). *How, then, can turbulence be calculated consistently when one transport variable calculates the effect of the larger eddies, while being mathematically coupled onto a transport variable that calculates the impact of the small eddies?* Not only are the dominant regions distinct, where the larger and the smallest eddies reside, but their timescales are significantly different as well, with $t_\ell \gg t_\eta$.

Finally, though L_ε was defined as a "dissipation length scale" for the ε PDE, Hanjalic recognized that the following three scales are different: ℓ (integral length, i.e., Hanjalic's "L"), Chou's λ, and L_ε (i.e., Hanjalic's Kolmogorov eddies). Indeed, Hanjalic asserted that (Hanjalic 1970) "Although these scales are not the same in general, it is reasonable to expect them to vary in a similar fashion. *Thus, considering the degree of approximation implied by the method so far, it may be assumed that $L \approx L_\varepsilon$.* It remains to specify a unique length scale which will represent both L and L_ε."

Thus, the development of the ε PDE assumes that $L \approx L_\varepsilon$, which means that a single length is supposed to "represent" the larger energy bearing eddies and the smaller dissipative eddies. This implies that the larger eddies $\approx$ smallest eddies, i.e., $\ell \approx \eta$; but in the real world, $\ell \gg \eta$.

4.7.2   Inconsistent SKE Closure

Consider the SKE formulation for calculating the turbulent kinematic viscosity, ν_t. Based on dimensional arguments, Taylor postulated in 1935 that

$$\varepsilon = C \frac{k^{3/2}}{\ell}. \tag{4.85A}$$

Based on experimental data, Wilcox recommends that $C = C_\mu$ (Wilcox 2006):

$$\varepsilon = C_\mu \frac{k^{3/2}}{\ell}. \tag{4.85B}$$

Solving for ℓ yields the integral eddy length scale in question:

$$\ell = C_\mu \frac{k^{3/2}}{\varepsilon}. \tag{4.85C}$$

To compute ν_t based on two-equation models, two variables are sought such that

$$\nu_t = \nu_t(k, x), \tag{4.86A}$$

where k and x are the desired transport variables. For the SKE, $x = \ell$, so using dimensional arguments,

$$\nu_t = k^{1/2}\ell. \tag{4.86B}$$

Substituting Taylor's ℓ relationship (Eq. 4.85C) into the above expression yields

$$\nu_t = k^{1/2}\ell = k^{1/2}\left(C_\mu \frac{k^{3/2}}{\varepsilon}\right) = C_\mu \frac{k^2}{\varepsilon}, \tag{4.86C}$$

which is the well-known Prandtl-Kolmogorov relationship. Therefore, the SKE ν_t is a function not only of k but also of ε. That is, ε is based on a relationship involving the larger eddies ℓ via $\varepsilon = C_\mu \frac{k^{3/2}}{\ell}$, and thus,

$$\nu_t = \nu_t(k, \varepsilon) = \nu_t(k, \ell) = C_\mu \frac{k^2}{\varepsilon} = k^{1/2}\ell. \tag{4.86D}$$

By contrast, the 2006 k-ω does not require an eddy length scale:

$$\nu_t = \nu_t(k, \widetilde{\omega}) = \frac{k}{\widetilde{\omega}}, \tag{4.86E}$$

where

$$\widetilde{\omega} = \max\left(\omega, C_{\lim}\sqrt{\frac{2S_{ij}S_{ij}}{\beta^*}}\right). \tag{4.86F}$$

In any case, the SKE relies on the *larger energetic* eddy scale ℓ to calculate the impact of the *smaller dissipating* eddies to achieve closure via ν_t, which puts the SKE in a rather tenuous situation, to say the least! For this reason, the SKE (as expressed in Launder and Sharma's classic 1974 paper) uses a modified Prandtl-Kolmogorov relationship for closure (Launder and Sharma 1974). This is an attempt to skew the impact of the larger eddies with a damping function that decreases in magnitude as Re_T decreases. In particular, the SKE Prandtl-Kolmogorov relationship is based on a function for Re_T (i.e., $C_\mu \equiv 0.09e^{-[3.4/(1+Re_T/50)^2]}$), which in turn is based on the *larger* eddies, with the explicit goal of dampening turbulence in regions near the wall, and thereby attempts to reflect the behavior of the smallest eddies. This Prandtl-Kolmogorov relationship issue has been noted previously, where it was reaffirmed that the relationship (Myong and Kasagi 1990) "...is approximately valid only at high turbulent Reynolds number flows *remote from the wall*."

The turbulent Reynolds number as used by the SKE is as follows (Jones and Launder 1972, 1973; Launder and Sharma 1974):

$$Re_T \equiv \left(\frac{\rho k^2}{\mu\varepsilon}\right)_{Jones-Launder} = \frac{\left(\frac{k^{3/2}}{\varepsilon}\right)(k^{1/2})\rho}{\mu} = \frac{(\ell_{SKE})(u_{SKE})\rho}{\mu}. \tag{4.87}$$

Therefore, the implied SKE Re_T-based definition for the large eddy length is

$$\ell_{SKE} = \frac{k^{3/2}}{\varepsilon}. \tag{4.88A}$$

Note that the SKE uses the Prandtl-Kolmogorov relationship for closure, such that

$$\nu_t = k^{1/2}\ell = C_\mu \frac{k^2}{\varepsilon}, \tag{4.89A}$$

which implies that

$$\ell_{Prandtl\text{-}Kolmogorov} = C_\mu \frac{k^{3/2}}{\varepsilon}. \tag{4.88B}$$

A comparison between Eqs. 4.88A and 4.88B shows that the large-scale length is inconsistent by a factor of C_μ. Furthermore, this inconsistency is reflected in the two damping SKE functions f_μ and C_2, because both use Re_T.

Note that Launder and Sharma use the following expression:

$$C_{\mu,SKE} = C_\mu f_\mu = 0.09\left\{e^{-[3.4/(1+Re_T/50)^2]}\right\}. \tag{4.90}$$

Because the Prandtl-Kolmogorov relationship is based on the larger eddies, the SKE modifies the relationship in the following manner:

$$\nu_{t,\text{SKE}} = C_{\mu,\text{SKE}} \frac{k^2}{\varepsilon} = \left\{ 0.09 e^{-\left[3.4/(1+Re_T/50)^2\right]} \right\} \frac{k^2}{\varepsilon}. \qquad (4.89\text{B})$$

To show its impact, the C_μ function is plotted in Fig. 4.5, where it is noted that C_μ remains constant at 0.09 for $Re_T > 1200$ (this refers to the turbulent *Reynolds number*, not the hydraulic *Reynolds number*). As Re_T approaches 0, C_μ approaches 0.003, or 3.3% of the peak value, and hence the large degree of overdamping near the wall (Zhao et al. 2017). Of course, the direct consequence of a smaller C_μ is that $\nu_{t,\text{SKE}}$ will be much smaller near the wall (which is generally good, but not so if the damping is excessive, as the SKE is notoriously known to do near the wall—recall that the smaller ν_t is, the lower the degree of turbulence in the local region; see Example 4.1 under "A note about turbulent viscosity vs. kinematic viscosity"). Furthermore, the relationship is based on Re_T, but not on y^+. In particular, what happens in a situation where $Re_T < 1200$ (but is turbulent), in the region where $7 \le y^+ \le 30$? Then, $C_\mu \to 0.003$ in the very region where turbulence is supposed to be the highest! And therefore, ν_t will be very small in such region per Eq. 4.89B.

The same issue applies to $C_2 = C_{\varepsilon 2} = 1.92\left(1.0 - 0.3e^{-Re_T^2}\right)$, though at a grander scale. The C_2 coefficient applies to the ε PDE term that represents the "destruction" rate for dissipation associated with eddy velocity fluctuation gradients and velocity fluctuation diffusion. The coefficient function is shown in Fig. 4.6. Note that the peak value is 1.92 for $Re_T > 2.5$ and rapidly decays to a minimum of 1.34 as Re_T approaches 0. Thus, its damping effect only occurs at very small Re_T, irrespective of the y^+ magnitude! That is, the damping function regime can apply to regions that may have significant turbulent eddy motion. Because part of the function domain can be outside of the viscous sublayer, its attempt to dampen eddies will adversely impact the velocity field. By the same token, to foment eddy behavior

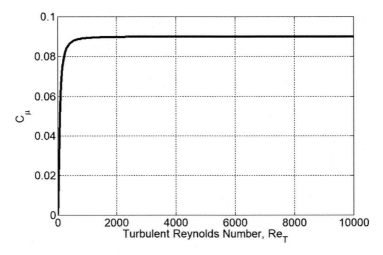

Fig. 4.5 C_μ as a function of Re_T

Fig. 4.6 $C_2 = C_{\varepsilon 2}$ vs. Re_T

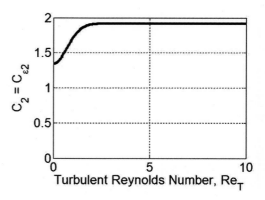

in regions that have no eddies will also generate issues. Not surprisingly, recent comparisons with data attribute overdamping of turbulence near the wall, thereby generating "unrealistic results" both in velocity distribution and heat transfer (Zhao et al. 2017).

Example 4.3 Consider internal flow in a pipe with $D = 0.0254$ m, wall roughness $= 2 \times 10^{-7}$ m, and $\bar{u} = 12.5$ m/s. The fluid, water in this case, is at 500 K and 7.5×10^6 Pa. Will using the SKE turbulence model result in simulation issues?

Solution For this situation, $\rho = 835.8$ kg/m^3 and $\mu = 1.19 \times 10^{-4}$ Pa-s. From the LIKE algorithm, $Re = 2.23 \times 10^6$, *so the flow is highly turbulent and rather isotropic, which is where SKE should perform its best.* For this case, the Kolmogorov and Taylor eddies are easily obtained from the LIKE algorithm as $\eta = 5.53 \times 10^{-6}$ m and $\lambda = 2.67 \times 10^{-4}$ m, respectively, while the viscous sublayer thickness is $\delta(y^+ = 7) = 2.24 \times 10^{-6}$ m and $\delta(y^+ = 1) = 3.20 \times 10^{-7}$ m. The LIKE algorithm also calculates ε at 3.10 m^2/s^3 and k at 0.155 m^2/s^2, so the SKE Re_T is easily calculated as

$$Re_{T,\text{SKE}} = \frac{\rho k^2}{\mu \varepsilon} = \frac{835.8 * (0.155)^2}{1.19x10^{-4} * 3.10} = 54,432.$$

From Figs. 4.5 and 4.6, it is clear that $C_2 = C_{\varepsilon 2}$ and C_μ are not damped because $Re_{T,\text{ SKE}}$ is too large; this comes as no surprise, as the SKE was built for high Re. However, the viscous sublayer is 2.5 and 118 times smaller than the Kolmogorov and Taylor eddies, respectively. Safely assuming that there are no Kolmogorov eddies within the viscous sublayer (refer to Sect. 3.7), then the *first* layer of Kolmogorov eddies would be at a distance y_η = viscous sublayer thickness + Kolmogorov eddy length = $\delta(y^+ = 7) + \eta = 2.24 \times 10^{-6}$ m + 5.53×10^{-6} m = 7.77×10^{-6} m. That is, this simplified analysis assumes that the viscous sublayer thickness and the typical Kolmogorov eddy length are added. This implies a dimensionless distance $y_\eta/[\delta(y^+ = 1)] = (7.77 \times 10^{-6}$ m)/$(3.20 \times 10^{-7}$ m) = 24 for the *first* layer of Kolmogorov eddies. By the same token, the *first* layer of Taylor eddies would be at a distance $y_\lambda = \delta(y^+ = 7) + \lambda = 2.24 \times 10^{-6}$ m +

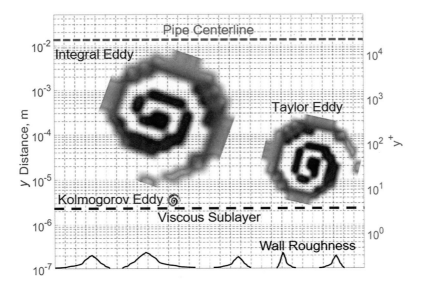

Fig. 4.7 Relative size of the viscous sublayer vs. the Kolmogorov, Taylor, and integral eddies

2.67×10^{-4} m $= 2.69 \times 10^{-4}$ m. This implies a dimensionless distance $y_\lambda/[\delta(y^+ = 1)] = (2.69 \times 10^{-4}$ m$)/(3.20 \times 10^{-7}$ m$) = 831$ for the *first* layer of Taylor eddies. As a check, note that at $y = D/2$, $y^+ = 39{,}666$ ($u_* = 0.45$ m/s from the LIKE algorithm). This further demonstrates that the damping functions are well outside of the region where eddies would first be expected near the wall—a grievous disconnect to say the least. In other words, the damping effect of both $C_2 = C_2(Re_T)$ and $C_\mu = C_\mu(Re_T)$ is nullified because Re_T is out of their applicability range. This is shown conceptually in Fig. 4.7, which compares the relative location (y and y^+) and size of the viscous sublayer vs. the Kolmogorov and Taylor eddies; note that the ordinate axis is based on the log of length, because of the large number of decades typically involved in turbulence.

Example 4.4 Consider Example 4.3, but modify $\bar{u}$ such that $Re_h = 3000$ (the hydraulic Reynolds number). Will there be any issues if the SKE is used?

Solution Backtracking $\bar{u}$ from Re, $\bar{u} = 0.0168$ m/s. From the LIKE algorithm,

$u_* = 1.24 \times 10^{-3}$ m/s,
y at $y^+ = 1 = 1.15 \times 10^{-4}$ m,
y at $y^+ = 7 = 8.02 \times 10^{-4}$ m,
$\eta = 4.23 \times 10^{-4}$ m,
$\lambda = 4.82 \times 10^{-3}$ m,
$k = 1.47 \times 10^{-6}$ m^2/s^2, and
$\varepsilon = 9.04 \times 10^{-8}$ m^2/s^3.

Calculation of the turbulence Reynolds number shows

$$Re_{\text{T,SKE}} = \frac{\rho k^2}{\mu \varepsilon} = \frac{835.8 * \left(1.47 \times 10^{-6}\right)^2}{1.19 \times 10^{-4} * 9.04 \times 10^{-8}} = 168,$$

which now has a significantly lower turbulence Reynolds number compared with the previous example. The SKE net damping behavior can be estimated from Fig. 4.5 or using

$$C_{\mu,\text{SKE}} = 0.09\left\{e^{-\left[3.4/(1+Re_{\text{T}}/50)^2\right]}\right\} = 0.09\left\{e^{-\left[3.4/(1+168/50)^2\right]}\right\} = 0.075.$$

Hence, the damping factor does little to reduce $C_{\mu,\text{SKE}}$, despite the flow occurring at a low Re! Furthermore, from Fig. 4.6, it is clear that $C_2 = C_{\varepsilon2}$ does not come into play because $Re_{T,\text{SKE}}$ is still too large despite Re_{h} being marginally turbulent. This certainly comes as an unpleasant surprise!

Example 4.5 Consider Example 4.3, with all parameters being the same, except that the velocity is unknown. Find $\bar{u}$ such that $Re_{\text{T,SKE}} = 2.2$ (recall that $C_2 = C_{\varepsilon2}$ is damped if $Re_{\text{T,SKE}} < 2.2$). What is the maximum y distance where $C_2 = C_{\varepsilon2}$ no longer provides damping?

Solution This problem can be solved in various ways. A not-so-elegant manner is to take the LIKE algorithm and supply it an estimated value for $\bar{u}$, which is then used to output k and ε so that $Re_{T,\text{SKE}}$ can be evaluated, and the iteration proceeds until $Re_{T,\text{SKE}} = 2.2$. Another approach is to find the number of n unknowns and seek the relevant n independent equations; this more elegant approach is left as an exercise in Sect. 4.9 (Problem 4.11). In any case, following the iterative method, it is found that $\bar{u} = 0.0012$ m/s when $Re_{T,\text{SKE}} = 2.2$. This shockingly low value ought not to come as a surprise, as $Re_{T,\text{SKE}} = 2.2$ translates onto a very, very small *hydraulic Re*, Re_{h}. In fact, $Re_{\text{h}} = 21.4$, which is actually very much in the *laminar* regime for a pipe! Again, because Re is so low, u_* is a paltry 1.91×10^{-5} m/s. But, of course, u_* is actually 0, as this flow is laminar (this is a pipe flow with $Re_{\text{h}} \ll 2200$). Said differently, it was not until the flow was *laminar* that the SKE damping function *finally* started "working"!

But, for the sake of running this problem to its logical end, to consider the implications of this "turbulent" laminar SKE application, the analysis will proceed. Thus, if the viscous sublayer extends to $y^+ = 7$, then y is calculated as

$$y = \frac{y^+\nu}{u_*} = \frac{7 * \left(\frac{1.19 \times 10^{-4}\,\text{Pa·s}}{835.8\,\text{kg/m}^3}\right)}{1.91 \times 10^{-5}\,\text{m/s}} = 0.052 \text{ m}.$$

That the pipe has $D = 0.0254$ m indicates that the entire pipe flow domain is impacted, and not for the better. In brief, it is ironic that the SKE's $C_2 = C_{\varepsilon2}$ damping function should become active only if the pipe flow were laminar; that is not a desirable turbulence model feature!

4.7.3 "Ironic" SKE Behavior Near the Wall

As pointed out by Wilcox (2006), it is indeed "ironic" that the SKE does not perform well near walls. The irony stems from the notion that the SKE calculates ε, which is associated with the decay of the *small* eddies. And yet, slightly away from the viscous sublayer, say $y^+ < 10$, it would be reasonable to expect that the SKE should perform its best in this region because this is exactly where larger eddies *do not exist* (refer to Fig. 3.10 in Chap. 3 or Fig. 4.7). Therefore, based on the decay of the smallest eddies, should not the SKE perform its calculational best for $y^+ < 10$? Quite the contrary, it is well-known that the SKE actually requires a wall function to improve its calculational behavior near walls.

Again, regarding the notion that the SKE has issues near the wall, consider experimental data for damping and production; in particular, the viscous sublayer decreases as Re increases, and the same trend applies for the region associated with $y^+ < 10$. Thus, as Re increases, the near-wall region decreases, so theoretically, the SKE should perform better and better near the wall. And yet, this region is exactly where poor SKE behavior is consistently found in the literature (Myong and Kasagi 1990; Speziale et al. 1992; Wilcox 2006; Bae 2016; Bae et al. 2016, 2017).

Finally, it is well-known that the SKE's 1974 damping function, f_μ, approaches a near-zero value (0.033); that is, $\left\{ e^{-\left[3.4/(1+0/50)^2 \right]} = 0.033 \right\}$ as y^+ approaches 0, which rapidly decreases k and ε near the wall (Myong and Kasagi 1990; Wilcox 2006). A more consistent formulation would be (Myong and Kasagi 1990)

$$f_\mu \to \frac{1}{y} \quad \text{as } y^+ \to 0. \tag{4.91}$$

The Myong-Kasagi model enables f_μ to increase near the wall, which is consistent with experimental data. Other researchers derive f_μ such that ν_t behaves as y^4 near the wall, with y being the distance normal to the wall (Lam and Bremhorst 1981). Others have employed similar near-wall functions with reasonable success (Shih and Hsu 1992). By contrast, the SKE tends to under predict k and ε for $y^+ < 10$ to 15 (Bernard 1986; Myong and Kasagi 1990). Curiously, the SKE over predicts ν_t by >50% near the pipe centerline at *high Re* (Myong and Kasagi 1990), despite this being the region where the flow tends to be more isotropic—the SKE's ideal domain. Again, this is another unexpected SKE behavior.

To reduce wall inconsistencies, the k PDE can be used to derive a boundary condition by considering the following at the wall: steady state, no convection, no production, and no turbulent viscous dissipation, respectively,

$$\frac{\partial k}{\partial t} + \overline{u}_j \frac{\partial k}{\partial x_j} = R_{ij}\frac{\partial \overline{u}_i}{\partial x_j} - \varepsilon + \frac{\partial}{\partial x_j}\left[\left(\nu + \frac{\nu_t}{\sigma_k}\right)\frac{\partial k}{\partial x_j}\right], \qquad (4.92A)$$

where $x_j = y$ (distance from the wall).

The PDE simplifies into the following second-order derivative boundary condition (Speziale et al. 1992; Grunloh 2016):

$$\nu\frac{\partial^2 k}{\partial y^2} = \varepsilon. \qquad (4.92B)$$

Unfortunately, the boundary condition introduces numerical stiffness. Ways to reduce the stiffness have been proposed but are generally not satisfactory (Speziale et al. 1992).

4.8   The 2003 SST Compared with the 2006 k-ω

The 2003 SST is produced by blending the SKE and 1988 k-ω models. In 2006, David Wilcox developed what would be his final k-ω model (unfortunately, he died on February 24, 2016; he truly led "An Improbable Life" (Wilcox 2007a, b)). In any case, much to Wilcox's delight, his 2006 model overcame the key weak points of his 1988 model. Had his 2006 model been developed in 1988, the motivating drive for the 2003 SST would not have existed. Nevertheless, a term-by-term comparison of the 2003 SST and 2006 k-ω models shows that both are much more similar than would be expected, as will be shown next.

First, compare the k PDE for both models. Beginning with the 2003 Menter SST,

$$\frac{\partial k}{\partial t} + \overline{u}_j \frac{\partial k}{\partial x_j} = \widetilde{P}_k - \beta^* k\omega + \frac{\partial}{\partial x_j}\left[\left(\nu + \sigma_k \nu_t\right)\frac{\partial k}{\partial x_j}\right], \qquad (4.93A)$$

where

$$P_k = R_{ij}\frac{\partial \overline{u}_i}{\partial x_j}, \qquad (4.93B)$$

and the turbulence kinetic energy production limiter is

$$\widetilde{P}_k = \min\left(P_k, 10\beta^* k\omega\right). \qquad (4.93C)$$

Now compare the above with the 2006 k-ω:

$$\frac{\partial k}{\partial t} + \bar{u}_j \frac{\partial k}{\partial x_j} = R_{ij} \frac{\partial \bar{u}_i}{\partial x_j} - \beta^* k\omega + \frac{\partial}{\partial x_j}\left[(\nu + \sigma^* \nu_t)\frac{\partial k}{\partial x_j}\right]. \tag{4.94}$$

The two k PDEs are fairly identical, with a couple of differences:

1. The k production terms are nearly the same, except that the 2003 SST uses a production limiter, whereas the 2006 k-ω does not.
2. The multiplier for the SST is $0.85 \le \sigma_k \le 1.0$, while the k-ω is fixed at $\sigma^* = 3/5$. This indicates that the SST model has a higher degree of turbulent diffusion.

Next, the ω PDEs are compared. Starting again with the 2003 Menter SST,

$$\begin{aligned}\frac{\partial \omega}{\partial t} + \bar{u}_j \frac{\partial \omega}{\partial x_j} =& \alpha \frac{\widetilde{P}_\omega}{\nu_t} - \beta\omega^2 + \frac{\partial}{\partial x_j}\left[(\nu + \sigma_\omega \nu_t)\frac{\partial \omega}{\partial x_j}\right] \\ &+ 2\sigma_{\omega 2}(1 - F_1)\frac{1}{\omega}\frac{\partial k}{\partial x_j}\frac{\partial \omega}{\partial x_j}\end{aligned} \tag{4.95A}$$

where

$$\alpha \frac{\widetilde{P}_\omega}{\nu_t} = \frac{\alpha}{\nu_t} R_{ij} \frac{\partial \bar{u}_i}{\partial x_j} = \alpha \frac{\omega}{k} R_{ij} \frac{\partial \bar{u}_i}{\partial x_j} \tag{4.95B}$$

and the turbulence frequency production limiter is

$$\widetilde{P}_\omega = \min\left(P_\omega, 10\beta^* k\omega\right). \tag{4.95C}$$

Now compare the above formulation with the 2006 k-ω:

$$\frac{\partial \omega}{\partial t} + \bar{u}_j \frac{\partial \omega}{\partial x_j} = \alpha \frac{\omega}{k} R_{ij} \frac{\partial \bar{u}_i}{\partial x_j} - \beta\omega^2 + \frac{\partial}{\partial x_j}\left[\left(\nu + \sigma\frac{k}{\omega}\right)\frac{\partial \omega}{\partial x_j}\right] + \frac{\sigma_d}{\omega}\frac{\partial k}{\partial x_j}\frac{\partial \omega}{\partial x_j}. \tag{4.96}$$

Again, the similarities are striking, though there are two major differences:

1. The ω production terms are nearly the same, except that the 2003 SST uses a production limiter, whereas the 2006 k-ω does not. In addition, the SST α ranges from 0.44 to 5/9, whereas the k-ω has a fixed α at 13/25. Despite this, the α values for both models are within 15% or less, and depending on the value for F_1, the α magnitudes can be identical.
2. The SST uses a blending function in the cross-diffusion term that makes $2\sigma_{\omega 2}(1 - F_1)$ range from 0 to 1.71 vs. the k-ω constant value of $\sigma_d = 0.0$ near the wall and $\sigma_d = 1/8$ for free shear flow. So, except for large F_1 in the boundary layer, the SST employs much more cross-diffusion away from the surface. This is perhaps one of the biggest differences, albeit that both models have essentially the same cross-diffusion derivatives and are multiplied by $1/\omega$.

Note that the SST uses σ_ω in the viscous term, while the k-ω uses σ. Nevertheless, $\sigma_\omega = \sigma$, so the terms are equal.

Next, it is noted that both models use stress limiters for ω. For the 2003 SST,

$$\widetilde{\omega} = \max\left(\omega, \frac{F_2 S}{a_1}\right) = \max\left(\omega, \frac{F_2\sqrt{2S_{ij}S_{ij}}}{a_1}\right) \tag{4.97}$$

with

$$\nu_t = \frac{k}{\widetilde{\omega}} = \frac{k}{\max\left(\omega, \frac{F_2\sqrt{2S_{ij}S_{ij}}}{a_1}\right)}. \tag{4.98}$$

It is noteworthy that the 2006 k-ω uses the following similar formulation:

$$\widetilde{\omega} = \max\left(\omega, C_{\lim}\sqrt{\frac{2S_{ij}S_{ij}}{\beta^*}}\right) \tag{4.99}$$

with

$$\nu_t = \frac{k}{\widetilde{\omega}} = \frac{k}{\max\left(\omega, C_{\lim}\sqrt{\frac{2\overline{S}_{ij}\overline{S}_{ij}}{\beta^*}}\right)}. \tag{4.100}$$

Note that the stress limiters for the SST and k-ω, respectively, are nearly identical:

$$\frac{1}{a_1} = 3.23 \tag{4.101}$$

and

$$C_{\lim}\sqrt{\frac{1}{\beta^*}} = \frac{7}{8}\sqrt{\frac{1}{(9/100)}} = 2.92. \tag{4.102}$$

Finally, the 2006 k-ω uses the f_b blending function to distinguish between free shear and near-wall flows, while the 2003 SST uses the F_1 "inter-model" blending function to calculate the free shear and near-wall flows (refer to Fig. 4.4). And as alluded earlier, the 2003 SST uses the P_k and P_ω production stress limiters, while the 2006 k-ω does not. All said, the 2003 SST and 2006 k-ω are much more alike than different, with most terms being the same, a few having slightly different coefficients, while a couple of terms are clearly different. Therefore, not surprisingly, a review of the literature shows that both models tend to generate similar results (Wilcox 2006; Fraczek and Wroblewski 2016). For example, researchers computing drag coefficients in labyrinth seals noted that the results for "the SST and k-omega turbulence model are almost the same" (Fraczek and Wroblewski 2016).

Finally, Wilcox noted that the 2006 k-ω "predicts reasonably close" when compared with experimental data and 1992 SST simulations for incompressible, transonic, supersonic, and hypersonic flows. To investigate this further, he increased C_{lim} to 1.0, so that $C_{\text{lim}}\sqrt{1/\beta^*} = 3.33$ (which is much closer to the SST value of $1/a_1 = 3.23$). This change increased the k-ω model's agreement with experimental data in the transonic range but at a penalty for reduced accuracy in sonic and hypersonic flows. Wilcox therefore recommends his original value for $C_{\text{lim}} = 7/8$ (Wilcox 2006). Thus, the 1992 SST (which is very similar to the 2003 SST) can outperform the 2006 k-ω in the transonic regime (Wilcox 2006), though it appears that the SST coefficients were fine-tuned for this particular regime.

4.9 Problems

4.1 Figure 4.6 indicates that C_2 remains fixed for $Re_T > 2.2$. Assume a cylindrical pipe with $D = 0.25$ m and $\nu = 4.5 \times 10^{-6}$ m^2/s. What is the equivalent Re_h? Discuss and justify the assumptions if simulations are of interest in the regime for $Re_T > 2.2$. (Hint: use the LIKE algorithm equations to estimate the missing quantities.)

4.2 Review (Menter 1992) and derive the 1992 SST model by combining the SKE and the 1988 k-ω models.

4.3 Show that $Re_T = \rho k^2/\mu\varepsilon$ (which is used in the SKE) is not consistent with the implied eddy length ℓ derived from the Prandtl-Kolmogorov relationship, $\nu_t = C_\mu \frac{k^2}{\varepsilon}$. (Hint: recall that the Prandtl-Kolmogorov relationship is $\nu_t = u_{\text{char}} x_{\text{char}} = k^{1/2}\ell$, whereby it is shown that ℓ is a function of ε. Use that ℓ relationship to derive $Re_T = x_{\text{char}} u_{\text{char}} \rho/\mu$ and compare with the SKE version.)

4.4 Transform the SKE model into an analogous formulation for the 1988 k-ω model by using a relationship that associates ε with ω (e.g., $\varepsilon = C_\mu \omega k$), where $C_\mu = \beta^*$.

4.5 Take the 2003 SST and modify its coefficients to be equivalent to the 2006 k-ω model. For example, let $1/a_1 = C_{\text{lim}}\sqrt{1/\beta^*} = 3.33$ and so forth. Use the newly derived model to simulate a system that consists of a smooth flat plate under isothermal boundary layer flow, is 0.1 m long and 0.05 m wide, and has air at 300 K and 1 atmosphere ($\nu = 1.58 \times 10^{-5}$ m^2/s and $U_s = 347.3$ m/s). The air flows from left to right along the 0.1 m plate at a constant velocity U_∞ based on a specified Ma. Conduct simulations with the revised SST model, and compare the results vs. the 2006 k-ω simulation using the same geometry, mesh, and initial and boundary conditions. How do the velocity solutions compare for $Ma = 0.25$, 0.5, 1.0, 5.0, and 10? Is the flow always turbulent?

4.6 Repeat Exercise 4.5, except that now the 2006 k-ω model is modified so that it more closely approaches the 2003 SST. For example, let $C_{\text{lim}}\sqrt{1/\beta^*} =$

$1/a_1 = 2.92$ and so forth. How do the velocity solutions compare for $Ma = 0.25, 0.5, 1.0, 5.0$, and 10? Is the flow always turbulent?

4.7 Build a two-equation turbulence model by choosing k and a (the eddy acceleration). Apply dimensional arguments to develop the transport PDEs (e.g., the Buckingham Pi theorem or similar arguments). What is the relevant expression for ν_t?

4.8 What is the thickness of the viscous sublayer at the point where C_μ in the SKE drops precipitously at $Re_T < 1200$? What is the size of the integral, Taylor, and Kolmogorov eddies at that point? Are there any potential issues?

4.9 What is the thickness of the viscous sublayer when C_2 in the SKE drops precipitously at $Re_T < 2.5$? What is the size of the integral, Taylor, and Kolmogorov eddies? Are there any potential issues?

4.10 Consider the MK k-ε model, where its authors added the Taylor length onto the integral length, ℓ. Starting with their length expression, derive a new MK formulation if the Kolmogorov length scale is added as well.

4.11 Consider Example 4.5. Find an analytical expression for the distance y from the wall such that the SKE $C_2 = C_{\varepsilon 2}$ is no longer dampening. This point can be assumed as $Re_{T,SKE} = 2.2$. Hint: the equations for y^+ and $Re_{T,SKE}$ will be needed, and some of the unknowns include k, ε, u_*, ℓ, and I_T.

4.12 Consider a cylindrical pipe. Convert Re_T onto the hydraulic Reynolds number, Re_h.

4.13 Would it be appropriate to replace ν with ν_t in order to define a new turbulence Re as $Re_T = k\ell/\nu_t$? Why or why not?

References

Abdollahzadeh, M., et al. (2017). Assessment of RANS turbulence models for numerical study of laminar-turbulent transition in convection heat transfer. *International Journal of Heat and Mass Transfer, 115*, 1288–1308.

Andersson, B., et al. (2012). *Computational fluid dynamic for engineers*. Cambridge: Cambridge University Press.

Argyropoulos, C. D., & Markatos, N. C. (2015). Recent advances on the numerical modelling of turbulent flows. *Applied Mathematical Modeling, 39*, 693–732.

Bae, Y. Y. (2016). A new formulation of variable turbulent Prandtl number for heat transfer to supercritical fluids. *International Journal of Heat and Mass Transfer, 92*, 792–806.

Bae, Y. Y., Kim, E. S., & Kim, M. (2016). Numerical simulation of supercritical fluids with property-dependent turbulent Prandtl number and variable damping function. *International Journal of Heat and Mass Transfer, 101*, 488–501.

Bae, Y. Y., Kim, E. S., & Kim, M. (2017). Assessment of low-Reynolds number k-ε turbulence models against highly buoyant flows. *International Journal of Heat and Mass Transfer, 108*, 529–536.

Bernard, P. S. (1986). Limitations of the near-wall k-ε turbulence model. *AIAA Journal, 24*(4), 619–622.

Bna, S., et al., Heat transfer numerical simulations with the four parameter k-ω-k_t-$ε_t$ model for low-Prandtl number liquid metals, XXX UIT Heat Transfer Conference, Bologna, June 2012.

Bredberg, J. (2000). *On the wall boundary condition for turbulence models*. Göteborg: Dept. of Thermo and Fluid Dynamics, Chalmers University of Technology.

Bulicek, M., & Malek, J. (2000). *Large data analysis for Kolmogorov's two-equation model of turbulence*. Prague: Charles University, Mathematical Institute, Czech Republic.

Chou, P. Y. (1940). On an extension of Reynolds' method of finding apparent stress and the nature of turbulence. *Chinese Journal of Physics, 4*, 1–33.

Chou, P. Y. (1945). On velocity correlations and the solutions of the equations of turbulent fluctuation. *Quarterly of Applied Mathematics, 3*, 38–54.

Chou, P. Y., & Chou, R. L. (1995). 50 Years of turbulence research in China. *Annual Review of Fluid Mechanics, 27*, 1–16.

Davydov, B. I. (1961). On the statistical dynamics of an incompressible turbulent fluid. *Doklady Akademiya Nauk SSSR, 136*, 47–50.

Diaz, D. d. O., & Hinz, D. F. (2015). Performance of eddy-viscosity turbulence models for predicting Swirling pipe-flow: simulations and Laser-Doppler velocimetry, arXiv:1507.04648v2.

Dong, J., Wang, X., & Tu, J. (2010). Numerical research about the internal flow of steam-jet vacuum pump evaluation of turbulence models and determination of the shock-mixing layer. *Physics Procedia, 32*, 614–622.

El-Behery, S. M., & Hamed, M. H. (2009). A comparative study of turbulence models performance for turbulent flow in a planar asymmetric diffuser. *World Academy of Science, Engineering and Technology, 53*, 769–780.

Fluent. (2012). "Turbulence", ANSYS, Lecture 7.

Fraczek, D., & Wroblewski, W. (2016). Validation of numerical models for flow simulation in labyrinth seals. *Journal of Physics, Conference Series, 760*, 012004.

Goldberg, U. C., & Batten, P. (2015). A wall-distance-free version of the SST turbulence model. *Engineering Applications of Computational Fluid Mechanics, 9*(1), 33–40.

Gorji, S., et al. (2014). A comparative study of turbulence models in a transient channel flow. *Computers & Fluids, 89*, 111–123.

Grunloh, T. P. (2016). A novel multi-scale domain overlapping CFD/STH coupling methodology for multi-dimensional flows relevant to nuclear applications, PhD dissertation, University of Michigan.

Hamba, F. (2005). Nonlocal analysis of the Reynolds stress in turbulent shear flow. *Physics of Fluids, 17*, 1–10.

Hamlington, P. E., & Dahm W. J. A. (2009). Reynolds stress closure including nonlocal and nonequilibrium effects in turbulent flows. 39th AIAA fluid dynamics conference, AIAA 2009–4162.

Hanjalic, K. (1970). Two-dimensional asymmetric turbulent flow in ducts, PhD dissertation, University of London.

Harlow, F. H., & Nakayama, P. I. (1967). Turbulence transport equations. *Physics of Fluids, 10*(11), 2323.

Harlow, F. H., & Nakayama, P. I. (1968). Transport of turbulence energy decay rate. *Los Alamos Scientific Laboratory, LA-3854*.

He, S., Kim, W. S., & Bae, J. H. (2008). Assessment of performance of turbulence models in predicting supercritical pressure heat transfer in a vertical tube. *International Journal of Heat and Mass Transfer, 51*, 4659–4675.

Hellsten, A. (1997). Some improvements in Menter's k-ⵁ SST turbulence model, AIAA, AIAA-98-2554.

Hellsten, A., & Laine, S. (1997). Extension of the k-ⵁ-SST turbulence model for flows over rough surfaces, AIAA, AIAA-97-3577.

Hinze, J. (1987). *Turbulence* (McGraw-Hill Classic Textbook Reissue Series) (p. 227). New York: McGraw-Hill.

Hirota, M., et al. (2017). Vortex stretching in a homogeneous isotropic turbulence. *Journal of Physics: Conference Series, 822*.

Hrenya, C. M., et al. (1995). Comparison of low Reynolds number k-ε turbulence models in predicting fully developed pipe flow. *Chemical Engineering Science, 50*(12), 1923–1941.

Jones, W. P., & Launder, B. E. (1972). The prediction of laminarization with a two-equation model of turbulence. *International Journal of Heat and Mass Transfer, 15*, 301–314.

Jones, W. P., & Launder, B. E. (1973). The calculation of low-Reynolds-number phenomena with a two-equation model of turbulence. International Journal of Heat and Mass Transfer, 16, 1119–1130.

Knopp, T., Eisfeld, B., & Calvo, J. B. (2009). A new extension for k–ω turbulence models to account for wall roughness. *International Journal of Heat and Fluid Flow, 30*(1), 54–65.

Kolmogorov, A. N. (1942). Equations of turbulent motion of an incompressible fluid, Izvestiya Akademiya Nauk SSSR, Seria Fizicheska VI, No. 1-2. (Translated by D. B. Spalding.)

Lam, C. K. G., & Bremhorst, K. (1981). A modified form of the k-ε model for predicting wall turbulence. *Transactions of the ASME, 103*, 456–460.

Launder, B. E., et al. (1973). Prediction of free shear flows a comparison of the performance of six turbulence models, Free Shear Flows Conference Proceedings, NASA Langley, Vol. 1. (Also as NASA Report SP-321.)

Launder, B. E., & Sharma, B. I. (1974). Application of the energy-dissipation model of turbulence to the calculation of flow near a spinning disc. *Letters in Heat and Mass Transfer, 1*, 131–137.

Mansour, N. N., Kim, J., & Moin, P. (1988). Reynolds-stress and dissipation-rate budgets in a turbulent channel flow. *Journal of Fluid Mechanics, 194, 15*.

Menter, F. R. (1992). Improved two-equation k-ω turbulence models for aerodynamic flows, NASA technical memorandum 103975, Ames Research Center.

Menter, F. R. (1993). Zonal two equation k-ω turbulence models for aerodynamic flows, AIAA 93-2906, 24th fluid dynamics conference.

Menter, F. R. (1994). Two-equation eddy-viscosity turbulence models for engineering applications. *AIAA, 32*(8), 1598.

Menter, F. R., Kuntz, M., & Langtry, R. (2003). Ten years of industrial experience with the SST turbulence model. *Turbulence, Heat and Mass Transfer, 4*, 1–8.

Mielke, A., & Naumann, J. (2015). Global-in-time existence of weak solutions to Kolmogorov's two-equation model of turbulence. *Comptes Rendus Mathematique Academie des Sciences, Paris, Series I, 353*, 321–326.

Millionshtchikov, M. (1941). On the theory of homogeneous isotropic turbulence. *Doklady Akademii Nauk SSSR, 32*(9), 615–618.

Myong, H. K., & Kasagi, N. (1988). A new proposal for a k-ε turbulence model and its evaluation: 1st report, development of the model. *Transactions of the Japan Society of Mechanical Engineers, Series B, 54*(507).

Myong, H. K., et al. (1989). Numerical prediction of turbulent pipe flow heat transfer for various Prandtl number fluids with the improved k-e turbulence model. *JSME International Journal, Series II, 32*(4).

Myong, H. K., & Kasagi, N. (1990). A new approach to the improvement of k-ε turbulence model for wall-bounded shear flows. *JSME International Journal, 33*(1), 63–72.

NASA1, The Wilcox k-omega turbulence model, Langley Research Center, https://turbmodels.larc.nasa.gov/wilcox.html. Accessed 13 Feb 2018.

NASA2, The Menter shear stress transport turbulence model, Langley Research Center, https://turbmodels.larc.nasa.gov/sst.html. Accessed 9 April 2018.

Otero R. G. J., Patel A., & Pecnik, R. (2018). A novel approach to accurately model heat transfer to supercritical fluids, The 6th International Symposium - Supercritical CO2 Power Cycles.

Pope, S. B. (1978). An explanation of the turbulent round-jet/plane-jet anomaly. *AIAA Journal, 16*, 279–281.

Prandtl, L. (1945). "Uber ein neues Formelsystem fur die ausgebildete Turbulenz", ("On a New Formulation for the Fully Developed Turbulence" or "On a New Equation System for Developed Turbulence"), Nacr. Akad. Wiss. Gottingen, Math-Phys. Kl.

Rodi, W. (1993). *Turbulence models and their application in hydraulics a state-of-the art review* (3rd ed.). Rotterdam/Brookfield: Balkema.

Rodi, W., & Mansour, N. N. (1993). Low Reynolds number k-ε modelling with the aid of direct simulation data. *Journal of Fluid Mechanics, 250*, 509–529.

Rodriguez, S. (2011). Swirling jets for the mitigation of hot spots and thermal stratification in the VHTR lower plenum, PhD dissertation, University of New Mexico.

Rotta, J. (1951). Statistische theorie nichthomogener turbulenz. *Zeitschrift fur Physik, 129*, 547–572.

Rotta, J. (1962). Turbulent boundary layers in incompressible flow. *Progress in Aeronautical Sciences, 2*, 1–219.

Saffman, P. G. (1970). A model for inhomogeneous turbulent flow. *Proceedings of the Royal Society of London. Series A, 317*, 417–433.

Sawko, R. (2012). "Mathematical and computational methods of non-Newtonian, multiphase flows", PhD dissertation, Cranfield University.

Schmitt, F. G. (2007). About Boussinesq's turbulent viscosity hypothesis: Historical remarks and a direct evaluation of its validity. *Comptes Rendus Mecanique, 335*, 617.

Schobeiri, M., & Abdelfattah, S. (2013). On the reliability of RANS and URANS numerical results for high-pressure turbine simulations: A benchmark experimental and numerical study on performance and interstage flow behavior of high-pressure turbines at design and off-design conditions using two different turbine designs. *ASME, Journal of Turbomachinery, 135*, 061012.

Shih, T. H., & Hsu, A. T. (1992). An improved k-ε model for near wall turbulence, NASA Lewis Research Center, N92-23349.

SimScale, Turbulence models K-epsilon. www.simscale.com/docs/content/simulation/model/turbulenceModel/kEpsilon.html. Accessed 18 April 2018.

Sondak, D. L. (1992). Wall functions for the k-ε turbulence model in generalized nonorthogonal curvilinear coordinates, PhD dissertation, Iowa State University.

Spalart, P. R. (2015). Philosophies and fallacies in turbulence modeling. *Progress in Aerospace Sciences, 74*, 1–15.

Spalding, D. B. (1991). Kolmogorov's two-equation model of turbulence. *Proceedings of the Royal Society of London A, 434*, 211–216.

Speziale, C. G., & Eringen, A. C. (1981). Nonlocal fluid mechanics description of wall turbulence. *Computers and Mathematics with Applications, 7*, 27–41.

Speziale, C. G., Abid, R., & Anderson, E. C. (1992). Critical evaluation of two-equation models for near-wall turbulence. *AIAA Journal, 30*(2), 324–331.

Spiegel, M. R. (1999). *Mathematical handbook of formulas and tables* (Schaum's Outlines) (2nd ed.). New York: McGraw-Hill Education.

Wilcox, D. C. (1988a). Reassessment of the scale-determining equation for advanced turbulence models. *AIAA Journal, 26*(11). (1988 *k-ω* model.).

Wilcox, D. C. (1988b). Multiscale model for turbulent flows. *AIAA Journal, 26*(11). (1988 *k-ω* model.).

Wilcox, D. C. (1993a). *Turbulence modeling for CFD* (1st ed.). La Cañada: DCW Industries, Inc. (1988 *k-ω* model.).

Wilcox, D. C. (1993b). Comparison of two-equation turbulence models for boundary layers with pressure gradient. *AIAA Journal, 31*, 1414–1421. (1988 *k-ω* model and a variant; low and high *Re* models.).

Wilcox, D. C. (1994). Simulation of transition with a two-equation turbulence model. *AIAA Journal, 32*(2). (1988 *k-ω* with transition from laminar to turbulent flows.).

Wilcox, D. C. (1998). Turbulence modeling for CFD, 2nd Ed., DCW Industries, Inc., La Cañada printed on 1998 and 2004. (1998 *k-ω* model.)

Wilcox, D. C. (2006). Turbulence modeling for CFD, 3rd Ed., DCW Industries, Inc., La Cãnada, printed on 2006 and 2010. (2006 k-*ω* model.)

Wilcox, D. C. (2007a). *An improbable life*. La Cañada: DCW Industries, Inc.

Wilcox, D. C. (2007b). Formulation of the k-ω turbulence model revisited, 45th AIAA Aerospace Sciences Meeting and Exhibit. *AIAA Journal 46*(11), 2823–2838. (2006 k-ω model.)

Wilcox, D. C. (2012). *Basic fluid mechanics* (5th ed.). La Cañada: DCW Industries, Inc.

Zeierman, S., & Wolfshtein, M. (1986). Turbulent time scale for turbulent-flow calculations. *AIAA Journal, 24*(10), 1606–1610.

Zhao, C. R., et al. (2017). Influence of various aspects of low Reynolds number k-ε turbulence models on predicting in-tube buoyancy affected heat transfer to supercritical pressure fluids. *Nuclear Engineering and Design, 313*, 401–413.

Chapter 5
LES and DNS Turbulence Modeling

Given the erratic track record of most turbulence models, new ideas are always welcome.

— David Wilcox 2006

Abstract This chapter is divided into two parts, LES and DNS. In the first part, the LES turbulence model is derived from first principles, and its terms are described in detail. The usage of LES filters is described, along with various recommendations. The LIKE algorithm is applied to show how to model large eddies properly by applying the appropriate node-to-node computational distances. LES-specific boundary and initial conditions are described, and dozens of practical recommendations are provided. In the second part, analogous discussions and recommendations for DNS are included as well.

5.1 LES Modeling (and Comparison with RANS and DNS)

The larger the eddy, the higher its nonisotropic nature and the more complex its behavior. The larger eddies obtain their kinetic energy from the bulk fluid energy, contain most of the turbulent kinetic energy (~80%), transfer kinetic energy to the smaller eddies by stretching and breaking them up ("cascading"), and are responsible for the majority of the diffusive processes involving mass, momentum, and energy. For these reasons, the simulation of large eddies is highly desirable. On the other hand, the smaller eddies take the kinetic energy from the larger eddies and transfer their energy back to the fluid through viscous shear. For high Re, the small-scale turbulent eddies are *statistically isotropic.* Therefore, they are "more universal" and more independent of the boundary conditions and the mean flow velocity than the larger eddies. Thus, simulation of the smaller eddies is also desirable.

So, why not simulate (resolve) the larger eddies and approximate (model) the behavior of the smaller eddies? Based on this premise, large eddy simulation (LES) models have been developed for several decades to capture these important eddy features (Smagorinsky 1963; Leonard 1974; Germano et al. 1991; Kim and Menon 1995; Nicoud and Ducros 1999; You and Moin 2007; Zhiyin 2015). Figure 5.1

© Springer Nature Switzerland AG 2019
S. Rodriguez, *Applied Computational Fluid Dynamics and Turbulence Modeling*,
https://doi.org/10.1007/978-3-030-28691-0_5

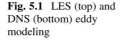

Fig. 5.1 LES (top) and
DNS (bottom) eddy
modeling

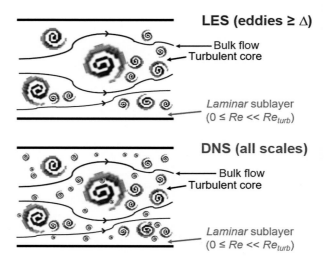

Fig. 5.2 LES (top) and
DNS (bottom) instantaneous
velocities

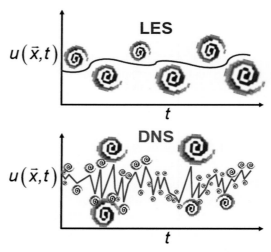

conceptually shows that LES resolves integral and Taylor eddies up to a user- or mesh-defined minimum eddy size Δ, while direct numerical simulation (DNS) resolves the integral, Taylor, and Kolmogorov eddies (i.e., DNS calculates all scales). In this context, the scale Δ determines the minimum size for which eddies will be resolved, thereby acting as a filter for the subgrid scale (SGS), whereby eddies smaller than Δ are modeled. And this is crucial, as the SGS model enables the decay of the turbulent kinetic energy at the appropriate spatial locations within the turbulent flow (Clark et al. 1979).

Starting in 1963 via a single publication, the LES approach has seen an exponential increase in publications over the past three decades (Bouffanais 2010; Zhiyin 2015).

Figure 5.2 shows the instantaneous velocity based on resolved LES and DNS calculations. Note that LES will capture a significant number of velocity fluctuations

Fig. 5.3 A comparison of
RANS, LES, and DNS
instantaneous velocities

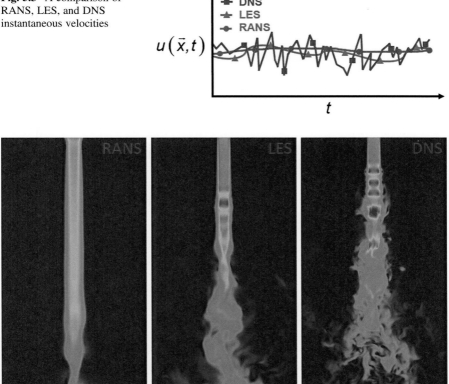

Fig. 5.4 Velocity distribution for a turbulent jet using RANS (LHS), LES (middle), and DNS (RHS)

associated with the larger eddies, while DNS will capture all the LES fluctuations, as well as the Taylor eddies that were cut off from LES (in the interest of a faster calculation), and all the Kolmogorov eddy fluctuations. In other words, DNS resolves the entire eddy spectrum: integral, Taylor, and Kolmogorov eddies. Hence, the DNS instantaneous velocity is more "jagged" and, of course, synonymous with the fluctuations seen in experimental data. *Note that Reynolds-averaged Navier-Stokes (RANS) will not calculate the dynamic eddy behavior but is instead a representative behavior of non-dynamic, time-averaged eddy behavior.* As such, the RANS instantaneous velocity as a function of time and space can never have the chaotic velocity wiggles that are resolved significantly by LES and more so by DNS, as shown in Fig. 5.3.

As an example of the different approaches, Fig. 5.4 compares the velocity distribution for a jet at $Re = 3220$ that was simulated using RANS, LES, and

DNS. A close inspection shows significant differences and similarities in velocity distribution. The main idea is that although all three provide useful details of the velocity field, RANS is faster than LES, which is faster than DNS; conversely, DNS is more detailed than LES, which is more detailed than RANS. It is also worthy of mention that theoretically speaking, DNS will calculate *all* turbulent systems and cases, while LES can do so for most situations, whereas RANS is much more limited in applicability. The choice of turbulence model is ultimately based on financial resources, time constraints, computational resources, and the necessary level of output required of the simulation. *Said more pragmatically, use the fastest model that gets reasonable accuracy and details for the system of interest.*

For the reasons discussed above, LES can be considered as an intermediate methodology between RANS and DNS, as a balance between output and computational effort. In general, LES is about an order or two of magnitude more time-intensive than RANS but is two to three (or more) orders cheaper than DNS. The reason for LES's computational cost is primarily due to the number of elements used (i.e., a simulation may require refinement up to the smaller eddies within the Taylor scale). On the other hand, RANS elements are much larger because they do not resolve eddy scales. Like RANS, the LES models typically employ the Boussinesq approximation, or a similar expression, for the smaller eddies in the subgrid scale.

LES is great for adverse pressure gradients, complex surfaces, and swirl but can be expensive in the boundary layer (Afgan 2007; Rodriguez 2011) and hence the use of the detached eddy simulation (DES). Due to its success in theoretical and, increasingly, in engineering calculations, there are dozens of LES models in the literature. The interested reader is encouraged to pursue this subject matter further (Kleissl and Parlange 2004; Vreman 2004; Lesieur et al. 2005; Bouffanais 2010; Nicoud et al. 2011; Yeon 2014; Zhiyin 2015). A partial list of such models includes:

- Standard Smagorinsky model (Smagorinsky 1963)
- Algebraic dynamic model (AKA "dynamic subgrid-scale model" or "dynamic Smagorinsky") (Germano et al. 1991; Lilly 1992; Kleissl and Parlange 2004)
- Localized dynamic model (Kim and Menon 1995)
- Wall-adapting local eddy-viscosity (WALE) model (Nicoud and Ducros 1999)
- Dynamic global-coefficient model (You and Moin 2007)
- RNG-LES model (CFD-Online 2018)
- Kinetic energy subgrid-scale model (KSGS) (Fuego 2016a, b)
- σ-SGS model (Nicoud et al. 2011)

5.1.1 How LES Works: A Brief Overview

As expected, the LES methodology divides the simulation into two areas. One portion calculates the velocity field of the larger eddies, thereby resolving their behavior explicitly, while the subgrid portion represents the smaller eddies, which are modeled (approximated). This is shown conceptually in Fig. 5.5.

Fig. 5.5 Size filtering of
LES model eddies

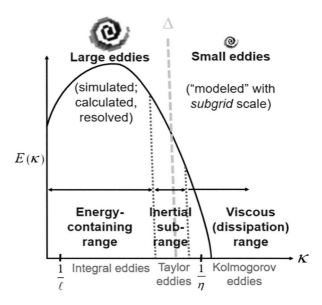

The computational approach is made by choosing a filtering length scale, Δ. In particular, if the eddy size is $\geq\Delta$, such eddy is resolved, but if the eddy is $<\Delta$, it is modeled. That is, the larger eddies are simulated (calculated, resolved), while the smaller eddies are "modeled" as a lumped, homogeneous-like group via an *SGS* model. Typically, the filtering strategy employs a filter function G that involves the parameter Δ. Then, as a general and well-proven LES approach, the large eddy filtered velocity is obtained by employing a "convolution filter" as follows (Leonard 1974; Shaanan et al. 1975; Clark et al. 1979):

$$\bar{u}_i\left(\vec{x},t\right) = \iiint G\left(\vec{x},\vec{x}'\right)u_i\left(\vec{x},t\right)d^3\vec{x}. \tag{5.1}$$

Some authors use the following notation to more explicitly indicate the existence of a mapping transformation based on parameter Δ across a distance $x - x'$,

$$G(x,x') \equiv G\left(\vec{x} - \vec{x}';\Delta\right). \tag{5.2}$$

Note that a convolution takes two functions and mathematically changes (maps) them onto a new relationship that is related to one of the original functions. In this case, a convolution takes the instantaneous velocity u_i and filters it as a function of eddy dimension Δ, to produce the filtered large eddy velocity $\bar{u}_i$ (i.e., this represents a "local," spatially averaged velocity region). *Thus, the direct consequence of the filtering operation is to eliminate the small fluctuations from u_i.* As an example, consider the most straightforward (and very popular) filter function G based on assuming a cubic volume with length dimension Δ. That is, consider the volume-averaged box filter (described later in this section). WLOG, the cube is taken in Cartesian coordinates centered about $\Delta/2$ for each of the three directions. Then,

$$\bar{u}_i\left(\vec{x},t\right) = \iiint G\left(\vec{x},\vec{x}'\right)u_i\left(\vec{x},t\right)d^3\vec{x}$$

$$= \int_{z=z_I-\Delta/2}^{z=z_I+\Delta/2} \int_{y=y_I-\Delta/2}^{y=y_I+\Delta/2} \int_{x=x_I-\Delta/2}^{x=x_I+\Delta/2} \frac{1}{\Delta^3} u_i(x_I - x_I', y_I - y_I', z_I - z_I', t)dx_I'dy_I'dz_I'$$

(5.3A)

where

$$\Delta = (\Delta x \Delta y \Delta z)^{1/3}.$$ (5.3B)

As explained later, $G = 0$ outside the cubic volume $(\Delta x \Delta y \Delta z)$ for the box filter; thus, any eddies smaller than Δ are filtered out; they are not included in $\bar{u}_i$. *Thus, only eddies within the integral volume are resolved, and anything lying outside of the integral volume is modeled.*

The small eddy velocity u_i' is related to the instantaneous velocity as follows,

$$u_i\left(\vec{x},t\right) = \bar{u}_i\left(\vec{x},t\right) + u_i'\left(\vec{x},t\right),$$ (5.4)

where $\bar{u}_i$ is the *resolved* velocity field for the larger eddies and u_i' is the subgrid velocity for the smaller, modeled eddies, as shown in Fig. 5.6. Note that although the

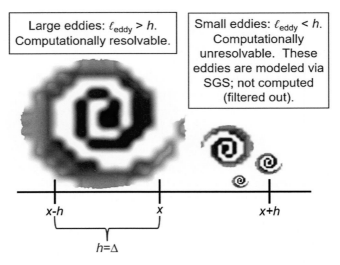

Fig. 5.6 Resolvable vs. filtered eddies

velocity-field superposition notation looks identical to the Reynolds decomposition for RANS, *it has a different meaning*. In particular, the LES velocity field is averaged in space (vs. RANS, which is averaged in time). Therefore, the entire LES velocity field is

$$u_i = \sum (\text{resolved} + \text{subgrid}) = \sum (\text{larger eddies} + \text{smaller eddies})$$
$$= \bar{u}_i + u_i'. \tag{5.5}$$

The filtering function G requires that (1) its volume integral is normalizable to 1.0, (2) the operator is linear (analogous to RANS Eq. 2.34L, Sect. 2.5), and (3) its derivative follows commutation (analogous to RANS Eq. 2.34E, Sect. 2.5). Note that the convolution filter allows the analyst to employ grid convergence index (GCI) and Richardson extrapolation, if so desired. If done correctly, the velocity solution should show filter length independence (convergence).

Many types of filters are used in the literature, and a general consensus is not yet fully formulated, primarily because the filters have properties that are not generically applicable (Deardorff 1970; Shaanan et al. 1975; Ferziger 1977; Galperin and Orszag 1993; Stefano and Vasilyev 2002; Lesieur et al. 2005). Said more ominously, the choice of filter will impact the solution, which is unfortunate; therefore, care must be taken (Stefano and Vasilyev 2002). For example, the SGS model output depends on whether the filter has a smooth, semi-sharp, or sharp cutoff. As a case in point, smooth filters permit the filtered region to be fuzzier, in the sense that the cutoff is not sharply defined between the larger eddies that are resolved and the smaller eddies that are modeled. By contrast, a sharp filter behaves more like a binary switch, being either "on" or "off" and thereby providing a well-defined separation cutoff in space (Stefano and Vasilyev 2002). Because turbulence is not smooth and fuzzy but is instead sharp and chaotic, the sharp filter resonates well (Stefano and Vasilyev 2002). More specifically, the sharp filter does a better job of capturing the energy cascade from the larger to smaller eddies; smooth filters are known to miscalculate the $-5/3$ slope in the inertial range. Smooth filters are known to dissipate significant fractions of the turbulent kinetic energy away from the larger scales and onto the smaller scales. Indeed, these issues can aggravate aliasing error (Stefano and Vasilyev 2002), which refers to the calculational error associated with the nonlinear convective term as represented by the node-based mesh. Some common filters are discussed next. Note that the filters listed below are in 3D; many references list the filters in 1D (Garnier et al. 2009).

The most straightforward filter is known as the volume-averaged box filter (AKA "running mean filter") (Deardorff 1970; Shaanan et al. 1975; Ferziger 1977; Clark et al. 1979; Fuego 2016b). The filter includes the element volume Δ^3 in its filtering criteria for cutoff. Because the eddy in question is either inside or outside of the volume, this filter is of the sharp type and is popular in CFD tools. The filter is

$$G\left(\vec{x} - \vec{x}';\Delta\right) = \begin{cases} \dfrac{1}{\Delta^3}, & |x - x'| < \dfrac{\Delta x}{2}, |y - y'| < \dfrac{\Delta y}{2}, |z - z'| < \dfrac{\Delta z}{2} \\ 0, & \text{otherwise} \end{cases} \quad (5.6)$$

The simplicity of the box filter has a cautionary issue, namely, that it is considered an "implicit filter," whereby the ratio of the test filter vs. the grid filter is a constant. Thus, issues can arise as to its ability to converge (and if the Δ becomes too small, then LES approaches DNS) (Zhiyin 2015).

A fairly sharp (but not as sharp as the volume-averaged box filter), is called the spectral filter (or sharp cutoff filter). The filter employs a sine function to filter out the smaller eddies (Ferziger 1977; Garnier et al. 2009),

$$G\left(\vec{x} - \vec{x}';\Delta\right) = \frac{\sin\left[\frac{(x-x')}{\Delta}\right]}{(x - x')} \frac{\sin\left[\frac{(y-y')}{\Delta}\right]}{(y - y')} \frac{\sin\left[\frac{(z-z')}{\Delta}\right]}{(z - z')}. \quad (5.7)$$

A Gaussian filter was developed to account for eddies based on a bell-shaped distribution (Leonard 1974; Shaanan et al. 1975; Ferziger 1977; Deardorff 1970; Mansour et al. 1977). The Gaussian filter is of the smooth type (Stefano and Vasilyev 2002),

$$G\left(\vec{x} - \vec{x}';\Delta\right) = \left(\frac{6}{\pi\Delta^2}\right)^{3/2} \exp\left[-\frac{6}{\Delta^2}|x - x'|^2\right] \exp\left[-\frac{6}{\Delta^2}|y - y'|^2\right] \exp\left[-\frac{6}{\Delta^2}|z - z'|^2\right]. \quad (5.8)$$

5.1.2   LES Mass and Momentum Conservation

So, now that $\bar{u}_i\left(\vec{x}, t\right)$ is known from Sect. 5.1.1, how is it applied to obtain the momentum terms? As was done with the derivation of RANS, assume an incompressible, Newtonian flow. Use SS mass conservation and the transient "laminar" (i.e., unmodified, unfiltered) NS equation. WLOG, simplify further by letting $\mu \neq \mu\left(\vec{x}\right)$. Then, the unfiltered conservation of mass PDE is

$$\frac{\partial u_i}{\partial x_i} = 0. \quad (5.9)$$

The unfiltered momentum conservation PDE is as follows:

$$\frac{\partial u_i}{\partial t} + \frac{\partial (u_i u_j)}{\partial x_j} = -\frac{1}{\rho}\frac{\partial \tau_{ij,\text{lam}}}{\partial x_j} - \frac{1}{\rho}\frac{\partial P}{\partial x_i} = \frac{\mu}{\rho}\frac{\partial}{\partial x_j}\left(\frac{\partial u_i}{\partial x_j} + \frac{\partial u_j}{\partial x_i}\right) - \frac{1}{\rho}\frac{\partial P}{\partial x_i}. \quad (5.10)$$

Note that in its full form,

$$\tau_{ij,\text{lam}} = -\mu\left(\frac{\partial u_i}{\partial x_j} + \frac{\partial u_j}{\partial x_i}\right), \quad (5.11\text{A})$$

which is not the same as

$$\tau_{ij,\text{lam}} = -\mu\frac{\partial u_i}{\partial x_j}. \quad (5.11\text{B})$$

Now, perform the LES *space*-filtering (bar or overbar) operation for the unfiltered NS equation to derive the LES momentum equation; do not apply $u_i = \bar{u}_i + u_i'$ yet. Then,

$$\frac{\partial \bar{u}_i}{\partial t} + \frac{\partial (\overline{u_i u_j})}{\partial x_j} = \frac{\mu}{\rho}\frac{\partial}{\partial x_j}\left(\frac{\partial \bar{u}_i}{\partial x_j} + \frac{\partial \bar{u}_j}{\partial x_i}\right) - \frac{1}{\rho}\frac{\partial \bar{P}}{\partial x_i}. \quad (5.12)$$

Note that the convective $\overline{u_i u_j}$ term is unknown, so some mathematics and physics are required. With this goal in mind, substitute $u_i = \bar{u}_i + u_i'$ into the convective term, and perform FOIL multiplication,

$$\overline{u_i u_j} = \overline{(\bar{u}_i + u_i')(\bar{u}_j + u_j')} = \overline{(\bar{u}_i\bar{u}_j + \bar{u}_i u_j' + u_i'\bar{u}_j + u_i'u_j')}$$
$$= \overline{\bar{u}_i\bar{u}_j} + \overline{\bar{u}_i u_j'} + \overline{u_i'\bar{u}_j} + \overline{u_i'u_j'}. \quad (5.13)$$

At this point, more succinct and universal notation can be used to represent the tensors that were derived from the convective term, namely,

$$\overline{u_i u_j} = \overline{\bar{u}_i\bar{u}_j} + \overline{\bar{u}_i u_j'} + \overline{u_i'\bar{u}_j} + \overline{u_i'u_j'} = \bar{u}_i\bar{u}_j + C_{ij} + L_{ij} + R_{ij} \quad (5.14\text{A})$$

where

$$C_{ij} = \overline{\bar{u}_i u_j'} + \overline{u_i'\bar{u}_j} = \text{cross-term stress (Clark et al.1979)}, \quad (5.14\text{B})$$

$$L_{ij} = \overline{\bar{u}_i\bar{u}_j} - \bar{u}_i\bar{u}_j$$
$$= \text{Leonard stress (Leonard 1974; Stefano and Vasilyev 2002)}, \quad (5.14\text{C})$$

and

$$R_{ij} = \overline{u_i' u_j'} = \text{SGS Reynolds stress (Clark et al.1979).} \qquad (5.14\text{D})$$

Thus, the LES space-filtered NS reduces to

$$\frac{\partial \overline{u}_i}{\partial t} + \frac{\partial \left(\overline{u}_i \overline{u}_j + C_{ij} + L_{ij} + R_{ij} \right)}{\partial x_j} = \frac{\mu}{\rho} \frac{\partial}{\partial x_j} \left(\frac{\partial \overline{u}_i}{\partial x_j} + \frac{\partial \overline{u}_j}{\partial x_i} \right) - \frac{1}{\rho} \frac{\partial \overline{P}}{\partial x_i}. \qquad (5.15)$$

Now simplify the expression to obtain the sought-after space-averaged *exact*, but unsolvable, LES PDE,

$$\frac{\partial \overline{u}_i}{\partial t} + \frac{\partial \left(\overline{u}_i \overline{u}_j \right)}{\partial x_j} = \frac{\mu}{\rho} \frac{\partial}{\partial x_j} \left[\left(\frac{\partial \overline{u}_i}{\partial x_j} + \frac{\partial \overline{u}_j}{\partial x_i} \right) \right] - \frac{\partial \left(C_{ij} + L_{ij} + R_{ij} \right)}{\partial x_j} - \frac{1}{\rho} \frac{\partial \overline{P}}{\partial x_i}. \qquad (5.16)$$

Finally, in shrewd anticipation of using various *inexact* closure *approximations*, the previous space-filtered NS form is conveniently reformulated as follows,

$$\frac{\partial \overline{u}_i}{\partial t} + \overline{u}_j \frac{\partial \overline{u}_i}{\partial x_j} = \frac{\mu}{\rho} \frac{\partial}{\partial x_j} \left[\underbrace{\left(\frac{\partial \overline{u}_i}{\partial x_j} + \frac{\partial \overline{u}_j}{\partial x_i} \right)}_{\text{Large resolved eddies}} - \underbrace{\tau_{ij}}_{\text{Small modeled eddies}} \right] - \frac{1}{\rho} \frac{\partial \overline{P}}{\partial x_i}, \qquad (5.17\ \text{A})$$

where

$$\tau_{ij} \equiv \overline{u_i u_j} - \overline{u}_i \overline{u}_j \equiv C_{ij} + L_{ij} + R_{ij} \equiv \text{SGS model.} \qquad (5.17\ \text{B})$$

Thus, once a complete expression for τ_{ij} is assumed, the LES PDE offers solvable, though *approximate,* solutions.

Note that unlike Reynolds averaging, $\overline{\overline{u}}_i \neq \overline{u}_i$ for the LES space-filtering (bar) operation; that is, the *space* filtering of a space-filtered velocity yields an entirely different velocity. By contrast, RANS *time* filtering of a time-filtered quantity yields the same quantity. Furthermore (Clark et al. 1979),

- $\overline{u_i u_j} \neq \overline{u}_i \overline{u}_j$.
- $\overline{\overline{u}_i u_j'} + \overline{u_i' \overline{u}_j} \approx 0$ is a poor approximation.
- $\overline{\overline{u}_i \overline{u}_j} \approx \overline{u}_i \overline{u}_j$ is not a good approximation.

And how in the world would one obtain $\overline{u_i' u_j'}$? So, some methodology has to be employed to approximate these terms.

The C_{ij} cross-term cannot be computed directly; it must be approximated. It is worthwhile noting that the C_{ij} cross-term decreases as *Re* increases (Clark et al. 1979), so higher-*Re* simulations might get away with ignoring the term. The cross-term can be modeled (approximated) quite reasonably as (Clark et al. 1979)

$$C_{ij} \approx \alpha \Delta^2 \left(\overline{u}_i \nabla^2 \overline{u}_j + \overline{u}_j \nabla^2 \overline{u}_i \right) \tag{5.18}$$

where α is a constant.

The L_{ij} Leonard stress can be solved *directly* by performing a second space averaging (Mansour et al. 1977), or it can be "approximated" as being on the order of the truncation error associated with second-order finite differences (Shaanan et al. 1975; Wilcox 2006) and is therefore ignored by some (Wilcox 2006). Furthermore, L_{ij} is at most a weak function of Re (Clark et al. 1979). On the other hand, it is best to not discount the L_{ij} term so readily, as it has a significant role associated with the transfer of eddy turbulent kinetic energy during cascading (Shaanan et al. 1975). For low- to intermediate-Re applications ($Re \leq 18,200$) (Clark et al. 1979), L_{ij} was approximated as (Leonard 1974),

$$L_{ij} \approx \frac{\gamma_L}{2} \nabla^2 \left(\overline{u}_i \overline{u}_j \right), \tag{5.19A}$$

where γ_L is calculated as the second moment of the filter function,

$$\gamma_L = \int\limits_{-\infty}^{\infty} x^2 dx \int\limits_{-\infty}^{\infty} G(x, y, z) dy dz. \tag{5.19B}$$

Finally, the Reynolds stress R_{ij} needs to be approximated, as no solvable expression is known for $\overline{u_i' u_j'}$. There are dozens of R_{ij} closure models in the literature. Of the turbulent kinematic viscosity models (i.e., eddy viscosity models), the simplest assume a constant eddy viscosity, while others are more complex, such as those that involve vorticity, turbulent kinetic energy, or mean strain rate (e.g., the standard Smagorinsky model). The first LES model originated over half a century ago and is called the standard Smagorinsky, after its originator, Joseph Smagorinsky (Smagorinsky 1963; Wilcox 2006; Zhiyin 2015). Not only does the Smagorinsky model provide sufficient dissipation for the modeled eddies, but its mechanism is backed by theory and comparisons with DNS (Deardorff 1971; Clark et al. 1979). For example, the magnitude of the Smagorinsky constant compares well with theory, the model is analogous to the Newtonian shear model, and comparisons with DNS are generally good. The model is expressed as follows:

$$R_{ij} \equiv 2\nu_t \overline{S}_{ij} = (C_S \Delta)^2 \sqrt{\overline{S}_{ij} \overline{S}_{ij}}. \tag{5.20}$$

The resolved strain rate is

$$\overline{S}_{ij} = \frac{1}{2} \left(\frac{\partial \overline{u}_i}{\partial x_j} + \frac{\partial \overline{u}_j}{\partial x_i} \right), \tag{5.21}$$

while the Smagorinsky eddy kinematic viscosity is

$$\nu_t = \frac{\mu_t}{\rho} \equiv (C_S \Delta)^2 \sqrt{\overline{S}_{ij} \overline{S}_{ij}}. \tag{5.22A}$$

In this context,

$$\Delta = (\Delta x \Delta y \Delta z)^{1/3} \tag{5.22B}$$

and

C_s = the Smagorinsky constant.

Some authors use $0.1 \leq C_s \leq 0.3$ (Tutar and Holdo 2001), while others use $0.1 \leq C_s \leq 0.24$ (Fuego 2016a). Additional guidelines regarding C_s are as follows,

- C_S shows some independence on Re and the energy/wavelength spectrum, especially in isotropic turbulence (Clark et al. 1979).
- C_S has a strong dependency on surface curvature and strain rate magnitude—unfortunately, there is no universal value that is valid for all scenarios.
- Some reasonable guidelines and values for C_S for various specific cases include:

  - $C_S = 0.1$ for internal flow in ducts (Rogallo and Moin 1984).
  - $C_S = 0.1$ for flows near the wall (Zhiyin 2015).
  - $C_S = 0.13$ in flows where the large-eddy velocity shear dominates the cascade energy transfer (Deardorff 1971).
  - $C_S = 0.15$ for flow around a sphere (Tutar and Holdo 2001).
  - The Fuego CFD code has $C_S = 0.17$ as a default value, which is the average of its minimum and maximum range (Fuego 2016a), in an attempt toward having a good, all around value. The value of 0.17 also corresponds to the estimated magnitude of Lilly (Lilly 1966; Leonard 1974).
  - $C_S = 0.18$ for isotropic turbulence (Zhiyin 2015).
  - $C_S = 0.21$ for flows where buoyancy dominates (Deardorff 1971; Rogallo and Moin 1984).
  - $C_S = 0.21$ for flows with low mean shear (Deardorff 1971; Rogallo and Moin 1984).
  - C_S should be decreased for situations where the mean strain rate increases.
  - If C_S is too large, eddies will undergo excessive damping.
  - Ensure that C_s is independent of Δ.

- The dynamic Smagorinsky model automatically calculates C_S in both space and time as the calculation proceeds, whereby $C_S = C_S\left(\vec{x}, t\right)$.

The standard Smagorinsky model is reasonable for large Re and isotropic flows but is not as good for low Re flows (it is too dissipative near the laminar to turbulent transition (Zhiyin 2015)). Flows with large deviations from isotropy (anisotropic) are not well-resolved, either.

Unfortunately, the *standard* Smagorinsky calculates a nonzero value for ν_t at the wall, so some LES models use a damping function such as van Driest. Nevertheless, this issue can be overcome by using sufficiently discretized resolution near the wall, without having to use damping functions (but at a higher computational cost) (Rodi et al. 1997; Tutar and Holdo 2001; Zhiyin 2015).

Given the choice between the 1963 Smagorinsky model and the dynamic Smagorinsky model, the latter is preferred because of its ability to calculate C_s as a function of space and time based on turbulence dynamics (Germano et al. 1991; Lilly 1992), and there is no need to specify C_s a priori based on flow type; a "constant" C_S can never fully represent a *dynamic* situation involving spatial and temporal changes.

Newer models with reduced dissipation near the wall and during transition have been developed recently (Vreman 2004), with excellent comparison with DNS for friction coefficients in channel flow. In addition, the σ-SGS model captures the cubic y dependence near the wall (Nicoud et al. 2011). Certainly, more validation is desirable.

5.1.3   Miscellaneous LES Modeling Recommendations

- Choose Δ and the average grid length h such that $\Delta/h < 3.3$ to reduce filtering and numerical issues (e.g., aliasing) (Stefano and Vasilyev 2002; Yeon 2014).
- The sharper the filter, the better the large-scale to small-scale energy dynamics (Stefano and Vasilyev 2002).
- Computer processing unit (CPU) time is strongly dependent on the minimum element size, Δ (and number of elements, of course).

  - The smaller Δ is, the longer it will take to run the simulation.
  - On the other hand, the analyst must demonstrate that the simulation is sufficiently discretized to capture the desired larger eddy behavior.

- For very large simulations, RANS can be submitted first to get faster, preliminary results.

  - Then, the LES calculations ought to be submitted as soon as possible, as they can take 10 to 100 times more computational time than RANS.

- A good point for an LES simulation is to let $\Delta \approx \lambda$ (i.e., simulate down to the Taylor eddies):

  - The logic behind this approach is that LES resolves the larger eddies, which carry about 80% of the total turbulent kinetic energy. Therefore, setting $\Delta \approx \lambda$ means that sufficient computational nodes are included in the region occupied by the integral eddies, as well as a significant fraction of the Taylor eddies that carry an important fraction of k (recall that the Taylor eddies span a wide spatial range).

- The integral and Taylor eddy dimensions can be easily estimated using the LIKE algorithm (refer to Sect. 3.4).

- The usage of higher-order elements such as hexahedrals is highly recommended (conversely, avoid lower-order tetrahedrals) (Tyacke et al. 2014).
- The addition of wall and damping functions tends to induce a counterintuitive and unpalatable flavor in LES purists. Namely, LES theory adheres to the notion that flow ought to be adequately calculated using a reasonable Δ that includes sufficient quantities of the larger and intermediate-sized eddies (i.e., a good fraction of the Taylor eddies) plus the behavioral contributions of the lumped, smallest eddies. So why "fudge" the simulation with extraneous functions? Particularly, any numerical accuracy that damping and wall functions may add to the simulation can doubtlessly be obtained using sufficient spatial resolution, higher-order elements, and good mesh metrics, especially near the wall.
- A wall function can be detrimental for LES simulations with separated flow (Rodi et al. 1997).
- As for what is sufficiently discretized for LES, the limit can be based on purely theoretical arguments: the first computational node ought to be the minimum of either the smallest eddies being resolved (say the Taylor eddies) *or* the size determined by the physics in question (e.g., the viscous layer extends to $y^+ = 5$, so one or two computational nodes in this region would be advisable).
- Large *Re* simulations should include a mechanism for calculating the Leonard stress (Shaanan et al. 1975).

Example 5.1 Helium at 400 K and 6 MPa is flowing inside a smooth cylindrical pipe with $D = 0.125$ m and $L = 5.0$ m. The system has an average velocity of 1.4 m/s. Suppose an analyst would like to use the LES model to simulate up to the Taylor eddy scale. What should the minimum node-to-node distance be? What is the value of y at $y^+ = 1.0$?

Solution At this temperature and pressure, the density is 7.08 kg/m^3, and the dynamic viscosity is 24.4E−6 Pa-s; thus, $\nu = \mu/\rho = 3.45$E−6 m^2/s. The characteristic length is the pipe diameter, D. Therefore, the fluid is turbulent:

$$Re = \frac{(0.125)(1.4)(7.08)}{24.4E - 6} = 50,779.$$

By running the LIKE MATLAB script, $\varepsilon = 3.65$E−3 m^2/s^3 and $k = 5.02$E−3 m^2/s^2. Alternatively, the necessary equations are:

- Characteristic length of the integral eddies (e.g., Eq. 3.17, $\ell \approx 0.07 D_h$)
- Turbulence intensity (e.g., Eq. 3.26C, $I = 0.16\, Re_h^{-1/8}$)
- k [e.g., Eq. 3.27, $k = \frac{3}{2}(\overline{u}I)^2$]
- ε (e.g., Eq. 3.28, $\varepsilon = C_\mu \frac{k^{3/2}}{\ell}$)

Then, the Taylor eddy length scale is readily calculated as

$$\lambda = \left(\frac{10k\nu}{\varepsilon}\right)^{1/2} = \left[\frac{10(5.02E-3)(3.45E-6)}{3.65E-3}\right]^{1/2} = 0.00687 \text{ m}.$$

But how does this scale compare with y^+? In this case, the LIKE algorithm calculates $u_* = 0.0706$ m/s and $\tau_w = 0.0353$ kg/m-s^2 (in anticipation of comparison with Example 5.2). Then, from $y^+ = \frac{yu_*}{\nu}$, this implies that $y = \frac{\nu\,y^+}{u_*}$. For $y^+ = 1.0$, then $y = \frac{\nu*1.0}{u_*}\Big|_{y^+=1.0} = \frac{3.45E-6*1.0}{0.0706} = 4.89E-5$ m.

Example 5.2 Redo Example 5.1, except that now the pipe is a bit rusty, having a surface roughness of 1.0E−4 m. Will this affect either the Taylor eddy size or the value of y at $y^+ = 1.0$?

Solution An inspection of the relevant equations shows that the Taylor eddy remains the same, as the equations in this book show no roughness dependency. It is possible that such dependency exists somewhere else in the literature, for clearly, very large roughness will generate its own significant Taylor eddy scale fluctuations, whereas small surface roughness will not affect the Taylor scale. In any case, the wall friction factor is larger (0.0059 vs 0.0051 from Example 5.1), so the wall shear τ_w is now larger by 15.7%. This means that u_* is larger as well, because $u_* = \sqrt{\frac{\tau_w}{\rho}}$. For this situation, $u_* = 0.0761$ m/s and $\tau_w = 0.041$ kg/m-s^2.

Therefore, $y_{\,y^+=1.0} = \frac{\nu*1.0}{u_*}\Big|_{y^+=1.0} = \frac{3.45E-6*1.0}{0.0761} = 4.53E-5$ m. In this case, the roughness generated extra shear at the wall, requiring a computational node that is much closer to the wall vs. the smooth surface.

5.2 DNS Modeling Recommendations

RANS calculates *time-averaged* turbulence effects, but no dynamics based on individual eddy behavior. By contrast, DNS is a turbulence approach that solves the unsteady Navier-Stokes equations such that *all* turbulence scales are resolved, and unlike LES, no subgrid model is employed (Wilcox 2006; Afgan 2007). As a result, DNS oftentimes requires multiple tens of millions to billions of computational nodes (Day et al. 2009). DNS is therefore used sparingly, especially in large systems at high *Re*; the higher the *Re*, the higher the required node count because eddies become smaller. Nevertheless, time favors this turbulence method, especially as computers and algorithms become faster. As of 2019, DNS is generally *at least* two to three orders of magnitude more expensive than LES.

DNS is unique among the turbulence models because it does not employ averaging (no $\bar{u}$, no u'); Boussinesq approximation (or nonlocal, nonequilibrium approaches); k, ε, ω, or ν_t; wall functions; curve fits; ad hoc models; and so forth and so on. *DNS is purely Navier-Stokes calculated for all time and spatial scales.* DNS uniquely solves Navier-Stokes to calculate all instantaneous, primitive-variable fluctuations.

Fortunately, as of 2019, DNS is no longer limited to low to moderate Re flows for reasonably sized domains. Still out of reach as of 2019 are very large systems, such as an entire nuclear plant, cruise ships, jet liners, very large geophysical systems, etc. Starting around the late 1990s and especially around 2005 or so, DNS became more commonplace for higher Re calculations of small- to mid-sized industrial applications (Huser and Biringen 1993; Dong and Karniadakis 2005; Kaneda and Ishihara 2006; Terentiev 2006; Stein 2009; Leonardi and Castro 2010; Kaneda and Ishihara 2006; Yeung et al. 2010; Naqavi et al. 2018; Elghobashi 2019; Chu et al. 2019).

For example, excellent results were obtained for swirling jets at $Re = 5000$ and swirl number $S = 0.79$ (Freitag and Klein 2005). A higher Re in the range of $12{,}000 \leq Re \leq 33{,}500$ and $S \leq 0.5$ was achieved for swirling jets a year later (Facciolo 2006). A plane wall jet at $Re = 7500$ compared favorably with theory and experimental data (Naqavi et al. 2018). Channel flow involving cubes as wall roughness was simulated up to $Re = 7000$ (Leonardi and Castro 2010). DNS has also been conducted on oscillating and stationary cylinders at $Re = 10{,}000$ (Dong and Karniadakis 2005). Recent DNS simulations confirm its ability to model bubbles, droplets, and porous media (Chu et al. 2019). Using spectral methods, a total of 4096^3 (6.87×10^{10} grid points) were used to explore isotropic turbulence in atmospheric flows at $Re_\lambda = 1000$ (Kaneda and Ishihara 2006; Yeung et al. 2010). Certainly, numerous examples can be cited that corroborate a favorable trend toward DNS.

Moreover, many recent studies can be cited in the literature whereby DNS calculations compare favorably with experimental data, to the point that many authors go as far as considering the output as good as experimental data (Moet et al. 2004; Freitag and Klein 2005; Duraisamy and Lele 2006; Afgan 2007; Bonaldo 2007; Busch et al. 2007; Walther et al. 2007; Taub et al. 2010). *In any case, if done properly, DNS is extremely accurate—as good, if not better, than experimental data.* For example, DNS can be better than experimental data because it can tract parameters that are difficult, and perhaps even impossible, to measure experimentally (e.g., P'). Furthermore, DNS provides much more detailed data than any experiment could ever achieve, e.g., its ability to employ millions to billions to trillions of computational nodes that behave as probes (Clark et al. 1979).

For the interested reader, many useful guidelines, too many to be cited here, can be found in the literature. A small sample includes (Moin and Mahesh 1998; Modi 1999; Wilcox 2006; Coleman and Sandberg 2010; Alfonsi 2011; Joslin 2012; Tryggvason and Buongiorno 2013; Argyropoulos and Markatos 2015; Wu 2017; Naqavi et al. 2018; Joshi and Nayak 2019).

The case for DNS is further strengthened in light of faster computational systems, such as multi-core and many core processors for increased computing performance (Alfonsi 2011), as well as nano- and quantum computers (Rudinger 2017). Figure 5.7 shows a quantum computer segment developed and manufactured recently at Sandia National Laboratories (Sandia 2011). Notice that despite its small length scale, which is on the order of 1500 nm total, the device has all the components necessary for a functional dual quantum dot structure. Such novel hardware, combined with quantum algorithms, will surely continue the DNS computational growth trend for

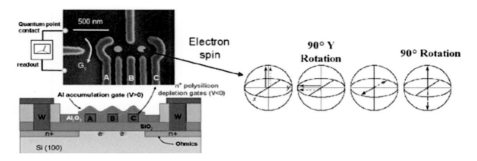

Fig. 5.7 A quantum computer segment. (Courtesy, Sandia National Laboratories, 2011)

many decades to come, if not centuries. In any case, as computational power increases, DNS will not only be used for turbulence research and small systems but for larger engineering designs as well; this trend is inevitable and was predicted long ago (Kim et al. 1987). This optimistic premise is supported by the strong potential from recent advances in quantum computers, topological quantum materials, and quantum algorithms (Lee et al. 2019; Singer 2019). As noted in Chap. 1, quantum algorithms already solve linear systems of equations (Singer 2019), which are essential for CFD solvers. Furthermore, it is expected that quantum algorithms will result in an *exponential decrease* of the time required to solve systems of linear equations (Singer 2019). Indeed, the literature as of 2019 indicates the potential for computational speed increases of at least a factor of 1000! And of course, the detailed DNS calculations will uncover fluid functionality that can be leveraged onto vastly improved engineered system behavior and performance.

But the ultimate grail is a quantum computer using symbolic computation and artificial intelligence algorithms to obtain *analytical* solutions for the Navier-Stokes PDE—to actually find unique mathematical solutions (expressions) for the PDE based on the particular BCs and ICs for any/all of the PDE terms.

5.2.1 DNS Numerical Methods

DNS *usually* requires higher-order numerical methods for spatial discretization (e.g., fourth, fifth, and sixth order), basically with the goal of reducing numerical errors such as truncation, diffusion, aliasing, and instabilities (Rai and Moin 1991; Huser and Biringen 1993; Drikakis and Geurts 2002; Sengupta and Bhaumik 2019). But there is another, more compelling reason: the small velocity fluctuations at the lower end of the Taylor scale and especially the Kolmogorov fluctuations can easily approach the magnitude of the numerical error and can thus be lost to the numerical "noise."

Generally, fourth-order methods are used in DNS, especially during its earlier years (Coleman and Sandberg 2010). More recently, this trend has seen some changes, including the usage of both second-order and optimized hybrid fourth-order methods that can be used successfully *if* error mitigation precautions are undertaken (Verstappen and Veldman 1997; Wilcox 2006). This is interesting, especially because most commercial CFD tools are second order in space. Certainly, higher-order methods have higher numerical accuracy than lower-order methods (Orszag and Israeli, 1974), and whether a method's order is even or odd has ramifications as well. On the other hand, higher-order methods are more prone to instabilities because they have less numerical dissipation and require more complex boundaries.

Many DNS solvers are *explicit* due to large memory constraints required by this approach. But, more recently, implicit solvers have become more common (Clark et al. 1979; Rodriguez 2000; Wilcox 2006; Coleman and Sandberg 2010; Alfonsi 2011).

5.2.2 DNS Spatial Domain

LES calculates the integral eddies and larger eddies up to some user-defined or mesh-defined minimum scale, such as the Taylor scale. On the other hand, DNS not only resolves all the LES eddies and the smaller Taylor eddies that were filtered out by LES, but it also resolves the Kolmogorov eddies. *Said more concisely, DNS must calculate all eddy scales.* Because DNS includes the Kolmogorov eddies, it is not uncommon for the first computational node to be at a small fraction of y^+, with $y^+ << 1.0$. For example, a channel flow at $Re = 3300$ included the first computational node at $y^+ = 0.05$, while its maximum spacing was set to $y^+ = 4$ (Kim et al. 1987). Of course, Kolmogorov eddy size is dependent on Re, and the larger Re is, the smaller y^+ will be. These are good guidelines, but to ensure more problem-specific node spacing, the user is encouraged to use the LIKE algorithm to determine node spacing; refer to Sects. 3.4 and 3.5. *In particular, η is a precarious function of y^+, so care must be taken to make the mesh consistent with the Kolmogorov eddies and not with a variable associated with the wall friction, u_*. Stated differently, what assurance is there that some fraction of y^+ is equal or smaller or in any way proportional to η? And if there is no proportionality, what assurance is there that the Kolmogorov eddy scale is modeled correctly using y^+ instead of η?*

The Kolmogorov eddy is the smallest eddy scale that is sustainable by the flow and is calculated as follows:

$$\eta = \left(\frac{\nu^3}{\varepsilon}\right)^{1/4}. \tag{5.23}$$

Therefore, the distance between the computational nodes must not exceed η. However, to fully capture the eddy's interior dynamics, a more rigorous restriction is imposed for the distance between the computational nodes, such that

$$\Delta x \leq \frac{\eta}{3} \text{ to } \frac{\eta}{2}. \tag{5.24}$$

As for the largest eddies, it is not appropriate to assume that the hydraulic diameter (or equivalent) is the limiting length scale. For example, because the larger integral eddies can stretch significantly (say at 45° from the main flow direction), the maximum bound is conservatively placed at

$$\Lambda = 2D_{h}. \tag{5.25}$$

This size restriction guides the minimum size that a mesh domain ought to be and still has enough space to capture all eddies. As a check that the DNS computational mesh is sufficiently large, the eddy fluctuations must be *uncorrelated* up to half the distance of the domain for the largest eddies.

5.2.3   DNS Time, Stability Criteria, and Computational Nodes

If the time step is limited by the Courant limit, then

$$\frac{u' \Delta t}{\Delta x} = \frac{u' \Delta t_{\text{Courant}}}{\eta} < 1.0. \tag{5.26}$$

On the other hand, the Kolmogorov eddy lifetime before collapsing into a tiny laminar sheet fragment is calculated as

$$\tau = \left(\frac{\nu}{\varepsilon}\right)^{1/2}. \tag{5.27}$$

Therefore, the smallest time step is the minimum of the following two:

$$\Delta t_{\text{DNS}} = \min\left(\tau, \Delta t_{\text{Courant}}\right). \tag{5.28}$$

If an explicit DNS simulation shows signs of numerical instability, the time step can be reduced further using a more restrictive criterion, one that is based on the wall friction velocity u_* and the channel characteristic length x_{char} (Kim et al. 1987; Wilcox 2006). Namely,

$$\Delta t \approx \frac{0.003}{\sqrt{Re_\tau}} \frac{x_{\text{char}}}{u_\tau}, \tag{5.29A}$$

where

$$Re_\tau = \frac{u_* x_{\text{char}}}{\nu}. \tag{5.29B}$$

A stability criterion that combines the 1D diffusive and convective limits, respectively, is as follows (Coleman and Sandberg 2010):

$$\left(1 - 4\frac{\nu\Delta t}{\Delta x^2}\right)^2 + \left(\frac{u\Delta t}{\Delta x}\right)^2 \leq 1. \tag{5.30}$$

As discussed in Sect. 3.5, the number of nodes required in 1D DNS calculations is estimated as

$$N_{1D} \approx Re_{\text{T}}^{3/4}, \tag{5.31A}$$

where

$$Re_{\text{T}} = \frac{\eta\sqrt{k}}{\nu}. \tag{5.31B}$$

For 3D DNS calculations, the number is significantly larger (Afgan 2007; Sodja 2007; Stein 2009; Taub et al. 2010):

$$N_{3D} \approx Re_{\text{T}}^{9/4} \text{ to } Re_{\text{T}}^{11/4}. \tag{5.32}$$

A more precise relationship is as follows (Wilcox 2006; Sodja 2007):

$$N_{3D} = (110\,Re_{\text{T}})^{9/4}. \tag{5.33}$$

As might be fully expected by now, the impressive 3D DNS calculations come at a high computational cost: the CPU needed to solve these problems is a strong function of the turbulent Re, so the required computational power increases stratospherically to the third power:

$$\text{CPU} \propto Re_T^3. \tag{5.34}$$

Example 5.3 Redo Example 5.1, except that it is now desired to run a DNS calculation. What are the minimum length and time scales based on the Kolmogorov eddies? What should the minimum computational domain length scale be (e.g., the minimum total mesh size)?

Solution Using the LIKE algorithm, the output is

Kolmogorov eddy size (η) = 3.26E−4 m
Kolmogorov eddy time = 3.07E−2 s

Of course, $\Delta x \leq \frac{\eta}{3}$ to $\frac{\eta}{2}$. As for the minimum domain that must be modeled,

$$\Lambda = 2D_h = 2 * 0.125 = 0.25 \text{ m}.$$

5.3 Special LES and DNS BCs and ICs

Descriptions of generic BCs and ICs suitable for ODEs, PDEs, and RANS turbulence models are discussed in Sects. 6.2, 6.2.1, 6.2.2, 6.2.3 and 6.2.4. However, LES and DNS methods require special treatment, as they must not generate spurious instabilities or introduce numerical errors that overshadow the eddy dynamics, especially those of the smaller eddies that are associated with very small velocity and length magnitudes. Therefore, LES- and DNS-specific BC and IC issues are discussed in what follows.

- *Inflow BCs*

Inflow BCs pose issues because it is not possible to implement a priori inflow distribution. For example, the BC can take the output from the computational domain, modify the results to reflect the inlet conditions, and then use the rescaled data for the next time step. An alternative approach is to let the flow reach a reasonable degree of turbulence, but such approach can extend the time domain to a prohibitive size. An elegant approach is to use a synthetic method, which supplies functions that randomly perturb the LES flow (Jarrin et al. 2006; Finn and Dogan 2019); such approach can be applied for DNS as well (Wu 2017). The synthetic method has become very rich in the literature, such that an entire book can easily be devoted to it; a great survey is provided by Wu (2017). Finally, a brute force approach is to take a relatively coarser "DNS" calculation that can be used as input for a finer DNS calculation, or perhaps a fine-scale LES simulation can be used as a starting point for the DNS input, and the process is iterated until convergence is reached.

- *Outflow BCs*

Well-posed outflow BCs permit eddies to exit the boundary seamlessly, without producing numerical errors, instabilities, or reflective waves. For this reason, researchers have developed non-reflecting, damping BCs that have a demonstrated ability to suppress spurious waves (Thompson 1987; Spalart 1990; Nordstrom et al. 1999).

- *Periodic BCs*

Periodic BCs are extremely useful for LES and DNS calculations. For example, recall that fully developed flow (FD) is relatively homogeneous along the perpendicular (spanwise) direction of the primary flow (Kim et al. 1987; Moin and Mahesh 1998; Leonardi and Castro 2010; Finn and Dogan 2019). This situation allows for significant reductions in geometry if a periodic BC is used to replace the repetitive flow pattern. (Refer to Sect. 2.6.2 for estimating the entrance length of turbulent flows.) Periodic BCs are also ideally suited when geometry is repetitive, such as turbomachinery components (e.g., blades) or nuclear reactor fuel lattices. This is an important point, as having to model the entire system can mean the addition of tens of millions to billions more computational cells.

- *No slip wall BCs*

Wall BCs with no slip (provided *Kn* is in the appropriate range) are considered fairly safe for LES and DNS applications (Coleman and Sandberg 2010; Leonardi and Castro 2010; Finn and Dogan 2019).

- *Initial Conditions (ICs)*

Regarding ICs, these can be obtained from coarser meshes and then superimposed on the LES or DNS grid (Coleman and Sandberg 2010; Finn and Dogan 2019). Then, the initial input is flushed out by allowing the calculation to run for several time periods, typically three or more flow throughs (Dong and Karniadakis 2005). Alternatively, ICs can be chosen such that they start the simulation using reasonable values, such as having the initial velocity equal to zero throughout the domain. Then, the simulation proceeds until it reaches its stationary limit (Day et al. 2009). At this point, the calculated turbulence data should be independent of the ICs. Another approach is to use Gaussian statistics with white noise ("artificial random force") to obtain a homogeneous initial velocity distribution (Jeng 1969; Stefano and Vasilyev 2002).

5.4   Problems

5.1 Under what circumstances would it be preferable to use LES instead of RANS? How might the results differ between the two models?

5.2 Under what circumstances would it be preferable to use DNS instead of LES or RANS? What fundamental differences would be expected between the LES and DNS output?

5.3 What would happen if a RANS model used node spacing down to the Kolmogorov scale? Would such results be defensible?

5.4 Does the LES model calculate k? Why or why not?

5.5 Water at 400 K and 6 MPa is flowing inside a smooth cylindrical pipe with $D = 0.1$ m and $L = 1.0$ m. The system has a mass flow rate of 50 kg/s. Suppose the analyst would like to use the LES model to simulate up to the Taylor eddy scale. What should the maximum node-to-node distance be? If the Taylor eddy is assumed as a spherical agglomeration, how many Taylor eddies would fit inside the pipe?

5.6 Redo Problem 5.5, except now suppose the analysist would like to use the DNS model to simulate up to the Kolmogorov eddy scale. What should the maximum node-to-node distance be? If the Kolmogorov eddy is assumed as a spherical agglomeration, how many Kolmogorov eddies would fit inside the pipe?

5.7 Water at room temperature and pressure flows at $U_\infty = 1500$ m/s around a pin with $D = 0.5$ mm. How many Kolmogorov eddies could fit at the tip of the pin?

5.8 Repeat Problem 5.7 using air. Any issues?

5.9 Consider a 3D smooth rectangular duct with $S = 0.05$ m and $L = 0.2$ m. The fluid is lead at 1000 K ($\rho = 10,161$ kg/m^3 and $\mu = 0.0013$ kg/m-s), with an average velocity of 2.5 m/s. Use your favorite LES model to simulate the system. What issues are present if uniform node-to-node spacing is set to 0.3ℓ? Does a swirl motion appear at the corners under sufficient spatial discretization?

5.10 Repeat Problem 5.9, but with uniform node spacing set to λ. Does a swirl motion appear at the corners? How does λ compare with $y^+ = 1$? Will node biasing help? (Consider an expansion ratio ≤ 1.5.)

5.11 Repeat Problem 5.9 but with node spacing set to η. Does a swirl motion appear at the corners? What issues are present under this spatial discretization? (i.e., issues other than a much longer computational time). What happens to the solution if the uniform node-to-node spacing is set to $\eta/4$?

5.12 Water is flowing in a smooth, cylindrical pipe with $D = 0.125$ m, at an average velocity of 7.5 m/s. The water is at 350 K and 8.0×10^5 Pa. Suppose that you want to use an LES model. What distance between the computational nodes will be sufficient to resolve the Taylor eddies? At what location should the first computational node be placed, and why?

5.13 Consider Problem 5.12. Suppose that you want to use DNS. What integral, Taylor, and Kolmogorov length, velocity, and time scales must be considered and resolved? Discuss various time step constraints. Suppose the initial fluid velocity is increased by a factor of 20, 40, and so on, until reaching a factor of 100. Plot the eddy length, velocity, and time scales for all three eddy types as a function of Re.

5.14 Consider a smooth flat plate under isothermal boundary layer flow. The smooth plate is 0.5 m long and 0.05 m wide, with air flowing parallel to the plate at 300 K and 1 atmosphere ($\nu = 1.58 \times 10^{-5}$ m^2/s and $U_s = 347.3$ m/s). The air flows from left to right along the 0.5 m plate at a constant velocity $U_\infty = 25$ m/s. Apply your CFD tool of choice to simulate this system using an LES model. Compare the flow distribution with your preferred RANS model. How are the velocity fields different?

References

Afgan, I. (2007, July). *Large eddy simulation of flow over cylindrical bodies using unstructured finite volume meshes*, PhD Diss., University of Manchester.

Alfonsi, G. (2011). Direct numerical simulation of turbulent flows. *ASME Applied Mechanics Reviews, 64*, 020802.

Argyropoulos, C. D., & Markatos, N. C. (2015). Recent advances on the numerical modelling of turbulent flows. *Applied Mathematical Modeling, 39*, 693.

Bonaldo, A. (2007). *Experimental characterisation of Swirl stabilized annular stratified flames*, PhD Diss., Cranfield University.

Bouffanais, R. (2010). Advances and challenges of applied large-eddy simulation. *Computers & Fluids, 39*, 735.

Busch, H., Ryan, K., & Sheard, G. J. (2007, Dec 2–7). *Strain-rate development between a co-rotating Lamb-Oseen Vortex pair of unequal strength*, 16th Australasian fluid mechanics conference, Gold Coast, Australia.

Chu, X., et al. (2019). Direct numerical simulation of convective heat transfer in porous media. *International Journal of Heat and Mass Transfer, 133*, 11.

CFD-Online, "RNG-LES Model", https://www.cfd-online.com/Wiki/RNG-LES_model. Accessed on 13 July 2018.

Clark, R. A., Ferziger, J. H., & Reynolds, W. C. (1979). Evaluation of subgrid-scale models using an accurately simulated turbulent flow. *Journal of Fluid Mechanics, 91*(Part 1), 1.

Coleman, G. N., & Sandberg, R. D. (2010). *A primer on direct numerical simulation of turbulence – Methods, procedures and guidelines*. Aerodynamics & Flight Mechanics Research Group, University of Southampton.

Day, M., et al. (2009). *Combined computational and experimental characterization of Lean premixed turbulent low Swirl laboratory flames*. Lawrence Berkeley National Laboratories, circa.

Deardorff, J. W. (1970). A numerical study of three-dimensional turbulent channel flow at large Reynolds numbers. *Journal of Fluid Mechanics, 41*(Part 2), 453.

Deardorff, J. W. (1971). On the magnitude of the subgrid scale eddy coefficient. *Journal of Computational Physics, 7*, 120.

Dong, S., & Karniadakis, G. E. (2005). DNS of flow past a stationary and oscillating cylinder at Re = 10000. *Journal of Fluids and Structures, 20*, 519.

Drikakis, D., & Geurts, B. J. (Eds.). (2002). *Turbulent flow computation*. Dordrecht: Kluwer Academic Publishers.

Duraisamy, K., & Lele, S. K. (2006). *DNS of temporal evolution of isolated vortices*, Center for Turbulence Research, Proceedings of the Summer Program.

Elghobashi. (2019). Direct numerical simulation of turbulent flows laden with droplets or bubbles. *Annual Review of Fluid Mechanics, 51*, 217.

Facciolo, L. (2006). *A study on axially rotating pipe and swirling jet flows*, Royal Institute of Technology, Department of Mechanics, S-100 44 Stockholm, Sweden, PhD Diss.

Ferziger, J. H. (1977). Large eddy numerical simulations of turbulent flows. *AIAA Journal, 15*(9), 1261.

Finn, J. R., & Dogan, O. N. (2019). *Analyzing the potential for erosion in a supercritical CO2 turbine nozzle with large eddy simulation*, Proc. of the ASME Turbo Expo, GT2019-91791.

Freitag, M., & Klein, M. (2005). Direct numerical simulation of a recirculating swirling flow. *Flow, Turbulence and Combustion, 75*, 51.

Fuego. (2016a). *SIERRA low Mach module: Fuego theory manual – Version 4.40*. Sandia National Laboratories.

Fuego. (2016b). *SIERRA low Mach module: Fuego user manual – Version 4.40*. Sandia National Laboratories.

Galperin, B., & Orszag, S. A. (Eds.). (1993). *Large eddy simulation of complex engineering and geophysical flows*. Cambridge, UK: Cambridge University Press.

Garnier, E., Adams, N., & Sagaut, P. (2009). *Large eddy simulation for compressible flows.* Netherlands: Springer.

Germano, M., et al. (1991). A dynamic subgrid scale eddy viscosity model. *Physics of Fluids A, 3* (7), 1760.

Huser, A., & Biringen, S. (1993). Direct numerical simulation of turbulent flow in a square duct. *Journal of Fluid Mechanics, 257*, 65.

Jarrin, N., et al. (2006). A synthetic-eddy-method for generating inflow conditions for large-eddy simulations. *International Journal of Heat and Fluid Flow, 27*, 585.

Jeng, D. T. (1969). Forced model equation for turbulence. *Physics of Fluids, 12*(10), 2006.

Joshi, J., & Nayak, A. (2019). *Advances of computational fluid Dynamics in nuclear reactor design and safety assessment* (1st ed.). Cambridge: Woodhead Publishing.

Joslin, R. D. (2012). *Discussion of DNS: Past, present, and future.* NASA, Langley Research Center, circa.

Kaneda, Y., & Ishihara, T. (2006). High-resolution direct numerical simulation of turbulence. *Journal of Turbulence, 7*(20), 20.

Kim, W., & Menon, S. (1995). *A new dynamic one-equation subgrid-scale model for large eddy simulation*, 33rd aerospace sciences meeting and exhibit, Reno, Nevada.

Kim, J., Moin, P., & Moser, R. (1987). Turbulence statistics in fully developed channel flow at low Reynolds number. *Journal of Fluid Mechanics, 177*, 133.

Kleissl, J., & Parlange, M. B. (2004). Field experimental study of dynamic Smagorinsky models in the atmospheric surface layer. *Journal of the Atmospheric Sciences, 61*, 2296.

Lee, S. R., et al. (2019). Topological quantum materials for realizing majorana quasiparticles. *Chemistry of Materials, 31*, 26. (Also available as SAND2018-11285J through Sandia National Laboratories).

Leonard, A. (1974). Energy cascade in large eddy simulations of turbulent fluid flow. *Advances in Geophysics, 18*, 237.

Leonardi, S., & Castro, I. P. (2010). Channel flow over large cube roughness: A direct numerical simulation study. *Journal of Fluid Mechanics, 651*, 519.

Lesieur, M., Metais, O., & Comte, P. (2005). *Large-eddy simulations of turbulence.* Cambridge: Cambridge University Press.

Lilly, D. K. (1966). *The representation of small-scale turbulence in numerical simulation experiments.* National Center for Atmospheric Research, NCAR Manuscript 281.

Lilly, D. K. (1992). A proposed modification of the Germano subgrid-scale closure method. *Physics of Fluids A, 4*(3), 633.

Mansour, N. N., et al. (1977). *Improved methods for large-eddy simulation of turbulence*, Proc. Penn State Symp, Turbulent Shear Flows.

Modi, A. (1999). *Direct numerical simulation of turbulent flows.* Penn State University.

Moet, H., et al. (2004). Wave propagation in Vortices and Vortex bursting. *Physics of Fluids*, 1–55.

Moin, P., & Mahesh, K. (1998). Direct numerical simulation: A tool in turbulence research. *Annual Review of Fluid Mechanics, 30*, 539.

Naqavi, I. Z., Tyacke, J. C., & Tucker, P. G. (2018). Direct numerical simulation of a wall jet: Flow physics. *Journal of Fluid Mechanics, 852*, 507.

Nicoud, F., & Ducros, F. (1999). Subgrid-scale modeling based on the square of the velocity gradient tensor. *Flow, Turbulence and Combustion, 62*, 183.

Nicoud, F., et al. (2011). Using singular values to build a subgrid-scale model for large eddy simulations. *Physics of Fluids, 23*, 085106.

Nordstrom, J., Nordin, N., & Henningson, D. (1999). The fringe region technique and the Fourier method used in the direct numerical simulation of spatially evolving viscous flows. *SIAM, Journal of Scientific Computing, 20*(4), 1365.

Orszag, S. A., & Israeli, M. (1974). Numerical simulation of viscous incompressible flows. *Annual Review of Fluid Mechanics, 6*, 681.

Rai, M. M., & Moin, P. (1991). Direct simulation of turbulent flow using finite difference schemes. *Journal of Computational Physics, 96*, 15. (Also available as Fluid Mechanics and Heat Transfer, AIAA Paper 89-0369, 1989).

Rodi, W., et al. (1997). Status of the large eddy simulation: Results of a workshop. *Transactions of the ASME, Journal of Fluids Engineering, 119*, 248.

Rodriguez, S. (2000). *A 4th order, implicit, adaptive mesh refinement algorithm for simulation of flame-vortex interactions*, Master Th., University of New Mexico.

Rodriguez, S. (2011, May). *Swirling jets for the mitigation of hot spots and thermal stratification in the VHTR lower plenum*, PhD Diss., University of New Mexico.

Rogallo, R. S., & Moin, P. (1984). Numerical simulation of turbulent flows. *Annual Review of Fluid Mechanics, 16*, 99.

Rudinger, K. (2017). *Quantum computing is and is not amazing*. Sandia National Laboratories, SAND2017-7067C.

Sandia. (2011) Quantum Information Science and Technology (QIST), https://www.sandia.gov/QIST. Sandia National Laboratories. Accessed on 3 May 2019.

Sengupta, T. K., & Bhaumik, S. (2019). *DNS of wall-bounded turbulent flows: A first principle approach*. Singapore: Springer.

Shaanan, S., Ferziger, J. H., & Reynolds, W. C. (1975). *Numerical simulation of turbulence in the presence of shear*, Report No. TF-6, Dept. of Mechanical Engineering, Stanford University.

Singer, N. (2019). *Quantum computing steps further ahead with new projects*. Sandia National Laboratories, Sandia LabNews.

Smagorinsky, J. (1963). General circulation experiments with the primitive equations I. The basic experiment. *Monthly Weather Report, 91*(3), 99.

Sodja, J. (2007). *Turbulence models in CFD*. University of Ljubljana.

Spalart, P. R. (1990). *Direct numerical study of crossflow instability*, laminar-turbulent transition, IUTAM Symposium.

Stefano, G., & Vasilyev, O. V. (2002). Sharp cutoff versus smooth filtering in large eddy simulation. *Physics of Fluids, 14*(1), 362.

Stein, O. (2009, March). *Large eddy simulation of combustion in swirling and opposed jet flows*, PhD Diss., Imperial College London.

Taub, G., et al. (2010, Jan 4–7). *A numerical investigation of swirling turbulent Buoyant jets at transient Reynolds numbers*, 48th AIAA aerospace sciences meeting, AIAA 2010-1362, Orlando, Florida.

Terentiev, L. (2006). *The turbulence closure model based on linear anisotropy invariant analysis*, Universitat Erlangen-Nurnberg, PhD Diss.

Thompson, K. W. (1987). Time dependent boundary conditions for hyperbolic systems. *Journal of Computational Physics, 68*, 1.

Tryggvason, G., & Buongiorno, J. (2013). *The role of direct numerical simulations in validation and verification*. University of Notre Dame and Massachusetts Institute of Technology, circa.

Tutar, M., & Holdo, A. E. (2001). Computational modeling of flow around a circular cylinder in sub-critical flow regime with various turbulence models. *International Journal for Numerical Methods in Fluids, 35*, 763.

Tyacke, J., et al. (2014). Large eddy simulation for turbines: Methodologies, cost and future outlooks. *ASME, Journal of Turbomachinery, 136*, 061009.

Verstappen, R. W. C. P., & Veldman, A. E. P. (1997). Direct numerical simulation of turbulence at lower costs. *Journal of Engineering Mathematics, 32*, 143.

Vreman, A. W. (2004). An eddy-viscosity subgrid-scale model for turbulent shear flow: Algebraic theory and applications. *Physics of Fluids, 16*(10), 3670.

Walther, J. H., et al. (2007). A numerical study of the stability of Helical Vortices using Vortex methods. *Journal of Physics: Conference Series, 75*, 012034.

Wilcox, D. C. (2006). *Turbulence modeling for CFD* (3rd ed.)., printed on 2006 and 2010.

Wu, X. (2017). Inflow turbulence generation methods. *Annual Review of Fluid Mechanics, 49*, 23.

Yeon, S. M. (2014). *Large-eddy simulation of sub-, critical and super-critical Reynolds number flow past a circular cylinder*, U. of Iowa, PhD Diss.

Yeung, P., et al. (2010). *Turbulence computations on a 4096^3 periodic domain: Passive scalars at high schmidt number and Lagrangian statistics conditioned on local flow structure*, 63rd Annual meeting of the APS Division of Fluid Dynamics, Vol. 55, No. 16, Long Beach, California.

You, D., & Moin, P. (2007). A dynamic global-coefficient subgrid-scale eddy-viscosity model for large-eddy simulation in complex geometries. *Physics of Fluids, 19*(6), 065110.

Zhiyin, Y. (2015). Large-eddy simulation: Past, present and the future. *Chinese Journal of Aeronautics, 28*(1), 11.

Chapter 6
Best Practices of the CFD Trade

"The result was very happy." Osborne Reynolds upon the initial formulation and validation of his dimensionless number, 1883.

Abstract A strong attempt is made to provide practical guidelines for CFD meshes. Dozens of mesh metrics are described in detail, and a mathematically-driven, physics-based set of "golden" mesh metrics is recommended. General CFD boundary and initial conditions are described, including boundary compatibility. Time step, stability, domain, and calculation speed-up guidelines are provided. Detailed guidelines for modeling laminar and turbulent natural circulation are discussed. The chapter concludes with dozens of data visualization recommendations for generating figures, movies, and other presentation media, with the goal of more effectively conveying the CFD results.

It was said in the 1970s with regard to finances that when "E. F. Hutton talks, people listen." Said in an overtly enthusiastic fashion, "If CFD calculations are done correctly, then nature listens!" The correct fluid dynamics and auxiliary equations, coded and applied correctly, will certainly mimic nature, and experimental data will inevitably follow pretest calculations. And under such careful modeling approach, on the uncommon instance when computational and experimental output do not match, there is a strong probability that the error source is experimental, such as faulty pressure gauges, incorrect experimental procedures, and so forth.

The sections that follow endeavor to provide guidelines that increase the likelihood that computational output will accurately reflect system behavior and experimental data. But the converse is also true and too common: when CFD is done incorrectly, without checks and balances, the GIGO acronym becomes valid—garbage in, garbage out. CFD is not to be treated as a black box. In this chapter, numerous guidelines and rules of thumb are provided, with the goal of increasing the computational accuracy of simulations.

© Springer Nature Switzerland AG 2019
S. Rodriguez, *Applied Computational Fluid Dynamics and Turbulence Modeling*,
https://doi.org/10.1007/978-3-030-28691-0_6

6.1   Toward Bullet-Proof Meshes

WLOG consider a 3D system. The mesh represents the system geometry that was parsed (divided) into computational elements or nodes, each reflecting the system behavior of the primitive variables and a numerous set of derived computational quantities. Each finite element or finite volume represents (reflects, maps) a small region of space for the system in question. Each element is in turn comprised of discrete computational points, or nodes, that have no volume. Or, if finite differences are used instead, then the system is directly represented with discrete computational nodes. In any case, the nodal agglomeration represents somewhat abstractly the physical behavior, whereby mathematical computational nodes are used to calculate mass, momentum, and energy to determine the overall system behavior. In the cumulative sense, the entire set of discrete, volumeless computational nodes represent the entire contiguous volume. In this virtual world, each computational node simultaneously behaves as if it were a thermocouple for recording temperature, as a pressure transducer for recording pressure, as a flow meter to obtain mass flow rate, and so forth.

   Without doubt, anyone can develop a mesh. But how is a *quality* mesh developed, and how can its "quality" pedigree be defined and measured? Moreover, what is "mesh quality," and is it not subjective? Fortunately, there are many guidelines that have well-served CFD modelers over the past few decades. In the words of P. M. Knupp (2007):

> Mesh quality concerns the characteristics of a mesh that permit a particular numerical PDE simulation to be efficiently performed, with fidelity to the underlying physics, and with the accuracy required for the problem.

Many mesh metrics are available to quantify and control the quality of a computational mesh. These are summarized in Table 6.1, where over two dozen mesh metrics for hexahedral elements are listed, including their definition and acceptable range. A metric's "acceptable range" is considered as a reasonable "rule of thumb" that generally ensures that a mesh will provide defensible output (or at least, minimizes additional, unacceptable errors!).

   To begin with, it is emphasized that a single mesh metric is akin to a rule of thumb and cannot ever, in of itself, guarantee that a mesh will be "bullet proof." Indeed, there might not even exist a unique set of mesh metrics that will always guarantee bullet proof meshes. Nevertheless, that is the grail being sought here (and a proposed such set will be presented later, along with some guidelines).

   Indeed, the shrewd analysis will demonstrate that the mesh in question satisfies a set of reasonably independent, complimentary mesh metrics, and not just a single metric. That is, multiple mesh aspects must be tested and improved as necessary. For example, having a good aspect ratio says nothing about element angle and vice versa. Thus, it is possible for a mesh to simultaneously have a great aspect ratio and a poor skew angle and so forth. In fact, as mentioned by Knupp, skew is "insensitive to length or aspect ratios," having sole dependence on the element's angles that are formed by the element faces (Knupp 2003). Thus, no *single* (unweighted) mesh

Table 6.1 Mesh metric definition and acceptable range for *hexahedral* elements

Metric	Definition	Acceptable range (units)	References
Angle (minimum)	The smallest element angle formed by the intersecting planes (dihedral angle). Too small of an angle increases numerical stiffness	45–90 (degrees)	Stimpson et al. (2007), Brewer and Marcum (2008), Zigh and Solis (2013)
Angle (maximum)	The largest element angle formed by the intersecting planes (dihedral angle). Too large of an angle increases numerical error	90–135 (degrees)	Stimpson et al. (2007), Brewer and Marcum (2008), Zigh and Solis (2013)
Aspect ratio	The ratio of maximum vs. minimum edge length. The aspect ratio seeks to ensure that quantities such as momentum and heat are transferred appropriately throughout the system	≤ 5 (unitless), can reach up to 10 if gradient is small (e.g., longer length parallel to flow direction, and smaller length perpendicular to the wall). The closer to 1.0, the better	Robinson (1987), Andersson et al. (2012), Cubit (2017)
Condition number	Jacobian matrix condition number based on the maximum value of the four element corners	1–4 (unitless)	Knupp (2000), Stimpson et al. (2007), Cubit (2017)
Distortion	The minimum of the Jacobian determinant times the ratio of the local (transformed) and global (actual) areas. This represents the element surface's deviation from a square	0.4–1.0 (length squared)	SDRC (1988), Lawry (2000), Stimpson et al. (2007), Cubit (2017)
Element area	Jacobian determinant based on the element's center	None (length squared)	Robinson (1987), Cubit (2017)
Element volume	Scaled to the Jacobian determinant magnitude at the element's center; minimum pointwise volume	None (length cubed)	Cubit (2017)
Expansion ratio	Element growth rate between adjacent elements	≤ 1.5 (unitless) The closer to 1.0, the better	Fluent (2012), Zigh and Solis (2013)
Jacobian	The Jacobian matrix relates how the computational variables map linearly onto their spatial location, e.g., the computational nodes. The matrix has geometrical information such as volume, shape, and orientation (Knupp 2001). The Jacobian determinant is calculated to gauge the relative stretching of the local	None (length squared)	Knupp (2000), Stimpson et al. (2007), Cubit (2017)

(continued)

Table 6.1 (continued)

Metric	Definition	Acceptable range (units)	References
	spacing in an element. It is also a measure of the orientation of the surface normals relative to each other. To obtain a relative, measurable metric, it is scaled vs. a perfect element; refer to "Scaled Jacobian" in this table		
Oddy	Oddy represents the largest metric tensor variation in the four corners of an element	0.0–0.5 (none)	Stimpson et al. (2007)
Orthogonal quality	The normalized dot product minimum of the element area vector and the centroid vector based on either the element's face or that of the adjacent element	0.15–1.0 (unitless). A value approaching 0 is unacceptable	Fluent (2018)
Quality index	A code-defined approach to factor the relative impact of a number of mesh metrics into a single metric (e.g., HyperMesh uses 12 different mesh metrics with user-defined weight factors)	The acceptable range is classified as "ideal" and "good," depending on the user-defined weight factors. Suspicious elements are flagged as "warn," while bad elements are tagged as "fail" and "worst"	HyperMesh (2018)
Relative size	J is the weighted Jacobian matrix determinant. Then, the relative size is the minimum of J and its inverse, J^{-1}	0.3–1.0 (unitless)	Knupp (2003), Cubit (2017)
Scaled Jacobian	The scaled Jacobian is obtained by taking the minimum Jacobian and then scaling it by dividing by the length of two element-edge vectors. Scaled Jacobians are used in many CFD codes when inverting system matrices. If the scaled Jacobian is less than 0.5, the calculation may abort; 1.0 refers to a cube (and hence is considered excellent). A value approaching zero implies a highly distorted (and undesirable element). Negative Jacobians refer to	≥ 0.5 (unitless) The closer to 1.0, the better	Knupp (2000), Stimpson et al. (2007), Cubit (2017)

(continued)

Table 6.1 (continued)

Metric	Definition	Acceptable range (units)	References
	inverted, concave, or bowed elements and should be avoided at all cost. Refer to "Jacobian" in this table		
Shape	2 divided by the magnitude of the condition number of the weighted Jacobian matrix	0.3–1.0 (unitless)	Knupp (2003), Stimpson et al. (2007), Cubit (2017)
Shape and size	The product of the shape and the relative size	0.2–1.0 (unitless)	Knupp (2003), Stimpson et al. (2007), Cubit (2017)
Shear	2 divided by the magnitude of the condition number of the Jacobian skew matrix	0.3–1.0 (unitless)	Knupp (2003); Stimpson et al. (2007), Cubit (2017)
Shear and size	The product of the shear and the relative size	0.2–1.0 (unitless)	Knupp (2003), Stimpson et al. (2007), Cubit (2017)
Skew	The maximum of |cos α|, where α represents the angle between the edges at the element's center. For example, a perfect element with 90° angles has cos (90°) = 0, while an element with 60° has a value of 0.5. Thus, the smaller the skew, the better	≤0.5 (unitless) The closer to 0.0, the better	Robinson (1987), Knupp (2003), Stimpson et al. (2007), Cubit (2017)
Skewness	Skewness compares the shape difference between a given element and that of a perfect hexahedral of the same volume. The larger skewness is, the larger the numerical error and the potential for instabilities	<0.9 (unitless); average mesh value should approach <1/3	Fluent (2009, 2012), Andersson et al. (2012)
Squish index	Calculates how much the faces of an element diverge from an ideal, orthogonal face	<0.9 (unitless)	Fluent (2009)
Stretch	$\sqrt{2}$ times the ratio of the element's minimum edge length and the maximum diagonal length	0.25–1.0 (unitless)	FIDAP (1999), Stimpson et al. (2007), Cubit (2017)
Taper	Maximum ratio of element lengths based on opposite sides	0.0–0.7 (unitless)	Robinson (1987), Stimpson et al. (2007), Cubit (2017)
Warp	Cosine of the smallest dihedral angle. That is, this	0.9–1.0 (unitless)	Cubit (2017)

(continued)

Table 6.1 (continued)

Metric	Definition	Acceptable range (units)	References
	represents the angle formed by the element planes that intersect diagonally		
Warpage	1 minus the cosine of the smallest dihedral angle. That is, the angle formed by the element planes that intersect diagonally	0.0–0.7 (unitless)	Stimpson et al. (2007)

Fig. 6.1 Skew vs. aspect ratio over the same region

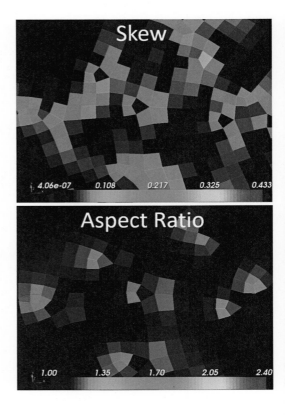

metric can capture all the geometric issues associated with elements, because element geometry involves several exclusive parameters such as length, angle, etc. This point is showcased in Fig. 6.1, which zooms into an airfoil region, showing both skew and aspect ratio. In this situation, though the mesh quality is reasonable, the point is that regions with good aspect ratios have skew magnitudes that are approaching the high limit, and regions with higher skew and higher aspect ratio are not always at the same location. Figure 6.2 shows a zoomed region of the airfoils colored by aspect ratio, with the highest magnitudes occurring where the geometry

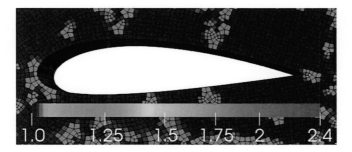

Fig. 6.2 Overlay of mesh colored with aspect ratio

has the sharpest changes. This situation is typical, as regions with the sharpest geometrical edges tend to have the worst mesh metrics. Therefore, those regions merit special attention, in addition to regions where mesh styles and boundaries intersect.

The first modern CFD meshes were boldly generated, without much regard for mesh metrics; who knew at the time that there would be mesh issues? Anyway, it was quickly discovered that mesh guidelines were required to generate defensible output. The first attempts involved a single metric, and in particular, the aspect ratio (Robinson 1987). By the mid-1980s, some researchers recommended aspect ratio, skew, and taper. As of 2019, no unanimous consensus exists as to what qualifies as a sufficient set of mesh metrics (though the literature does express some ideas). To make matters worse, it is also clear that many mesh metrics are not independent of each other (i.e., some are redundant).

Other metrics may include distortion, which is the minimum of the Jacobian determinant times the ratio of the local (transformed) and global (actual) areas. Said in simpler, more geometric terms, this represents the element surface's deviation from a square. Some software packages offer a combination of quality metrics that are weighted and factored into a single metric, such as HyperMesh's "quality index." This software uses a combination of 12 different mesh metrics, including aspect ratio, skew, Jacobian, warpage, and angle (HyperMesh 2018). On the other hand, Fluent recommends skewness, aspect ratio, and squish (a measure for how much the faces diverge from an ideal, orthogonal face) (Fluent 2009). From the point of view of synthesis, this brief mesh metrics overview shows a commonality for certain mesh metrics, such as aspect ratio and skew, as well as a commonality in geometric parameters that should be qualified. In effect, by the start of the twenty-first century, the general notion was, and continues to be, that mesh metrics need to show that elements are not unduly deformed geometrically (Fluent 2009, 2012).

Therefore, excellent, general mesh guidelines should include the following considerations:

- Stretching (length issues)
- Distortion (angle issues)
- Transitioning (distance and propagation issues between adjacent elements)
- Adequate computational variable mapping onto node distribution

Consequently, a complete set of mesh metrics ought to consider length ratios, element angles, distance between adjacent nodes (i.e., a growth ratio), and an approach that gauges the computational variable's mapping onto the node distribution. For these reasons, ideal mesh metric sets (or equivalently similar criterion) ought to include such metrics and justification. One such set, and highly recommended, is the following "grail" mesh metric group:

- Aspect ratio ≤5 (the closer to 1.0, the better; considers length ratios)
- Skew ≤0.5 (the closer to 0.0, the better; factors in element angles)
- Expansion ratio ≤1.5 (the closer to 1.0, the better; gauges node-to-node distance growth between adjacent nodes/elements)
- Scaled Jacobian ≥0.5 (the closer to 1.0, the better; a measure of the computational variable mapping onto the node distribution)

As shown in Table 6.1, there is a bewildering number of mesh metrics, whose intent is to provide guidelines for the generation of quality meshes. Thus, Table 6.1 is not meant to intimidate users but is instead intended to offer many options, especially because CFD tools tend to be associated with diverse metrics, and the user is encouraged to experiment with different guidelines. For illustration purposes, Table 6.2 shows various mesh metric output for a compact heat exchanger with airfoil surfaces. Notice that a more concise mathematical expression for mesh metrics can be pinpointed if the average, minimum, maximum, and standard

Table 6.2 Cubit meshing tool output showing mesh metrics for a compact heat exchanger with airfoil surfaces

Mesh metric	Average	Standard deviation	Minimum/element number	Maximum/element number
Aspect ratio	1.093E+00	9.885E−02	1.000E+00 (95714)	2.405E+00 (51721)
Skew	5.867E−02	5.856E−02	4.058E−07 (553135)	4.335E−01 (368)
Taper	4.620E−02	6.404E−02	3.399E−06 (497420)	4.025E−01 (22026)
Element volume	2.081E−13	3.856E−14	5.344E−14 (107610)	5.336E−13 (10085)
Stretch	9.228E−01	6.818E−02	3.568E−01 (55611)	9.973E−01 (66323)
Diagonal ratio	9.626E−01	3.645E−02	6.949E−01 (56257)	1.000E+00 (664913)
Dimension	3.394E−05	2.307E−06	1.892E−05 (163499)	4.389E−05 (10085)
Condition number	1.024E+00	5.678E−02	1.000E+00 (513435)	1.765E+00 (502167)
Jacobian	1.965E−13	4.031E−14	3.543E−14 (107610)	5.141E−13 (10085)
Scaled Jacobian	9.837E−01	3.562E−02	6.910E−01 (742)	1.000E+00 (384691)
Shear	9.837E−01	3.562E−02	6.910E−01 (742)	1.000E+00 (384691)
Shape	9.793E−01	4.231E−02	6.212E−01 (107610)	1.000E+00 (513435)
Relative size	8.062E−01	1.898E−01	6.596E−02 (107610)	1.000E+00 (92062)
Shear and size	7.974E−01	1.976E−01	5.785E−02 (107476)	9.999E−01 (92062)
Shape and size	7.957E−01	2.004E−01	4.097E−02 (107610)	9.998E−01 (45726)
Distortion	9.392E−01	8.792E−02	4.862E−01 (75966)	1.000E+00 (50308)

Table 6.3 Tabulation of key mesh metrics for laminar and turbulent flows (this table is intended as an aide, to be filled by analyst prior to submitting CFD simulations)

Parameter	Coarse (0.5×)	Medium (×)	Fine (2×)	Very fine (4×)
Number of computational nodes				
Average aspect ratio (desired range: 1.0–5.0)				
Maximum aspect ratio (desired range: 1.0–5.0)				
Average skew (desired range: 0.0–0.5)				
Maximum skew (desired range: 0.0–0.5)				
Minimum scaled Jacobian (desired range: > 0.5)				
Average node spacing growth rate				
$Re =$ (If turbulent, the desired node-to–node distance is $\sim\lambda$ for LES and $\sim\eta$ for DNS)				
Kolmogorov eddy size $=$				
Taylor eddy size $=$				
Integral eddy size $=$				
At $y^+ = 1$, $y =$				
At $y^+ = 7$, $y =$				
At $y^+ = 30$, $y =$				
First computational node for calculation is at $y =$				

deviations are calculated by the meshing package. Though 16 metrics are shown in the table, just 4 are needed per the mesh set recommended above, 3 are highlighted in the table, and the fourth was applied when the mesh was generated, namely, that the expansion ratio ≤1.5. For convenience, Table 6.3 lists key mesh and flow metrics that ought to be calculated *prior* to beginning the CFD analysis; it is intended as a general guide.

While no mesh metric could ever be "perfect" or "universal," mesh metric guidelines should be viewed as more than "rules of thumb" that are intended to save the analyst much grief. The metrics generally have a strong mathematical basis that is strongly founded upon computational principles (Knupp 2003), and as such, are intended to increase fidelity in the computational output. Certainly, some meshes will produce reasonable results even when metrics are ignored, but more often than not, ignoring the guidelines will result in poor output. Some of the most curious results that have been pinned down as a direct consequence of poor meshes include stainless steel that ignited at 400 K, flows exceeding the speed of light, regions with no flow that suddenly accelerate out of nowhere, levitating flows that only Houdini could explain, temperatures below absolute zero, negative densities, and countless other nonsense. If this discussion has not yet generated a healthy dose of caution

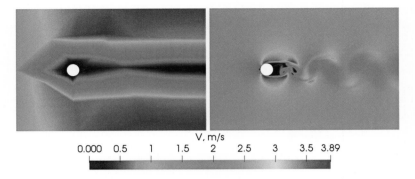

V, m/s

0.000 0.5   1   1.5   2   2.5   3   3.5 3.89

Fig. 6.3 Divergent solutions using noncompliant mesh metrics (LHS) vs. compliant (RHS)

when developing meshes, then consider Fig. 6.3, which shows how drastically solutions can diverge if mesh metric guidelines are not followed. In this situation, flow around a cylinder is considered, whereby the cylinder has $D = 0.1016$ m, $U_\infty = 2.44$ m/s, and $\nu = 4.95 \times 10^{-4}$ m^2/s. The fluid flows from left to right, at $Re = 500$. Two calculations were run, with everything being the same, except that the mesh on the RHS was developed with good mesh metrics (e.g., aspect ratio ≤ 5, skew ≤ 0.5, an expansion ratio ≤ 1.5, and scaled Jacobian >0.5), while the mesh on the LHS was purposely developed with poor mesh metrics (e.g., aspect ratio ≥ 50, skew ≥ 5, expansion ratio ≥ 15, and scaled Jacobian approaching 0). Of course, the image on the RHS is consistent with experimental data.

6.1.1 Additional Mesh Guidelines

- Highly distorted elements will inevitably produce highly distorted output (for hexahedral elements, such elements tend to deviate from a cubical geometry).
- Regions with large gradients should have a higher element density (e.g., the boundary layer near the wall, free surfaces).
- The element face should be as close to perpendicular to the wall. That is, this is one of the worst regions to have skewed elements. For this reason, tetrahedral elements are not recommended at the wall.
- Similarly to the criterion above, the element faces should be close to perpendicular to the main flow direction.
- The longer side of an element should be oriented in the direction of the flow, with the smaller side perpendicular to the large velocity gradient (e.g., a boundary layer, etc.).
- Numerical error is only compounded when elements have multiple, independent poor mesh metrics (e.g., an element having both large aspect ratio and large skew).

- Always ensure that the grid is refined spatially (coarse, medium, and fine meshes).
- The usage of grid convergence index (GCI) and/or Richardson extrapolation greatly increases confidence in the simulation output.

6.1.2   Computational Node Spacing for RANS Models

The spatial distribution of computational nodes in Reynolds-averaged Navier-Stokes (RANS) models is crucial for generating defensible solutions. This is especially so for phenomena-rich calculations involving complex flows, such as strong wall shear, swirl, rotational surfaces, backflows, turbomachinery, drag, and lift.

If no wall function is used, then it is important that RANS CFD meshes have the first computational node at $y^+ = 1$ (Fluent 2012). Generally speaking, this is the case for most RANS models, but there are exceptions. For example, Wilcox recommended that the first computational node be placed at $y^+ = 5$ if using his 2006 k-ω turbulence model (Wilcox 2006). For turbulent, low-Re flows, the first node can be as far as $y^+ = 4$ (Andersson et al. 2012). Mesh biasing can be used to reduce the number of computational nodes, as shown in Fig. 6.4. If biasing is used, it is crucial that the growth rate between nodes not exceed 1.5 (this is the expansion ratio metric; refer to Table 6.1). But, just how large should the node spacing be, especially near the wall? Certainly, RANS is not concerned with individual eddy behavior (though there are a few exceptions, such as the Myong-Kasagi RANS in relationship to Taylor eddies). Therefore, it is not necessary that RANS node spacing be less than or equal to the size of the Taylor and Kolmogorov eddies; in fact, doing so will be detrimental for various reasons. On the other hand, including an additional computational node or two at $y^+ \leq 30$ will help capture the complex turbulence motions occurring in the buffer layer, as explained in Sect. 3.6. In fact, conservative researchers recommend as many as five to ten computational nodes in the region

Fig. 6.4 Biased node distribution based on y^+ location

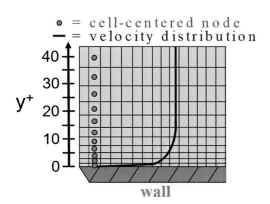

bounded by $y^+ < 20$ (Andersson et al. 2012). And this is certainly consistent with the use of node biasing when the growth factor does not exceed 1.5.

So, the above guidelines specify a minimum spacing limit for RANS. But, what should the maximum spacing limit be? The maximum nodal distance should be decreed by the node spacing needed to achieve sufficient spatial discretization. For example, if cutting the node distance in half only reduces the computational error by less than some reasonable metric acceptable to the analyst (say <1–2%), then the current maximum node spacing is satisfactory.

As noted in Sect. 3.6, the viscous sublayer and the buffer layer flow-physics forge crucial velocity gradients in these narrow regions. For instance, drag and lift are significant at and near the wall. Therefore, being able to model the viscous sublayer in the range of $0 \le y^+ \le 5$ (to as high as 8) is of outmost importance to capture wall effects, and so is the region in the range of $5 \le y^+ \le 30$, where most of the nonisotropic effects occur, where eddy production and decay predominate, and where sizable eddy fluctuations occur.

Therefore, it is not surprising that calculations that fail to include at least one or two computational nodes in the viscous sublayer and at least two in the buffer layer generally fail to deliver, especially if complex flows are involved. Certainly, following these guidelines will result in a large mesh that will require much computational time. There ought to be some give and take in a balance between accuracy and deadlines. Therefore, the analyst in encouraged to run coarser meshes to at least get some preliminary results and then submit the finer meshes as early as possible, to pin down the impact of the mesh discretizations.

6.1.3 Wall Functions

Wall functions are used to more accurately calculate certain behaviors at the wall, such as shear stress, wall friction, wall heat flux, wall temperature, and so forth. And for some turbulence models, they serve as the only way of ensuring that the velocity near the wall is calculated adequately (e.g., SKE); this practice is common, especially when the SKE is used in low-*Re* flows. In addition, wall functions can serve as a means to decrease the number of computational nodes in RANS models, as well as large eddy simulation (LES) and detached eddy simulation (DES) models. In addition, wall functions have been shown to improve calculations for low-*Pr* fluids such as liquid metals (Bna et al. 2012).

Ultimately, though, it is up to the turbulence model developer to demonstrate that as y^+ approaches the viscous sublayer, the turbulence model performance approaches $u^+ = y^+$ without the need to enforce (fudge) the behavior. And such is the case for certain turbulence models, including the 2006 k-ω; indeed, the usage of the 2006 k-ω with a wall function would corrupt the output, as the 2006 k-ω inherently has such asymptotic behavior. Furthermore, wall function usage is counter to the LES modeling philosophy and can have unintended consequences for separated flows (Rodi et al. 1997). Other concerns are listed in Sect. 5.1.3. Of course, wall functions will not work if the flow is detached (separated), because the wall function assumes a linear or log velocity distribution. In such cases, SKE should be avoided, and the 2003 SST, 2006 k-ω, LES, or direct numerical simulation (DNS) ought to be used.

If despite these cautionary notes, a user still decides that a linear wall function must be used, then the first computational node can be at $y^+ < 5$, to as low as 1. If a log wall function is used, then the first computational node can be at $30 < y^+ < 500$, where the lower limit applies for low Re calculations and the higher limit is for higher Re (Zigh and Solis 2013); however, this guideline can be nonconservative, as y^+ in the vicinity of 30 can have large velocity gradients and fluctuations.

6.1.4   Computational Domain Size

Another critical aspect of simulations is to determine the computational space that is sufficiently large; if the computational domain is too small, important phenomena will be missed, but if the domain is too large, the calculation will be needlessly slow. For example, it is important in CFD calculations that the mesh is sufficiently large to capture the domain of the primary flow (clearly), but also keep in mind recirculation, entrainment, any secondary flows, and so forth. For illustration purposes, the top of Fig. 6.5 shows a computational region for an expansion where the recirculation pattern is truncated, while the bottom section shows a domain that is sufficiently large, such that it captures the recirculation pattern. Figure 6.6 shows a similar situation for a jet model. In particular, the LHS conceptually shows what happens if the mesh domain is not large enough to capture the primary jet-flow pattern as it spreads, while the RHS shows a more satisfactory computational domain that is sufficiently large to include entrainment.

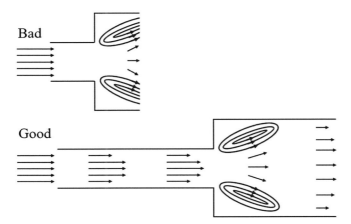

Fig. 6.5 Good and bad mesh domains for a flow expansion

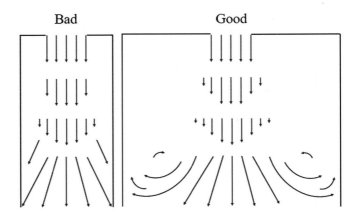

Fig. 6.6 Good and bad mesh domains for a jet

Generally, a visualization tool will rather quickly and intuitively show whether a mesh has an adequate domain or not. Fortunately, there are explicit guidelines that can serve as general guiding principles, without having to submit the calculation first. For example, some rules of thumb for flow around a body with hydraulic diameter D_h are as follows (Tutar and Holdo 2001):

- Distance between the inflow boundary and the body centerline $= 7D_h$
- Distance between the body centerline and the perpendicular side boundaries $= 7D_h$
- Distance between the body centerline and the flow exit boundary $= 15D_h$

Similar guidelines can be found elsewhere for other systems (Leonardi and Castro 2010), implying some sort of domain-size universality.

Finally, the P and u gradients normal to the BC should be relatively small. If not, then it will be necessary to further extend the mesh.

6.2   Boundary and Initial Condition Recommendations

A key to successful CFD modeling includes the proper selection of boundary conditions (BCs) and initial conditions (ICs). For instance, though a simulation with an incorrect BC may run to completion, it will nevertheless yield an incorrect solution. Recall that BCs can generate an infinite family of solution curves, but only the appropriate BCs will provide the correct solution for a given, uniquely specified system. In other words, BCs determine unique solutions for the ordinary and partial differential equations (ODEs and PDEs) found in conservation equations, such as mass, momentum, and energy. Under ideal conditions, a bad BC will immediately cause a code abort and thereby raise red flags that something is wrong. But, under many situations (Murphy's law), the code will be happy to proceed toward the generation of useless data. Besides generating bad output, incompatible BCs will result in other unintended consequences, including solution instabilities and significantly longer computational times. Thus, extra caution is desirable when selecting BCs.

For mature CFD codes, one of the most significant sources of boundary errors arises from the user, via poor meshes and/or BC selection. For these reasons and others, it is incumbent that analysts spend a reasonable amount of time reflecting on BC selection as the input model is developed and tested. Thereafter, the analyst ought to consider running additional simulations with alternative BCs, to explore the system behavior.

BCs are covered in Sects. 6.2.1, 6.2.2 and 6.2.3, while Sect. 6.2.4 discusses IC recommendations.

6.2.1 General BCs

BCs enable unique solutions to be obtained from ODEs and PDEs. BCs can also be viewed as solution drivers because they force the calculation into a unique solution. The number of required BCs for ODEs and PDEs is based on the largest order for each spatial coordinate being solved. For example, a second order ODE in 3D space will require two BCs for each of the three spatial coordinates, for a total of six. Note that BCs are associated with the system behavior at the spatial boundaries, that is, at the system edges, while ICs are specified for the entire domain at a given time point.

Example 6.1 Consider the following ODE, $\frac{d^2T}{dx^2} + \cos(3\pi x)\frac{dT}{dx} - 10T = 0$, where $T = T(x)$ and $0 \leq x \leq L$. How many BCs are needed?

Solution Two BCs are needed to appropriately solve this system because the highest spatial derivative is of order two and only one space coordinate is considered (x-direction). As an example of such system, one possibility is that it has fixed-temperature conditions, e.g., $T(x = 0) = 450$ K on the LHS and $T(x = L) = 600$ K on the RHS.

Example 6.2 Consider a system in 3D Cartesian space, whereby $T = T(x,y,z)$, $0 \leq x \leq L$, $0 \leq y \leq M$, and $0 \leq z \leq N$. The PDE is $\frac{\partial^2 T}{\partial x^2} + \frac{\partial^2 T}{\partial y^2} + \frac{\partial^2 T}{\partial z^2} = 0$. How many BCs are needed?

Solution This system is second order for each of the three space derivatives, and there are three directions. Therefore, six BCs are needed: two in the x-direction, two in the y-direction, and another two to take care of the z-direction.

More formally, BC types for a given spatial domain of length $[a,b]$ are classified as follows:

- Dirichlet (first type). This BC type associates the primitive variables (e.g., u, v, w, T, P, ρ, etc.) with a fixed value or scalar field. For example, $w(a) = 5.5$ m/s.

- Neumann (second type). This BC type associates a derivative with zero, meaning that the spatial gradient is zero. For example, if no heat transfer occurs, then $\frac{dT}{\partial x} = 0$, so the system is adiabatic, which means that $T_{\text{wall}} = T_{\text{node adjacent to wall}}$. In theory, it is impossible to have a perfectly adiabatic boundary, but for engineering purposes, this is a great approximation if the system is reasonably insulated at the boundary.
- Robin (third type). This BC is a superposition (generalization in this case) for the terms associated with a Dirichlet and Neumann BC. For instance, a convective/conductive BC can be represented as $k\frac{dT}{dx} + hT = C_l$.

6.2.2   BC Types in CFD Codes

There are many types of BCs used in commercial CFD codes. *However, it is up to the analyst to justify the usage of all input model BCs and to perform sensitivity studies to quantify their impact on the solution. This is the case because BC input is one of the top sources of modeling errors. Furthermore, boundaries are one of the worst places to have poor mesh metrics.* For this reason, it is highly recommended that analysts use aspect ratios near 1.0 for elements normal to boundaries. Likewise, having skew less than 0.5 is highly recommended and more so if skew approaches 0. A selection of major BC types is discussed next. Special LES and DNS BCs, as well as ICs, are found in Sect. 5.3.

- *Symmetry BC*

A symmetry boundary is ideal for use when the flow field has geometric symmetry *and* the flow is symmetric. (Asymmetric flow can occur in situations with symmetric geometry. For example, boundaries can have different BCs and thus result in asymmetric flow). Where appropriate, use symmetry to reduce the element count. Symmetry BCs are ideal for DNS analysis and are also suitable in the spanwise direction of boundary layer flows, far away from the wall. For example, a free-stream boundary allows the user to model fluid conditions far away from the wall and its boundary layer.

- *Periodic BC*

Periodic boundaries are ideal for repetitive geometry, such as occurs in the periodic flow lattice of a nuclear reactor core and turbines (Rodriguez and Turner 2012; Finn and Dogan 2019; Joshi and Nayak 2019), and are ideal in Fourier space applications (Orszag and Israeli 1974).

- *Inflow and Outflow BC*

An inflow (or outflow) BC allows the user to impose velocity or mass flow rate at the inlet (or outlet). The quantity can be fixed or variable and can be rather complex, involving functions, subroutines, tables, and so forth. More often than not, inflows

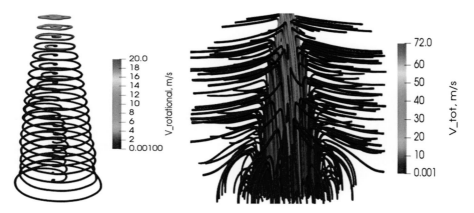

Fig. 6.7 LHS: azimuthal swirl velocity. RHS: total swirl velocity

and outflows should be normal to the BC surface and its elements, but there are exceptions, such as a swirl BC (Rodriguez 2011).

- *Circular Swirl BC*

Some CFD tools already have swirl boundaries, so their application is convenient and straightforward. Otherwise, a *circular* surface boundary with swirl can be approximated as follows (Rodriguez and El-Genk 2010; Rodriguez 2011) and can be included into the CFD code as a user defined function or subroutine. Only the average swirl velocity V_{ave} and swirl angle θ (in degrees) are needed to calculate the u_o, v_o, and w_o velocity magnitudes; the desired average swirl velocity can be easily obtained if the mass flow rate, fluid density, and jet diameter are known. Basically, the total velocity field (RHS of Fig. 6.7) is the superposition of the azimuthal rotation in the x-y plane (u and v; refer to the LHS of Fig. 6.7) plus the axial motion in the z-direction (e.g., the w velocity component). In particular,

$$\vec{V}(x,y,z) = u(x,y,z)\,\vec{i} + \text{v}(x,y,z)\,\vec{j} + w(x,y,z)\,\vec{k}. \tag{6.1}$$

Then, the geometric swirl number S can be approximated as

$$S \approx \frac{2}{3}\tan(\theta). \tag{6.2}$$

At this point, the u_o, v_o, and w_o velocity magnitudes can be calculated as follows:

$$u_0 = \frac{V_{ave}}{\sqrt{2}\sqrt{1 + \frac{4}{9S^2}}}, \tag{6.3A}$$

$$v_0 = -\frac{V_{ave}}{\sqrt{2}\sqrt{1 + \frac{4}{9S^2}}}, \tag{6.3B}$$

and

$$w_0 = \sqrt{V_{ave}^2 - 2u_0^2}. \tag{6.3C}$$

Knowing the velocity magnitudes, then the velocity distributions as a function of x, y, and z can be expressed as follows:

$$u(y) = u_0 \sin\left[2\pi\left(\frac{y - y_{min}}{y_{max} - y_{min}}\right)\right], \tag{6.4A}$$

$$v(x) = v_0 \sin\left[2\pi\left(\frac{x - x_{min}}{x_{max} - x_{min}}\right)\right], \tag{6.4B}$$

and

$$w = w_0. \tag{6.4C}$$

x_{min}, x_{max}, y_{min}, and y_{max} correspond to the minimum and maximum location where the circular swirl boundary occurs. The sine functions distribute the flow at the desired swirl angle and with the correct velocity magnitudes in all three directions.

In this situation, the azimuthal (rotational) velocity is

$$V_\theta = \sqrt{2}u_0. \tag{6.5}$$

Finally, the average swirl velocity can be back-calculated using

$$V_{ave} = \sqrt{u_0^2 + v_0^2 + w_0^2}. \tag{6.6}$$

- *Open BC*

An open boundary allows the fluid to enter or exit the domain, without impacting the interior solution (hopefully!). In theory, such is the case, whereby the momentum distribution within the system is not impacted. In practice, these boundaries can generate spurious reflective waves. Therefore, careful inspection should be undertaken to ensure a defensible solution. The reader is encouraged to consult the literature in this specialized research topic, which includes many areas regarding reflectiveless BCs and wave-mitigation techniques (Thompson 1987; Spalart 1990; Nordstrom, Nordin, and Henningson 1999).

- *Wall BC with No Slip*

Wall BCs are used to define mathematically how the flow is shaped by surfaces. This includes characteristics such as wall roughness, with the "smooth wall" being the general CFD code default. Wall BCs are generally "no slip," meaning that the fluid is attached to the wall (Orszag and Israeli 1974). Therefore, the fluid velocity at the wall matches the wall velocity, which is normally 0; the same situation applies to temperature T.

- *Wall BC with Slip*

Slip is present in situations with small Knudsen number (Kn), $0.01 \leq Kn \leq 0.3$, which is usually in the realm of micro and nanoflows. In such cases, $u_{\text{fluid}} \neq u_{\text{wall}}$ and $T_{\text{fluid}} \neq T_{\text{wall}}$. Rather, the fluid velocity at the wall is greater than zero when slip occurs! It is thought that slip is present as a result of thin, trapped gas sheets that behave as a lubricant between a liquid and the wall surface roughness (Tabeling 2009; Bolaños and Vernescu 2017). If slip is present, a reasonable wall BC for liquid flow is (Tabeling 2009)

$$u = L_{\text{N}} \frac{\partial u}{\partial z} \tag{6.7A}$$

where

$$L_{\text{N}} \approx \frac{\nu}{u_{\text{therm}}} \approx \frac{\nu}{u_{\text{s}}} = \text{Navier length.} \tag{6.7B}$$

In this context,

$\nu =$ kinematic viscosity of the gas trapped between the liquid and the wall

and

$u_{\text{therm}} \sim u_{\text{s}}$ (sound velocity).

Kn is defined as

$$Kn = \frac{\lambda}{\ell_{\text{char}}} \tag{6.8}$$

where

$\lambda =$ mean free path, which is the average distance traveled by particles between collisions and $\ell_{\text{char}} =$ characteristic length of the flow channel.

For liquids, λ is approximately of the magnitude of the liquid's molecular size. For water, $\lambda \sim 0.02$ nm (Wang et al. 2014; Bolaños and Vernescu 2017). Recently, researchers have considered the impact of surface roughness on slip (Bolaños and Vernescu 2017).

Example 6.3 Find the Navier length for water at 30 °C and 1 atmosphere in a tree root that can be approximated as a cylinder. How small can the hydraulic diameter be such that slip occurs?

Solution Water has the following physical properties: $U_s = 1507$ m/s and $\nu = 0.801 \times 10^{-6}$ m^2/s. Hence, $L_N \sim \frac{\nu}{u_s} = \frac{0.801 \times 10^{-6} \text{ m}^2/\text{s}}{1507 \text{ m/s}} = 5.32 \times 10^{-10}$ m. Solving for the smallest hydraulic diameter implies that $Kn < 0.3$. Therefore, $\ell_{\text{char}} = \frac{\lambda}{Kn} \sim \frac{0.2 \times 10^{-9} \text{ m}}{0.3} = 6.67 \times 10^{-10}$ m, which is clearly in the nanoregion.

6.2.3   Compatible Versus Incompatible BCs

When various BCs are used, the following Western classic comes to mind, "The Good, the Bad, and the Ugly." In particular, certain BC combinations for inlets and outlets are generally defensible and reliable, while others are notoriously untrustworthy. This is summarized in Table 6.4.

For convenience, the total static pressure is defined as

$$P_{\text{total}} = P_{\text{static}} + \frac{\rho u^2}{2}. \tag{6.9}$$

Finally, a set of BC conditions must be self-consistent, and along with the PDEs/ODEs being solved, the system must not form an ill-posed problem (Rempfer 2006). This can occur when the BC fails to provide additional independent information, such as when $n + 1$ unknowns are solved with n equations. Additionally, BC

Table 6.4 Inlet and outlet BC combinations

The good...	The bad...	And the ugly
The specification of u at the inlet and static P at the outlet yields reasonable results. These BCs are highly compatible	Usage of total P at the inlet and static P at the outlet can result in numerical instabilities	The specification of *both* inlet and outlet u will result in incorrect velocity distributions. In any case, what is driving the momentum here?
Because mass flow rate is proportional to u, the specification of mass flow rate and static P at the outlet yields reasonable results for incompressible situations	–	The usage of total P at the inlet, while using an outflow BC at the outlet can result in flow instabilities and incorrect momentum calculation
P can be specified at the inlet, so long as the code calculates the exit conditions	–	Usage of mass flow rate at the inlet and an outflow BC at the outlet is not recommended. This is particularly so when ρ is not constant, i.e., when the flow is compressible

intersections, such as the interphase where curves and surfaces meet, can be prone to numerical issues if the intersecting BCs are not compatible. The output from these regions should always be checked with a visualization software.

6.2.4 Initial Conditions

Initial conditions (ICs) are needed to solve initial value problems that march in time as the solution progresses. Analogous to BCs, the number of required ICs for ODEs and PDEs is based on the maximum order of the time derivative. For example, consider a chunk of ice. The IC could state how much ice there is at time zero, so that its mass as a function of time can be determined. Of course, other factors are important, such as how much heat the ice is absorbing; this could be modeled using a BC such as a heat flux or a more detailed Robin BC.

The mathematical utility of an IC is that it specifies the initial quantity (value) of a given parameter, usually at time 0. For example, $T(t = 0) = 310$ K specifies the temperature distribution at time zero for the domain in question.

Often, it is better to set the initial velocities to 0 and just let the code calculate consistent values based on the BCs. Otherwise, there may be inconsistencies between the pressure and velocity field as specified by the ICs and BCs, and these can cause the code to run longer before converging and may even result in solution divergence! In other words, failure to include ICs that are consistent with the values represented by the BCs may cause numerical issues.

Example 6.4 Consider the following PDE:

$$\frac{\partial u}{\partial t} + u\frac{\partial u}{\partial x} + v\frac{\partial u}{\partial x} + w\frac{\partial u}{\partial x} = \frac{\mu}{\rho}\left(\frac{\partial^2 u}{\partial x^2} + \frac{\partial^2 u}{\partial y^2} + \frac{\partial^2 u}{\partial z^2}\right) - \frac{1}{\rho}\frac{\partial P}{\partial x}.$$ How many ICs are required?

Solution This is the Navier-Stokes equation, which has a first-order partial derivative for time. Therefore, only one IC is needed.

Example 6.5 Consider the following Navier-Stokes PDE:

$$u\frac{\partial u}{\partial x} + v\frac{\partial u}{\partial x} + w\frac{\partial u}{\partial x} = \frac{\mu}{\rho}\left(\frac{\partial^2 u}{\partial x^2} + \frac{\partial^2 u}{\partial y^2} + \frac{\partial^2 u}{\partial z^2}\right) - \frac{1}{\rho}\frac{\partial P}{\partial x}.$$ How many ICs are required?

Solution This equation is the steady-state (SS) version of Navier-Stokes, so the time partial derivative is zero, and therefore no ICs are needed.

Example 6.6 Consider the following PDE:

$$\frac{\partial^2 u}{\partial t^2} = c^2\left(\frac{\partial^2 u}{\partial x^2} + \frac{\partial^2 u}{\partial y^2} + \frac{\partial^2 u}{\partial z^2}\right).$$ How many ICs are required?

Solution This is the wave equation with a second order time derivative. Therefore, two ICs are needed.

While not needed for LES or DNS, RANS methods require IC input for various turbulence quantities, depending on the model being used:

- k for the k PDE (e.g., Prandtl k, k-ε, v2-f, SST, and k-ω models)
- ε for the ε PDE (e.g., SKE, realizable k-ε, RNG k-ε, SST, and Myong-Kasagi k-ε models)
- ω for the ω PDE (e.g., SST, Kolmogorov k-ω, 1988 k-ω, 1998 k-ω, and 2006 k-ω models)
- ν_t for Prandtl-Kolmogorov closure models (e.g., SKE) or similar closure models (see Tables 4.2 and 4.3)

These input parameters have a significant impact near open BCs, but not as much far away from inlets and outlets. The best approach is to obtain k, ε, ω, and ν_t from experimental data. But such data is usually not available. The next best approach is to estimate input values using the LIKE algorithm and associated equations; refer to Sects. 3.4 and 3.5. The approach of last resort is to use the CFD code default values. But such values are generalized and are therefore not likely to be suitable for the specific situation of interest.

6.3   RANS Modeling Recommendations

Provided the caveats and limitations are understood, algebraic (zero-equation) models can be used when very fast simulations are desirable. Of these, the Prandtl mixing length, Baldwin-Lomax, and Cebeci-Smith are worthwhile investigating as well (Wilcox 2006).

For one-equation RANS models, the Spalart-Allmaras is highly recommended.

Regarding two-equation RANS, if there is a choice between the 1988, 1998, and 2006 k-ω models, the latter is much superior and is therefore the preferred version. Unfortunately, the coefficients for the Kolmogorov k-ω model are not well documented, and the production term is missing. Therefore, the original Kolmogorov k-ω is not recommended, though some acceptable fixes have been documented in the literature; refer to Sect. 4.6.3.1.

If only k-ε models are available, the Myong-Kasagi is a fairly recent, promising contender with much potential, but does not (yet) have as much usage as the SKE. Yet, because of its promising characteristics (see Sect. 4.6.3.6), it is highly recommended. On the one hand, the realizable and RNG k-ε models resolve various issues associated with the SKE. Unfortunately, issues associated with the ε PDE regarding eddy scales do impact, *to various degrees, all k-ε models*, as discussed in Sect. 4.7 and its subsections; this includes ε maldistribution for Myong-Kasagi and similar models. *However, for the reasons stated in Sects. 4.7, 4.7.1, 4.7.2 and 4.7.3, the SKE is unequivocally not recommended.*

In summary, for great, all-around RANS models, Prandtl mixing length, Baldwin-Lomax, Cebeci-Smith, Spalart-Allmaras, 2006 k-ω, 2003 SST, and Myong-Kasagi k-ε are recommended, while the SKE is not; refer to Chap. 4 for additional details and reasoning behind such choices. The v2-f model is quite useful but encounters stability issues. Table 6.5 summarizes some of the pros and cons of various turbulence models.

Table 6.5 Guidelines toward turbulence model selection

Turbulence model	Pros	Cons
Zero-equation (algebraic models; mixing length hypothesis)	Fastest models, great for getting analytical solutions from PDEs (e.g., Prandtl mixing-length model). Because this is the simplest form of turbulence model, it is the easiest to implement and very robust numerically. The Baldwin-Lomax is good for turbomachinery and aerospace applications, as well as attached, thin, boundary layers	Very limited applicability and successful mostly for very simple flows or specific domains. No transport of turbulent scales (e.g., v or ℓ). Do not use near solid boundaries, unless using a damping function (Van Driest)
One-equation (k-algebraic model)	Fast. Computes one turbulent length scale. Spalart-Allmaras is good for turbomachinery and aerospace applications. The Prandtl k PDE is useful for fast calculations, test cases, and research. Shows surprisingly great results for heat transfer in supercritical fluids (Otero et al. 2018)	Has no transport of the length scale. More limited in terms of physics compared to two-equation models (e.g., k-ε, k-ω)
Two-equation models	Compute the velocity turbulence scale and some other key scale, such as length, time, frequency, etc. Generally provide good results for many flows, with varying degrees of success. Reasonably fast	Shear stress diffusion and nonhomogeneity are not calculated. Cannot compute eddy dynamics
Reynolds stress model (RSM)	Zero-, one-, and two-equation RANS models assume isotropic viscosity via the Boussinesq approximation. RSM is anisotropic and Boussinesq-less. Great for strong swirl, adverse pressure gradients, and anisotropic turbulence	RSM uses six PDEs to solve each of the six independent components in the stress tensor. Its theoretical basis is great, but successes are limited, as the model has many coefficients that require justification. Computationally costly and may experience instability issues
Standard k-ε (SKE)	The most widely used model prior to ~2005. Good for isotropic (high Re) flows, simple flows, plane and radial jets (but NOT round jets), and plumes	Poor results for round jets (round jet anomaly), far wakes, strongly curved surfaces, intermediate to high swirl, flow separation, sudden acceleration, and low-Re. Too dissipative. Has serious eddy scale issues and closure inconsistencies
Renormalization group (RNG) k-ε	Improves the standard k-ε by including a term that reduces dissipation. This improves low Re, separation and swirling flows	Not good for round jets and plumes. Not as stable as the standard k-ε. Has serious eddy scale issues and closure inconsistencies
Realizable k-ε (RKE)	The realizability constraints only yield positive normal stresses, which is an improvement over the SKE. This upgrade improves separated and swirling flows, boundary	Not as stable as the SKE. Has serious eddy scale issues and closure inconsistencies

(continued)

Table 6.5 (continued)

Turbulence model	Pros	Cons
	flows, strong streamline curvature, and round jets. The model is better than RNG for separated flows and secondary flows. RKE also solves the round jet anomaly	
v2-f	Has the same k-ε PDEs as SKE, but dissipation is different. Models the wall region without using wall or damping functions. Good for strong swirl	Some misses at low swirl. Notoriously unstable (tends to abort more often than its peers). Has serious eddy scale issues and closure inconsistencies
2006 k-ω	Great for adverse pressure gradients, separated flows, low to high swirl, turbulent heat transfer, low-Re, and aerospace applications. Solves the round jet anomaly. Uses no wall functions. Wilcox recommends that the first node be at $y^+ < 5$ (vs. 1 for SKE). *Best all-around RANS model*	Requires a fine mesh near the wall, as it does not use wall functions (which is actually a pro)
2003 SST	Great for adverse pressure gradients, separated flows, low to high swirl, turbulent heat transfer, low-Re, and aerospace applications. *Great all-around RANS model*	Because of its reliance on the SKE for high Re, the SST inherited the SKE's eddy scale issues and closure inconsistencies

6.4   Miscellaneous Do's and Don'ts of the Trade

6.4.1   Well-Posed Solutions

Certain criteria must be met to solve fluid dynamics PDEs. This includes the mathematical notion that the problem is "well-posed," meaning that the following three conditions are satisfied: (1) the solution exists (which sounds obvious, but not all problems have solutions, as will be shown later!), (2) the solution is unique, and (3) the solution depends continuously on its boundary and initial conditions. For tough engineering problems that seem to exhibit more than their fair share as "code breakers," it is advisable to question if the problem at hand is well-posed. Are the boundaries consistent? Does a solution exist?

Example 6.7 A well-funded university wishes to perform wind tunnel experiments on a 3D-printed dimpled airfoil at a wide Ma range, from subsonic ($Ma < 1$) to supersonic ($Ma \geq 1.0$). It is desired to run posttest CFD analysis thereafter. Does a numerical solution exist at $Ma = 1$? Determine if this problem is well-posed.

Solution To simplify the analysis, assume that the flow is modeled as SS, inviscid, irrotational, and compressible. In addition, simplify the problem further by considering a 2D Cartesian geometry, with no external heat sources. Because the flow is SS and inviscid, the transient and viscous terms drop, but the convective term remains. For this highly simplified situation, the momentum and energy conservation equations reduce to $\left(1 - Ma^2\right)\frac{\partial u}{\partial x} + \frac{\partial v}{\partial y} = 0$ after several pages of elegant mathematical procedures involving the Prandtl-Glauert rule for linearizing compressible, isentropic flow (Hanson 2012; Pritamashutosh 2014).

Note that many terms in the conservation equations are zero for this idealized situation. In particular, because the flow is SS,

$$\frac{\partial}{\partial t} = 0.$$

Because the flow is inviscid,

$$\nabla^2 V = 0.$$

But most importantly, the flow is compressible ($Ma > 0.3$; refer to Sect. 2.2), and therefore,

$$\vec{\nabla} \cdot \vec{V} \neq 0.$$

And because there are no spatial density gradients for this isentropic, SS system, the conservation of mass PDE is reduced to

$$\frac{\cancel{\partial \rho}}{\cancel{\partial t}} = -\rho\left(\frac{\partial u}{\partial x} + \frac{\partial v}{\partial y} + \frac{\cancel{\partial w}}{\cancel{\partial z}}\right) - \left(u\frac{\cancel{\partial \rho}}{\cancel{\partial x}} + v\frac{\cancel{\partial \rho}}{\cancel{\partial y}} + w\frac{\cancel{\partial \rho}}{\cancel{\partial z}}\right).$$

That is,

$$\frac{\partial u}{\partial x} + \frac{\partial v}{\partial y} = 0.$$

However, the above simplified PDE will not be used further in this analysis because the momentum-energy PDE already provides a useful Ma dependency, $\left(1 - Ma^2\right)\frac{\partial u}{\partial x} + \frac{\partial v}{\partial y} = 0$, that can be exploited to solve this problem, as will be shown later. Therefore, another expression is needed to reach the same number of unknowns as equations.

Now, because the flow is irrotational, the cross product of the velocity is zero. Namely, from the definition of an irrotational flow,

$$\vec{\nabla} \times \vec{V} \equiv \vec{0} = \begin{vmatrix} \vec{i} & \vec{j} & \vec{k} \\ \dfrac{\partial}{\partial x} & \dfrac{\partial}{\partial y} & \dfrac{\partial}{\partial z} \\ u & v & w \end{vmatrix} = \begin{vmatrix} \vec{i} & \vec{j} & \vec{k} \\ \dfrac{\partial}{\partial x} & \dfrac{\partial}{\partial y} & 0 \\ u & v & 0 \end{vmatrix} = \dfrac{\partial v}{\partial x}\vec{k} - \dfrac{\partial u}{\partial y}\vec{k}.$$

Next, use the dot product to multiply the above expression by $\vec{k}$ (because $\vec{k} \cdot \vec{k} = 1$), thereby reducing the PDE to

$$\frac{\partial v}{\partial x} - \frac{\partial u}{\partial y} = 0.$$

By this point, the momentum-energy and irrotational PDEs are sufficient to solve this problem, especially once the elegant eigenvalue method is applied. This closure allows the investigation of any unruly behavior for this seemingly straightforward system of PDE equations. In particular, the two PDEs conform to the following PDE generic classification (DuChateau and Zachmann 2011):

$$\begin{cases} a_1 \dfrac{\partial u}{\partial x} + b_1 \dfrac{\partial u}{\partial y} + c_1 \dfrac{\partial v}{\partial x} + d_1 \dfrac{\partial v}{\partial y} = e_1 \\ a_2 \dfrac{\partial u}{\partial x} + b_2 \dfrac{\partial u}{\partial y} + c_2 \dfrac{\partial v}{\partial x} + d_2 \dfrac{\partial v}{\partial y} = e_2 \end{cases}.$$

Therefore, the two PDEs can be put into matrix form, in anticipation of calculating their eigenvalues, which will allow one to determine if the flow is hyperbolic, parabolic, elliptic, or mixed. If there are two real and distinct eigenvalues, then the solution is hyperbolic. If there is only a single real eigenvalue, the solution is parabolic. Finally, if the eigenvalues are imaginary, then the solution is elliptic.

The two PDEs can be expressed in general matrix-vector format as

$$M \frac{\partial \Psi}{\partial x} + N \frac{\partial \Psi}{\partial y} = E,$$

such that the M and N matrices are associated with the x and y partial derivatives, respectively,

$$M = \begin{bmatrix} a_1 & c_1 \\ a_2 & c_2 \end{bmatrix}$$

and

$$N = \begin{bmatrix} b_1 & d_1 \\ b_2 & d_2 \end{bmatrix}.$$

In this context, the two vectors are

$$\Psi = \left\{ \begin{matrix} u \\ v \end{matrix} \right\}$$

and

$$E = \left\{ \begin{matrix} e_1 \\ e_2 \end{matrix} \right\}.$$

Therefore, the equations can be expressed as follows:

$$M \frac{\partial \Psi}{\partial x} + N \frac{\partial \Psi}{\partial y} = \begin{bmatrix} a_1 & c_1 \\ a_2 & c_2 \end{bmatrix} \left\{ \begin{matrix} \dfrac{\partial u}{\partial x} \\ \dfrac{\partial v}{\partial x} \end{matrix} \right\} + \begin{bmatrix} b_1 & d_1 \\ b_2 & d_2 \end{bmatrix} \left\{ \begin{matrix} \dfrac{\partial u}{\partial y} \\ \dfrac{\partial v}{\partial y} \end{matrix} \right\} = \left\{ \begin{matrix} e_1 \\ e_2 \end{matrix} \right\}.$$

By comparing the PDE generic classification formula with the energy-momentum and irrotational PDEs, the coefficients for the M *and* N matrices are readily shown as

$$M = \begin{bmatrix} 1 - Ma^2 & 0 \\ 0 & 1 \end{bmatrix}$$

and

$$N = \begin{bmatrix} 0 & 1 \\ -1 & 0 \end{bmatrix}.$$

Therefore, the desired system of equations is

$$\begin{bmatrix} 1 - Ma^2 & 0 \\ 0 & 1 \end{bmatrix} \left\{ \begin{matrix} \dfrac{\partial u}{\partial x} \\ \dfrac{\partial v}{\partial x} \end{matrix} \right\} + \begin{bmatrix} 0 & 1 \\ -1 & 0 \end{bmatrix} \left\{ \begin{matrix} \dfrac{\partial u}{\partial y} \\ \dfrac{\partial v}{\partial y} \end{matrix} \right\} = \left\{ \begin{matrix} 0 \\ 0 \end{matrix} \right\}.$$

Fortunately, M is just a 2×2 matrix, so obtaining its inverse is straightforward (Kreyzig 1979); the inversion is performed in preparation for obtaining the system eigenvalues:

$$M^{-1} = \begin{bmatrix} \dfrac{1}{1 - Ma^2} & 0 \\ 0 & 1 \end{bmatrix}.$$

For convenience, let

$$\Phi \equiv M^{-1}N.$$

Then,

$$\Phi = \begin{bmatrix} \dfrac{1}{1 - Ma^2} & 0 \\ 0 & 1 \end{bmatrix} \begin{bmatrix} 0 & 1 \\ -1 & 0 \end{bmatrix} = \begin{bmatrix} 0 & \dfrac{1}{1 - Ma^2} \\ -1 & 0 \end{bmatrix}.$$

Finally, the λ eigenvalues are found by solving the following determinant (Kreyzig 1979),

$|\Phi - \lambda I| = 0$, where I is the identity matrix.

That is,

$$\left| \begin{bmatrix} 0 & \dfrac{1}{1 - Ma^2} \\ -1 & 0 \end{bmatrix} - \lambda \begin{bmatrix} 1 & 0 \\ 0 & 1 \end{bmatrix} \right| = \begin{vmatrix} -\lambda & \dfrac{1}{1 - Ma^2} \\ -1 & -\lambda \end{vmatrix} = 0 = \lambda^2 + \dfrac{1}{1 - Ma^2}.$$

It is evident that the eigenvalues are solely a function of Ma (which is why the coupled momentum-energy PDE was selected in the first place!). The resultant quadratic equation can now be solved for λ:

$$\lambda = \pm \sqrt{\frac{1}{Ma^2 - 1}}.$$

So, how does this system of equations behave? Suppose $Ma < 1$. Then, λ has two imaginary eigenvalues, so the behavior is elliptic. Now suppose $Ma > 1$. Then, λ has two real and distinct eigenvalues, so the behavior is hyperbolic. Thus, there are no issues at this point. But can this system ever be parabolic? No, because the solution expression demands two distinct roots (one positive and one negative for real solutions or the complex conjugate for imaginary solutions). So, the only way to have a *single root* in this case is for the square root expression to be equal to 0, which immediately results in a mathematical contradiction. In any case, because of its hyperbolic and elliptic behavior, this problem has mixed behavior. But, what happens when the hapless CFD engineer is asked to solve the problem for $Ma = 1$? Will a solution exist? Will such CFD calculation cease to abort if the time step is reduced, or more elements are added, or even if another CFD tool is used?

6.4.2 Time Steps, Stability, and CFL

Most modern commercial CFD tools are fully implicit. This means that their numerical methods are unconditionally stable. That is, the method *should* be stable for all time steps. However, this does not mean that very large time steps are encouraged. In fact, too large a time step in an implicit algorithm will increase the numerical error as a result of truncation. By contrast, explicit and semi-explicit methods are easier to program, but should never use a time step larger than the Courant limit; failure to do so will result in numerical instability, large parameter oscillations per time step, nonsensical output, a severe cut in the time step whenever the code automatically attempts to adjust the time step, and inevitable code aborts. Thus, whether a numerical method is implicit or explicit, it is always a good idea to calculate the Courant number, either as a guide to limit truncation error or to avoid instabilities, respectively. The Courant number is also referred as the CFL number, based on the last-name initials of its developers (Courant et al. 1967).

To determine the CFL limit for a 1D PDE, consider the following equation:

$$\frac{\partial \phi}{\partial t} + c \frac{\partial \phi}{\partial x} = 0, \tag{6.10}$$

where:

ϕ = a scalar (e.g., ρ, T, u, v, w, etc.)
c = parameter for the given PDE (e.g., u, etc.)

Suppose $c = u$, meaning that the PDE is a 1D *laminar*, inviscid flow momentum equation. In this case, the 1D CFL limit can be expressed as

$$\mathrm{CFL} = c \frac{\Delta t}{\Delta x} = u \frac{\Delta t}{\Delta x} \leq C_{\max}, \tag{6.11}$$

where:

Δt = time step
Δx = distance between computational nodes in the x direction
$C_{\max}$ is the maximum size of the CFL number, depending on the computational situation, as will be explained shortly

In 2D, the *laminar* CFL is expanded as follows:

$$\mathrm{CFL} = \frac{u \Delta t}{\Delta x} + \frac{v \Delta t}{\Delta y} \leq C_{\max}, \tag{6.12}$$

while in 3D, the *laminar* CFL is

$$\text{CFL} = \frac{u\Delta t}{\Delta x} + \frac{v\Delta t}{\Delta y} + \frac{w\Delta t}{\Delta z} \leq C_{\text{max}}. \tag{6.13}$$

CFL values greater than 1, and up to 5 or so, are acceptable for implicit solvers (Andersson et al. 2012). Some code developers push the envelope even further, using CFL as large as 10, usually as a quick turnaround for testing new code. (Though this approach will likely generate very large truncation errors.) Nevertheless, used with caution, CFL values greater than 1 for implicit codes are desirable, especially for those seeking faster numerical solutions. However, it is up to the analyst to show that temporal discretization is satisfied; that is, the solution converges as Δt is reduced. This can be shown as follows: once the solution converges *spatially* using a given time step Δt_1, a second simulation is run using the same mesh, but with a time step of $\Delta t_1/2$. If the solution does not change appreciably (say <1%), temporal discretization has been reached, at least reasonably so. For the truly obsessed (or diligent!), yet a third simulation can be performed, with the *spatial* distance being cut once more in half ($\Delta x_{\text{converged}}/2$), and the time step is cut in half from the previous simulation (i.e., the time step is now at $\Delta t_1/4$). This can be plotted to show temporal convergence vs. a desired variable, e.g., u, P, etc.

In contrast with implicit solvers, a CFL value of 1.0 or less is necessary for explicit solvers, lest the solution becomes unstable. For an *explicit* application for the *laminar, compressible, inviscid* Navier-Stokes in a 2D Cartesian system (Anderson et al. 1984), the maximum recommended time step is

$$\Delta t_{\text{max}} \leq \frac{1}{\frac{|u|}{\Delta x} + \frac{|v|}{\Delta y} + u_s \sqrt{\frac{1}{(\Delta x)^2} + \frac{1}{(\Delta y)^2}}} \tag{6.14}$$

where

u_s = sound speed.

For convenience, the above expression will be referred as "ATP," in honor of the referenced authors. The ATP expression can be extended onto a *laminar* 3D format as

$$\Delta t_{\text{max}} \leq \frac{1}{\frac{|u|}{\Delta x} + \frac{|v|}{\Delta y} + \frac{|w|}{\Delta z} + u_s \sqrt{\frac{1}{(\Delta x)^2} + \frac{1}{(\Delta y)^2} + \frac{1}{(\Delta z)^2}}}, \tag{6.15}$$

as well as simplified onto the *laminar* 1D expression,

$$\Delta t_{\text{max}} \leq \frac{1}{\frac{|u|}{\Delta x} + u_s \sqrt{\frac{1}{(\Delta x)^2}}}. \tag{6.16}$$

Note that the ATP expressions are for explicit numerics (not implicit), so CFL $\leq$ 1.0. Note as well that such expressions do not depend on viscosity, as Δt_{max} is limited in the above expressions to situations where the viscous effects are negligible compared with the inertial term. As a further note of caution, stability analysis not only depends on the PDE in question but also on the type of numerical discretization used (Courant et al. 1967). Thus, stability analysis tends to be rather ad hoc for specific applications and not necessarily generalizable. Fortunately, there are exceptions, because the physical PDEs tend to follow similar expressions (see Problems 6.11 and 6.12); this is the focus of much research (Courant et al. 1967; Anderson et al. 1984). *Nevertheless, the ATP expressions can serve as guidelines, especially when no other obvious Δt stability criterion/guidance exists. But, there is one final cautionary note: the above expressions are not suitable for turbulent flows!* This situation will be treated though simple approximations that are discussed later in this section.

Continuing on, a close inspection of the CFL criterion and the ATP expressions shows that these limits are specific to convection. So, what happens if stability is dependent on the viscous dissipation, meaning that only so much viscous momentum can be transferred per unit time to guarantee stability? Consider a system where Navier-Stokes has the following form:

$$\rho\frac{\partial u}{\partial t} + \rho\left(u\frac{\partial u}{\partial x} + v\frac{\partial u}{\partial y} + w\frac{\partial u}{\partial z}\right) = \mu\left(\frac{\partial^2 u}{\partial x^2} + \frac{\partial^2 u}{\partial y^2} + \frac{\partial^2 u}{\partial z^2}\right) - \frac{\partial P}{\partial x} + \rho g. \quad (6.17\text{A})$$

Suppose that the viscous (i.e., molecular diffusion) term is larger in magnitude than the convective term and that the pressure and body forces are small. Then, for a 1D *laminar* system,

$$\frac{\partial u}{\partial t} = \nu\frac{\partial^2 u}{\partial x^2}. \quad (6.17\text{B})$$

The above equation can be approximated using forward in time, centered in space (FTCS) finite differences that are second order in space and first order in time. Then, the von Neumann stability analysis method (or some such method) can be applied to determine stability limits. In this case, the numerical error E in both space and time can be modeled as a Fourier series,

$$E(x, t) = \sum_{n=1}^{M} e^{\alpha t} e^{ikx}. \quad (6.18)$$

And, after about one page of elegant algebra, a stable time step is obtained if the following condition is satisfied:

$$\Delta t_{max} \leq \frac{1}{2}\frac{\Delta x^2}{\nu}. \quad (6.19)$$

Again, note that the above stability expression applies for *laminar* flows, not turbulent flows.

Finally, to ensure stability for explicit methods, the minimum time step based on the convective (CFL or ATP) and diffusive bounds should be used, where diffusive refers to molecular diffusion based on viscous effects:

$$\Delta t_{min} \leq \min \left(\Delta t_{\text{convective}}, \Delta t_{\text{diffusive}} \right). \tag{6.20}$$

In preparation for obtaining stability criterion for turbulent flows, note that the diffusion stability expression that was derived using von Neumann analysis can also be derived by using *scaling* analysis. Thus, the transient viscous PDE is transformed according to its key parameters and variables. Namely, the PDE scales (transforms) approximately (to within an order of magnitude) as follows:

$$\frac{\partial u}{\partial t} = \nu \frac{\partial^2 u}{\partial x^2} \quad \leftrightarrow \quad \frac{u}{t} \sim \nu \frac{u}{x^2}. \tag{6.21}$$

Note that scaling is generally easily palatable for engineers but is known to give mathematicians extreme heartburn! In any case, for small time steps and small changes in space, then $t \sim \Delta t$ and $x \sim \Delta x$, respectively; it is assumed that the error in these approximations is small because the changes are linear, i.e., taken near a well-known point and are not deviated much from it. Therefore, substituting the approximations, scaling allows the following approximation:

$$\Delta t_{\text{max,scale}} \sim \frac{\Delta x^2}{\nu}. \tag{6.22}$$

Then, because diffusing more than half of the available mass per unit time step will lead to instabilities, the maximum allowable time step is cut in half and is therefore imposed as a limiting bound, yielding

$$\Delta t_{\text{max,scale}} \leq \frac{1}{2} \frac{\Delta x^2}{\nu}. \tag{6.23}$$

Thus, the same bounding equation is derived for laminar flows, whether scaling arguments are used, or a more rigorous method is applied (i.e., von Neumann stability analysis). In any case, the solution lends some corroborating credibility to the scaling method, and it is therefore now extended onto *turbulent* flows. That is, how can the maximum time step be estimated for a turbulent flow? Lacking much guidance from the literature, the analysis begins by noting that for turbulent flows, $\nu_t \gg \nu$, so the viscous term can be ignored (at least as a first-order approximation). Now, assume a situation where the convective term is sufficiently smaller than the turbulence diffusion term, and considering only the x-direction:

$$\frac{\partial \overline{u}}{\partial t} = 2 \frac{\partial \left(\nu_t \frac{\partial \overline{u}}{\partial x} \right)}{\partial x}. \tag{6.24}$$

Using scaling analysis, the PDE is transformed into:

$$\frac{\overline{u}}{t} \sim 2\nu_t \frac{\overline{u}}{x^2}. \tag{6.25}$$

Finally, in analogy with the laminar convective scaling completed earlier, the limiting bound for turbulent diffusion ought to permit at most half of the available mass to transfer per unit time step. Further, the approximation will consider only small time steps and spatial changes, so that the scaling assumptions are reasonably physical. Hence,

$$\Delta t_{\text{turbulent}} \leq \frac{\Delta x^2}{\nu_t}. \tag{6.26}$$

6.4.3   A Few More Tips

- *Computational time reduction:*
  - Calculational time is *substantially* cut if the problem can be reduced from 3D to 2D, or even 1D. This is perhaps one of the greatest computational time-reducing tricks.
  - Certainly, turbulence is a 3D phenomenon, but it can still be simulated in a 2D domain when the flow tends to be symmetric about the primary flow direction. Furthermore, the largest turbulent fluctuations tend to occur along the primary flow direction.
  - Where appropriate, use symmetry to reduce the element count.
  - Consider removing any system geometry that clearly is not needed (e.g., why include a 2-cm bolt in a 1000 m^3 tank when only CFD is required?). Recall that the finest structures will likely have the smallest node-to-node spacing, and this will drive the computational time step.
  - Reduce the system size, especially in regions that are not as important. For example, regions of interest may already be FD, so modeling the flow as it becomes FD is not necessary. In such cases, an FD flow BC will significantly reduce the computational domain.
  - For implicit calculations, let CFL > 1, to perhaps 2, and as high as 5. This is highly practical for exploratory calculations; but for the final calculation, reduce the CFL to 1 or so, so that truncation error is minimized.
  - Use node biasing to increase the distance between nodes in regions with small velocity gradients. But ensure that the expansion ratio is less than 1.5.

- The proper orthogonal decomposition (POD) method (or similar methodology) (Ly and Tran 2001; Willcox and Peraire 2002) is ideal for situations involving multiple simulations that require slight input variations (e.g., uncertainty quantification, sensitivity studies, minor design changes). In such cases, the first CFD simulation involves no computational time savings, but thereafter, adequate approximations for simulations that took days to weeks to complete would only require a few minutes to complete (Grunloh 2019).
  - Adaptive time stepping can reduce the computational time by a factor of about seven for transients involving sharp, fast pulses (e.g., water hammer) (Flownex 2019).
- In general, higher-order numerical methods provide higher accuracy and therefore require fewer computational nodes (Orszag and Israeli 1974). However, higher-order methods tend to be more unstable because they have less numerical dissipation. Furthermore, the boundary conditions become more complex. For example, fourth-order spatial methods can require so-called ghost boundaries.
- Use iterative solvers (e.g., Gauss-Siedel, Jacobi, and successive over-relaxation) for very large matrices (e.g., systems with a large number of computational nodes, say in the millions to billions, with some systems approaching trillions as of 2019).
- Use preconditioners to transform an unruly matrix so that it is more manageable, thereby making it more amenable toward numerical iteration.
- Preconditioners can be useful under the following circumstances:
  - Unstructured meshes (meshes with irregular patterns)
  - Ill-posed matrices, matrices with large Jacobians, or systems with large condition numbers
  - Multiphase flows
  - Flows with widely varying Ma
  - Meshes with large span (range) with regard to aspect ratio

6.5   Natural Circulation Modeling

Many systems undergo natural convection, which is also known as free convection and natural circulation. This phenomenon typically occurs as a result of temperature gradients that induce density changes in fluids, thereby causing the fluid to flow. Other natural circulation-inducing phenomena include magnetic circulation and species concentration, such as changes in salinity, and so forth. The key point is that under natural circulation, there is no forced circulation, such as occurs from pumps (whether positive displacement or centrifugal), injectors, mechanical devices, and the like.

Natural circulation systems include passive cooling of nuclear reactors under normal and accident conditions (Fernandez-Cosials et al. 2017), bodies of water (ponds, lakes, oceans), weather patterns (thunderstorms, wind, tornadoes, hurricanes), storage tanks, heat exchangers, micro devices, heat pipes, heat sinks, buildings (internal and external circulation flow patterns), etc. Natural circulation flows can be laminar, transitional, or turbulent. And to add more complexity, these flows can be purely natural or involve a degree of extraneous forced flow, in which case the flow is considered as "mixed" circulation. Because natural circulation flows tend to have lower velocities than forced flows and require coupled physics (e.g., energy, magneto hydrodynamics, etc.), they exhibit special numerical challenges.

6.5.1   Natural Circulation Approximation

Consider a Cartesian system under natural circulation, subject to conservation of mass, momentum, and energy, as shown in Fig. 6.8. If the system is 2D (which is a reasonable approximation due to symmetry), then conservation of momentum under *laminar* natural circulation is

$$\rho \frac{\partial u}{\partial t} + \rho \left(u \frac{\partial u}{\partial x} + v \frac{\partial u}{\partial y} \right) = \mu \frac{\partial^2 u}{\partial y^2} - \frac{\partial P}{\partial x} + \rho g \qquad (6.27)$$

where:

u = fluid velocity in the x-direction
v = fluid velocity in the y-direction

Fig. 6.8 Thermal boundary layer generated by a heated wall on the RHS

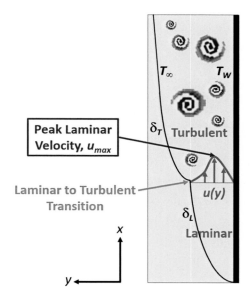

ρ = fluid density
μ = fluid dynamic viscosity
g = gravitational constant in the x-direction

Note that the viscous term in the x-direction is negligible because the u velocity gradient WRT to x is a relatively smaller quantity.

The conservation of energy equation is

$$\frac{\partial T}{\partial t} + \vec{V} \cdot \vec{\nabla} T = \frac{\dot{Q}_0'''}{\rho C_p} + \alpha \nabla^2 T, \tag{6.28}$$

where:

T = fluid temperature
$\vec{V}$ = velocity vector
$\dot{Q}_0'''$ = volumetric heat source
t = time

The fluid thermal diffusivity is defined as

$$\alpha = \frac{k}{\rho C_p}, \tag{6.29}$$

where:

k = fluid thermal conductivity
ρ = fluid density
C_p = fluid heat capacity at constant pressure

The energy equation simplifies under the assumption of a 2D system with no heat source:

$$\frac{\partial T}{\partial t} + u \frac{\partial T}{\partial x} + v \frac{\partial T}{\partial y} = \frac{k}{\rho C_p} \frac{\partial^2 T}{\partial y^2}. \tag{6.30}$$

At this point, the dimensionless Grashof (Gr) number can be defined, which is analogous to Re. In particular, Re represents the degree of laminarity or turbulence under *forced* circulation, while Gr times the Prandtl number (Pr) is a measure of the degree of laminarity or turbulence under *natural* circulation ($GrPr$ = Raleigh number = Ra). For example, for a vertical plate, $GrPr < 10^9$ implies laminar natural circulation, while $GrPr > 10^9$ implies turbulent natural circulation (Holman 1990). In this context, Gr and Pr are defined as

$$Gr = \frac{\beta g h^3 \Delta T}{\nu^2} \tag{6.31}$$

and

$$Pr = \frac{C_p \mu}{k}, \tag{6.32}$$

where h is the wall height.

Here, the volume expansion coefficient (AKA volume expansivity) in units of the inverse of the absolute temperature is defined as

$$\beta \equiv -\frac{1}{\rho}\left(\frac{\partial \rho}{\partial T}\right)_P \approx -\frac{1}{\rho}\frac{(\rho - \rho_\infty)}{(T - T_\infty)} = \frac{1}{\rho}\frac{(\rho_\infty - \rho)}{(T - T_\infty)}. \tag{6.33}$$

Solving for an expression that will later be substituted onto the pressure term in the momentum equation:

$$\rho_\infty - \rho = \rho\beta(T - T_\infty). \tag{6.34}$$

The remainder of the parameters is

$$\Delta T = (T_w - T_\infty), \tag{6.35}$$

T_w = temperature of the wall in contact with the fluid (heats-up the fluid),
T_∞ = fluid temperature far away from the wall,

   and

$$\nu = \frac{\mu}{\rho} = \text{fluid kinematic viscosity.}$$

Note that it is best to find the physical properties at the bulk temperature (AKA mixing cup temperature or flow average temperature), instead of T_∞. The bulk temperature in a given fluid region could be measured if such fluid were thoroughly mixed, resulting in a single, "average" temperature, and hence its utility in calculating physical properties. For example, for a cylindrical tube,

$$T_b = \frac{\int_0^{r_0} (\rho u C_P T)r\,dr}{\int_0^{r_0} (\rho u C_P)r\,dr}. \tag{6.36}$$

The integrals can be obtained with a CFD tool once the simulation has been conducted. Approximations are available in the literature for cylindrical tubes with a uniform heat flux (Lin et al. 2012), though, as a first cut "quick and dirty" approximation, let

$$T_b \approx \frac{T_w + T_\infty}{2} = T_{\text{film}}. \tag{6.37}$$

In any case, whether a rough approximation or a more precise bulk temperature is calculated, the physical properties are determined using the bulk temperature:

$$\nu = \nu(T_b); \beta = \beta(T_b), \text{etc.} \tag{6.38}$$

Continuing on, there is a pressure gradient caused by the fluid's weight per unit length of the fluid:

$$\frac{\partial P}{\partial x} = -\rho_\infty g. \tag{6.39}$$

At this point, it is convenient to substitute the two expressions for pressure and density into the momentum equation (White 1991), as follows:

$$\rho \left(\frac{\partial u}{\partial t} + u \frac{\partial u}{\partial x} + v \frac{\partial u}{\partial y} \right) = \mu \frac{\partial^2 u}{\partial y^2} + g(\rho_\infty - \rho) = \mu \frac{\partial^2 u}{\partial y^2} + \rho g [\beta(T - T_\infty)]. \tag{6.40}$$

Furthermore, the transient term can be dropped if SS is assumed, and dividing by ρ,

$$u \frac{\partial u}{\partial x} + v \frac{\partial u}{\partial y} = \nu \frac{\partial^2 u}{\partial y^2} + \frac{g(\rho_\infty - \rho)}{\rho} = \nu \frac{\partial^2 u}{\partial y^2} + g[\beta(T - T_\infty)]. \tag{6.41}$$

But the above PDE cannot be solved exactly. Fortunately, there are many ways to obtain approximate solutions for this intractable PDE, such as back-of-the-envelope energy balances (White 1991), numerical methods (Ostrach 1953), and extended numerical solutions resulting in a curve-fit solution (Rodriguez and Ames 2015).

A very simplified method for estimating the natural circulation velocity begins by assuming that there is a macroscopic (lumped) balance between the potential and kinetic energy (*PE* and *KE*, respectively) of the fluid (White 1991):

$$\text{PE} = \frac{1}{2} g h \Delta \rho \tag{6.42}$$

and

$$KE = \frac{1}{2} \rho V_{\text{NC}}^2, \tag{6.43}$$

where V_{NC} is the average natural convection velocity. If *PE* is approximately balanced by *KE*, then

$$\frac{1}{2}gh\Delta\rho \approx \frac{1}{2}\rho V_{NC}^2. \qquad (6.44A)$$

Solving for the characteristic fluid velocity,

$$V_{NC} \approx \left(gh\frac{\Delta\rho}{\rho}\right)^{1/2} \approx (gh\beta\Delta T)^{1/2}, \qquad (6.44B)$$

where it is assumed that

$$\frac{\Delta\rho}{\rho} \approx \beta\Delta T. \qquad (6.44C)$$

Various researchers have employed more sophisticated techniques, including polynomial velocity distributions to approximate the PDE solution (Blasius 1908; Holman 1990; Haberman 2004). For example, Holman assumed a cubic polynomial velocity distribution as a function of y and four unknown constants (Holman 1990):

$$u(y) = u(y; c_1, c_2, c_3, c_4) = u\left(c_1 + c_2 y + c_3 y^2 + c_4 y^3\right). \qquad (6.45)$$

It is then possible to solve the unknowns subject to the problem's specific BCs. In this case, the *laminar* velocity distribution u is assumed to be solely a function of y, with no x dependency. This is not a bad approximation, as most of the velocity changes occur in y, as shown in Fig. 6.8. Then, upon applying the BCs, and after a few pages of algebra, the desired velocity distribution is obtained for a *laminar*, natural circulation flow with $Pr \approx 1$ (Holman 1990):

$$u(y) = \frac{\beta\delta^2 g(T_w - T_\infty)}{4\nu} \frac{y}{\delta}\left(1 - \frac{y}{\delta}\right)^2. \qquad (6.46)$$

The laminar velocity solution can be further exploited by taking the derivative of u WRT y and setting it to 0. That is, the peak laminar u in the thermal boundary layer at a given location y can be obtained as follows:

$$\frac{du}{dy} = 0 = \left[\frac{\beta\delta^2 g(T_w - T_\infty)}{4\nu}\right]\left(\frac{1}{\delta} - 4\frac{y}{\delta^2} + 3\frac{y^2}{\delta^3}\right). \qquad (6.47)$$

After about one page of algebra, the desired location for the peak NC velocity is

$$y = \frac{\delta}{3}. \qquad (6.48)$$

In other words, for any location x, there is a maximum peak laminar velocity u, which is always located at $y = \delta/3$. This is a direct consequence of the velocity distribution being parabolic; refer to Fig. 6.8. Therefore, the maximum velocity $u(x)$ for laminar flows with $Pr \approx 1$ is now derived as

$$u_{max}\left(y = \frac{\delta}{3}\right) = \frac{\beta\delta^2 g(T_w - T_\infty)}{4\nu}\left(\frac{4}{27}\right) = \frac{\beta\delta^2 g(T_w - T_\infty)}{27\nu}. \tag{6.49}$$

A more general laminar velocity expression, valid for $0.001 \leq Pr \leq 1000$ (Rodriguez and Ames 2015), is cited here as follows:

$$u_{max}(x) = 2\left(0.5Pr^{-0.11} - 0.24\right)\frac{\nu_\infty}{x}\sqrt{Gr_x}. \tag{6.50}$$

The above expression is based on an extension of Ostrach's ground-breaking work for natural circulation (Ostrach 1953).

An expression for the thermal boundary layer thickness for *laminar* natural circulation is shown as a function of fluid properties and distance along the plate (Holman 1990):

$$\delta = \delta(x) = Cx^{1/4}, \tag{6.51A}$$

where Holman's constant is primarily a function of the fluid physical properties α, β, and ν, while the term in brackets is similar to Gr:

$$C = 3.93\frac{\left(\frac{20}{21} + \frac{\nu}{\alpha}\right)^{1/4}\left(\frac{\alpha}{\nu}\right)^{1/2}}{\left[\frac{\beta g(T_w - T_\infty)}{\nu^2}\right]^{1/4}}. \tag{6.51B}$$

The Gr-like term is the buoyancy parameter times the temperature difference (White 1991) and is very useful for comparing the relative potential of fluids to undergo natural circulation (White 1991).

The laminar boundary layer thickness for *forced* flow parallel to a wall was derived by Blasius over a century ago and compares well with experimental data (Blasius 1908):

$$\delta = \delta(x) = \frac{5x}{\sqrt{Re_x}}. \tag{6.52}$$

6.5.2   Additional Natural Circulation Modeling Guidelines

- Knowing the peak velocity allows the analyst to estimate a reasonable time step. For example, for laminar flows, the CFL limit can be estimated by using the peak velocity discussed in Sect. 6.4.2.

- In general, a reasonable time step for natural circulation flows is typically on the order of tens to hundreds of times larger than forced convection, because natural circulation flows tend to be that much slower.
- Pr has a significant impact on the modeling of natural circulation flows (Ostrach 1953; Holman 1990; Yokomine et al. 2007; Rodriguez and Ames 2015); some guidelines are as follows:

  – Low Pr materials (e.g., liquid metals) cannot be adequately modeled using a fixed turbulent Pr, as the results can diverge from Nusselt number (Nu) experimental data by as much as 50%, and this divergence trend only increases further as Re increases. Nevertheless, this situation can be fixed by using wall functions suitable for low-Pr fluids, along with low-Pr turbulence models such as the k-ω-k_t-ε_t (Bna et al. 2012). Note that Pr_t is generally a function of Re and Pr for low-Pr fluids (Jischa and Rieke 1979; Chen et al. 2013):

$$Pr_t = 0.9 + \frac{182.4}{PrRe^{0.888}}. \tag{6.53}$$

  – For large Pr materials (e.g., oils), the turbulent Pr_t is generally a function of both Re and Pr as well (Hasan 2007; Yokomine et al. 2007):

$$Pr_t = 6.374\,Re - 0.238Pr - 0.161. \tag{6.54}$$

- For codes that allow the user to use different (separate) solvers for mass, momentum, and energy (e.g., Fuego (2016)), it is preferable to choose the same solver for all three PDEs and to use the same convergence criteria. Failure to do so may result in inconsistent solutions and code aborts.
- Because the flow motion depends on density differences that occur near the wall, it is critically important that the mesh near the wall be sufficiently discretized and has good mesh metrics. For turbulent natural circulation flows, $y^+ = 1$ is an ideal starting point. The criteria for laminar, natural circulation flows is not as restrictive vs. turbulent natural circulation flow and is resolvable with meshes having 10–100 times fewer nodes across the boundary layer.
- Natural circulation flows can involve many transitions, including chaotic shifts, oscillations, reversals, bifurcations, and unexpected turbulence behaviors. Therefore, changes in key parameters may result in significantly divergent flows (Gleick 1988; Strogatz 1994).

  – Because of the above situation, small changes in initial conditions may result; under the right conditions, chaotic flows and bifurcations will occur as well (Strogatz 1994).
  – In fact, the famous system of three coupled PDE equations discovered by Lorenz that so beautifully capture the butterfly-like Lorenz attractor was derived directly from conservation of energy and the Navier Stokes equations (Lorenz 1963). This confirms that the seemingly deterministic energy and momentum PDEs have lots of hidden, intrinsic chaotic structure!

6.6 Data Visualization Tips

As discussed in Chap. 1, an extreme advantage of CFD and multiphysics over experiments is that a very large number of computational nodes can be used to model a system, with each node providing information as though it were a thermocouple, a pressure transducer, a flow meter, and so forth. This is certainly not possible experimentally. And of course, this discussion does not yet include CFD's ability to compute properties that are extremely difficult to measure experimentally, that cannot be measured due to physical constraints and the laws of nature, that are too expensive to measure, or that are simply unsafe to measure. More mundane issues include instrumentation that results in unintended parameter changes, such as a thermocouple being an unwanted heat source or sink, or interfering with the flow by blocking or diverting it, and so forth. Or, perhaps the instrumentation was not calibrated, or failed to work properly, or was not incorporated properly. Certainly, good experimentalists will take care of these and other issues, so they could never happen, and experimental data is always perfect, right?

And hence the potential benefits of CFD, if done properly. Needless to say, the human eye will not respond well to reams of CFD numerical data, so what are some ways to most effectively summarize millions or billions of computational data points? Because data visualization is a science and an art form that is embedded with diverse and conflicting human factors, there is no ultimate consensus as to how to generate excellent images, figures, charts, etc. (Sanders and McCormick 1987). But, speaking in general terms, computational output imagery should be focused, clear, legible, self-contained, and show a compelling point or story; having eye appeal ("eye-candy") is a definite bonus. That said, there are general, useful tips that blend various human factors, data display, and art forms. Some of these are suggested, while others are highly recommended.

- More often than not, it is recommended that the system geometry (or a cut-out section) be included in a figure, so use overlay as much as possible. Overlay refers to the superposition of two or more images, and in this context, this is the overlaying of key surfaces and some form of parameter color rendering. For example, the upper image on the RHS of Fig. 6.9 shows a velocity distribution for flow around six cylinders, but it is difficult to "see" the flow distribution. By overlaying the six tubes with translucent coloring (e.g., "volume" rendering in ParaView), the flow pattern comes alive and makes more intuitive sense, as shown in the lower LHS image. The overlay of streamlines and cylinders is shown in the lower RHS of Fig. 6.9 as an improved display. If necessary for a clearer view of the velocity distribution, set the solid body's opacity to 5–20%.
- Visualization can be used to show how spatial convergence is coming along and might even point to meshing issues. This is achieved by overlaying the computational mesh onto the parameter color display (whether it is T, u, v, w, etc.). In particular, meshes that are sufficiently discretized will show that the element coloring by parameter is independent of the mesh grid pattern—that is, the parameter values (by color) must not follow (hug) the computational elements.

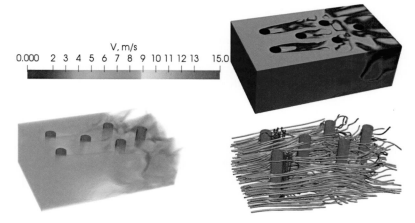

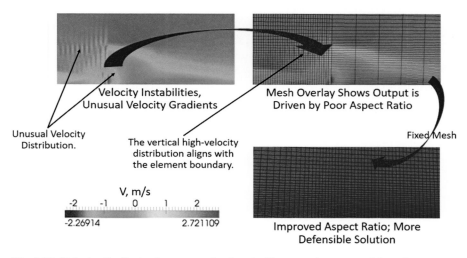

Fig. 6.9 Velocity distribution around vertical cylinders using various overlay schemes

Fig. 6.10 Velocity distribution in a rectangular domain. Upper two images: mesh has a large aspect ratio and large expansion ratio. Bottom: mesh with better mesh metrics

For example, notice that the upper LHS of Fig. 6.10 has various unusual vertical velocity streaks; this image provides few clues as to the source of the problem. However, once the mesh grid is overlaid onto the colored velocity distribution (upper RHS), it is immediately evident that the velocity distribution follows (is dependent upon) the element boundaries. Moreover, the overlay shows that the mesh has a large aspect ratio on the RHS of the domain and a large expansion ratio at the interface between the small and large elements (central region). Once better mesh metrics are applied to the model, the velocity distribution is shown to be independent of the element boundaries (see the lower RHS), where the baffling velocity distribution vanishes.

- It is often useful to include multiple sets of parameters on a single image (e.g., single figure). For example, if the image is split into two screens, then T can be shown on one side and perhaps u on the other and so forth. This juxtaposition of computational parameters can provide many useful engineering insights and behavior correlations.
- A significant portion of the population is color blind, so do not count on color to provide the entire exposition of data (Sanders and McCormick 1987). To avoid this issue, not only consider using assorted colors for the curves but also use symbols (e.g., triangles, circles, etc.) and diverse types of curves (e.g., dashed, dotted, etc.). The curves and symbols should be much larger than size 1 and more likely should be on the order of 3 to 7.
- Consider using log scales (both for curves and coloring by parameter) when there is a broad range in data parameter space.
- Use arrows and brief descriptions to identify key changes in parameters. Many programs can do this, including MATLAB and ParaView. However, the process can also be accomplished by copying the figure onto PowerPoint, overlaying arrows and comments on the slide, and then generating the final image through a screen shot, e.g., Any Capture software.
- Note that screen shot software generate images at various resolutions, such as .tif, .png, .bmp, and so forth. Always select the highest possible resolution.
- Certainly, too many arrows, comments, and "bells and whistles" can be distracting and even detrimental. Again, this is an art and a science. The point is to *focus* on the narrative that a figure should convey. If too much information is required, consider using several figures instead of one, employing a unified theme per figure.
- It is wise to spend an extra 5 minutes per figure to check for errors and pesky, ubiquitous typos, as well as legibility. To make matters worse, this is an area where spell checkers are not typically employed, especially for binary files. Therefore, check the units. Make sure the font is still readable in the document (not just on the image that was created!). The same applies to the curves; are they distinguishable? If possible, have a colleague inspect your figure(s); do they make sense? Does your image convey the story you need to express?
- For extremely appealing, artful images, consider using opposing colors, such as they appear on color wheels. For example, blue and orange go well together (i.e., when included in a figure where they are adjacent to each other), and so does green and red; ditto cyan and light green; etc.
- In contrast to the previous point, avoid using, on a side by side basis, colors that are adjacent on the color wheel. For example, blue and purple do not go well near each other, ditto yellow and light orange or light blue and green, contrast is important.
- Do not use yellow on white! Images require sharp contrast.
- Remember that the data will likely be viewed by important people with diverse backgrounds, including engineers of many types, managers, administrators, students, lawyers, and financiers with deep pockets. Therefore, terms that make sense to a civil engineer may not mean much to a nuclear engineer and vice versa.

A well-thought-out figure must cross language barriers and ought to appeal to a wider audience.

- Avoid using acronyms in figures, and always define them in the associated documents. (As a side note, some acronyms have diverse meanings, depending on the field.).
- A figure should be "self-contained," meaning that in of itself, it must contain all the information necessary for the audience to understand its message. For example, do not assume that a deeply buried paragraph in the report justifies the missing information that rightfully belongs in the figure as well. This includes well-described legends, titles, captions, etc.
- Images should have font size that is at least 20 or higher. Do not ruin a great computational effort with legends that are unreadable (or not present, as is sometimes the case).
- The legend should clearly label all curves and allow the reader to fully understand the parameter range under consideration.
- Indeed, "a picture is worth a thousand words." Make it count!

6.7 Problems

6.1 Is it ever acceptable to use only one mesh metric? Why or why not?

6.2 What is an acceptable minimum set of independent mesh metrics, and why?

6.3 Can a mesh metric replace the aspect ratio? Why or why not?

6.4 Explain the CFL number, and what are acceptable CFL values? Why is the CFL magnitude different for explicit and implicit solvers?

6.5 For an explicit calculation, is it sufficient to only check the CFL criteria? If not, what else should be checked? (Hint: which momentum term is associated with CFL?)

6.6 Choose an article from the journal of your choice and compare its figures with the guidelines listed in Sect. 6.6. What was done right and what could be improved?

6.7 Show that for the *laminar* region with the following velocity distribution, $U(y) = \frac{\beta \delta^2 g(T_w - T_\infty)}{4\nu} \frac{y}{\delta}\left(1 - \frac{y}{\delta}\right)^2$, the peak velocity is $u_{max} = \frac{\beta \delta^2 g(T_w - T_\infty)}{27\nu}$. Plot the peak velocity vs. the boundary layer thickness. The fluid is nitrous oxide at $T_\infty = 332$ K and $P = 1$ MPa. $T_w = 500$ K. Hint: would computing Gr help?

6.8 A system in 3D Cartesian space has $V = V(x,y,z)$, and its domain is $0 \le x \le L$, $0 \le y \le M$, and $0 \le z \le N$. The PDE is $\frac{\partial^2 u}{\partial x^2} + \frac{\partial^2 v}{\partial y^2} + \frac{\partial^2 w}{\partial z^2} = 0$. How many BCs are needed, and what types of BCs would be needed to solve the equation?

6.9 A system in 3D Cartesian space has $V = V(x,y,z)$. The domain is bounded by $0 \le x \le L$, $0 \le y \le M$, and $0 \le z \le N$. The mass conservation PDE is

$\frac{\partial \rho}{\partial t} = -\rho \left(\frac{\partial u}{\partial x} + \frac{\partial v}{\partial y} + \frac{\partial w}{\partial z} \right) - \left(u \frac{\partial \rho}{\partial x} + v \frac{\partial \rho}{\partial y} + w \frac{\partial \rho}{\partial z} \right).$ How many BCs and ICs are needed to solve the equation?

6.10 Water at 400 K and 6 MPa is flowing inside a cylindrical duct ($D = 0.1$ m) at a mass flow rate of 50 kg/s. If an implicit solver is used with CFL = 5 and the first computational node is at $y^+ = 1$, what is the expected time step?

6.11 Consider the following Cartesian 3D energy equation:

$$\frac{\partial T}{\partial t} + u \frac{\partial T}{\partial x} + v \frac{\partial T}{\partial y} + w \frac{\partial T}{\partial z} = \frac{k}{\rho C_p} \left(\frac{\partial^2 T}{\partial x^2} + \frac{\partial^2 T}{\partial y^2} + \frac{\partial^2 T}{\partial z^2} \right) + \frac{\dot{Q}_0'''}{\rho C_p}.$$

Assume no heat source and SS. What is the heat convection stability criteria? Hint: note the similarity between the energy and the u-momentum conservation equation, with no external pressure source, no body force (e.g., negligible g), and the viscous term is much smaller than the convective term:

$$\rho \frac{\partial u}{\partial t} + \rho \left(u \frac{\partial u}{\partial x} + v \frac{\partial u}{\partial y} + w \frac{\partial u}{\partial z} \right) = \mu \left(\cancel{\frac{\partial^2 u}{\partial x^2}} + \frac{\partial^2 u}{\partial y^2} + \frac{\partial^2 u}{\partial z^2} \right) - \cancel{\frac{\partial p}{\partial x}} + \cancel{\rho g}.$$

6.12 Consider the same situation for the energy equation as in Problem 6.11, except that now, the diffusive term is much larger than the convective term. What is the diffusive stability limit?

6.13 Which metric might be better, the condition number or the scaled Jacobian? Discuss pros and cons. What is the mathematical/computational basis for these parameters? For example, consider (Knupp 2003), and recall that mesh metrics are generally based on strong mathematical/computational foundations.

6.14 Consider a lead bismuth eutectic (LBE) at 700 K and 1 atmosphere flowing at mass flow rate of 100 kg/s. $Re = 5000$ inside a pipe. Assume that the first computational node is placed at $y^+ = 1$. Furthermore, assume that a node-to-node growth rate of 1.5 is used. How many computational nodes will be included in the viscous sublayer and the buffer layer? How does this nodalization distance compare with λ and η?

6.15 Consider a smooth flat plate under isothermal boundary-layer flow. The plate is 0.1 m long and 0.05 m wide, with air flowing parallel to it at 300 K and 1 atmosphere ($v = 1.58 \times 10^{-5}$ m²/s and $U_s = 347.3$ m/s). The air flows from left to right along the 0.1 m plate at a constant velocity $U_\infty = 15.8$ m/s. Will a turbulence model be needed? Develop a mesh with aspect ratios ≥ 50 and skew ≥ 5. Use biasing near the wall, with an expansion ratio ≥ 15. Use your CFD tool of choice and compare the solution with another mesh that has aspect ratio ≤ 5, skew ≤ 0.5, and an expansion ratio ≤ 1.5. What happened here?

6.16 An engineer is tasked with designing a vertical nuclear reactor where the heat source is based on uranium fuel enclosed within aluminum flat plates that are 2.5 m wide, resulting in a wall temperature at 850 K. The liquid bismuth coolant enters the bottom at 600 K and 1 atmosphere and flows at a mass flow rate of 50 kg/s as a result of natural circulation. What is the required minimum plate height h such that the fluid marginally enters the turbulent regime, $GrPr = 1.0E9$? Estimate the average velocity and the peak turbulent velocity at that point. What are the integral, Taylor, and Kolmogorov length scales? Which turbulence models are advisable, and what is the value of y for $y^+ = 1$ and 30?

References

Anderson, D. A., Tannehill, J. C., & Pletcher, R. H. (1984). *Computational fluid mechanics and heat transfer*. New York: Hemisphere Publishing Corporation.

Andersson, B., et al. (2012). *Computational fluid dynamic for engineers*. Cambridge University Press.

Blasius, H. (1908). Grenzschichten in Flussigkeiten mit kleiner Reibung. *Zeitschrift für Angewandte Mathematik und Physik, 56*(1). Also available in English as, "The Boundary Layers in Fluids with Little Friction", National Advisory Committee for Aeronautics, Technical Memorandum 1256, 1950.

Bna, S., et al. (2012, June). *Heat transfer numerical simulations with the four parameter k-ω-k_t-ε_t model for low-Prandtl Number liquid metals*, XXX UIT Heat Transfer Conference, Bologna.

Bolaños, S. J., & Vernescu, B. (2017). Derivation of the Navier slip and slip length for viscous flows over a rough boundary. *Physics of Fluids, 29*, 057103.

Brewer, M. L., & Marcum, D. (2008). *Proceedings of the 16th international meshing roundtable*. Springer.

Chen, F., et al. (2013). Investigation on the applicability of Turbulent-Prandtl-Number models for liquid Lead-Bismuth Eutectic. *Nuclear Engineering and Design, 257*, 128.

Courant, R., Friedrichs, K., & Lewy, H. (1967, March). On the partial difference equations of mathematical physics. *IBM Journal*. (English version; the German version first appeared in Mathematische Annalen, Vol. 100, in 1928).

Cubit. (2017). *CUBIT 15.3 user documentation*, SAND2017-6895 W, Sandia National Laboratories. Also available at https://cubit.sandia.gov/public/15.3/help_manual/WebHelp/cubithelp.htm. Accessed on 27 June 2018.

DuChateau, & Zachmann. (2011). *Partial differential equations* (3rd ed.). New York: Schaum's Outlines, McGraw-Hill.

Fernandez-Cosials, K., et al. (2017). Three-dimensional simulation of a LBLOCA in an AP1000 containment building. *Energy Procedia, 127*, 235–241.

FIDAP Version 8.52, Theory and user's manual, Fluent, Inc., 1999. (FIDAP uses the FIMESH meshing software.)

Finn, J. R., & Dogan, O. N. (2019). *Analyzing the potential for erosion in a supercritical CO2 turbine nozzle with large eddy simulation*, Proceedings of the ASME Turbo Expo, GT2019-91791.

Flownex. (2019, May 15). Transient simulations run 7 times faster in upcoming 2019 Flownex release, e-mail correspondence.

Fluent. (2009). ANSYS FLUENT 12.0 user's guide.

Fluent. (2012). Best practice guidelines, Lecture 10, ANSYS.

Fluent. (2018). ANSYS FLUENT 16.2.3 User's Guide. Also available at https://www.sharcnet.ca/Software/Ansys/16.2.3. Accessed on 27 June 2018.

Fuego (2016). SIERRA low Mach module: Fuego user manual – Version 4.40. Albuquerque: Sandia National Laboratories.

Gleick, J. (1988). *Chaos*. New York: Penguin Books.

Grunloh, T. (2019). *Accelerate CFD high fidelity fluid dynamics in a fraction of the time*. Illinois Rocstar.

Haberman, R. (2004). *Applied partial differential equations with fourier series and boundary value problems* (4th ed.). Upper Saddle River: Pearson Prentice Hall.

Hanson, R. (2012). *Aerodynamics*. University of Toronto Institute for Aerospace Studies, AER307.

Hasan, B. O. (2007). Turbulent Prandtl Number and its use in prediction of heat transfer coefficient for liquids. *Nahrain University, College of Engineering Journal (NUCEJ), 10*(1), 53.

Holman, J. (1990). *Heat transfer* (7th ed.). New York: McGraw-Hill.

HyperMesh. Element quality and checks, https://altairhyperworks.com/product/HyperMesh. Can also access at https://www.scribd.com/doc/6675303/Hypermesh-Quality-Tutorials. Accessed on 26 June 2018.

Jischa, M., & Rieke, H. B. (1979). About the prediction of turbulent Prandtl and Schmidt Numbers from modeled transport equations. *International Journal of Heat and Mass Transfer, 22*, 1547.

Joshi, J., & Nayak, A. (2019). *Fluid dynamics in nuclear reactor design and safety assessment* (1st ed.). Duxford: Woodhead Publishing.

Knupp, P. M. (2000). Achieving finite element mesh quality via optimization of the Jacobian matrix norm and associated quantities. *International Journal for Numerical Methods in Engineering, 48*, 1165.

Knupp, P. M. (2001). Algebraic mesh quality metrics. *SIAM Journal on Scientific Computing, 23* (1), 193.

Knupp, P. M. (2003). Algebraic mesh quality metrics for unstructured initial meshes. *Finite Elements in Analysis and Design, 39*, 217.

Knupp, P. M. (2007). *Remarks on mesh quality*, 45th AIAA aerospace sciences meeting and exhibit. Also available as Sandia National Laboratories SAND2007-8128C.

Kreyzig, E. (1979). *Advanced engineering mathematics* (4th ed.). New York: Wiley.

Lawry, M. H. (2000). *I-DEAS student guide*, Structural Dynamics Research Corporation, Structural Dynamics Research Corporation, Can also be found under https://www.scribd.com/doc/46303903/I-DEAS-Student-Guide.

Leonardi, S., & Castro, I. P. (2010). Channel flow over large cube roughness: A direct numerical simulation study. *Journal of Fluid Mechanics, 651*, 519.

Lin, M., Wang, Q. W., & Guo, Z. X. (2012). A simple method for predicting bulk temperature from tube wall temperature with uniform outside wall heat flux. *International Communications in Heat and Mass Transfer, 39*, 582.

Lorenz, E. (1963). Deterministic nonperiodic flow. *Journal of the Atmospheric Sciences, 20*, 130.

Ly, H. V., & Tran, H. T. (2001). Modeling and control of physical processes using proper orthogonal decomposition. *Mathematical and Computer Modelling, 33*, 223.

Nordstrom, J., Nordin, N., & Henningson, D. (1999). The fringe region technique and the Fourier method used in the direct numerical simulation of spatially evolving viscous flows. SIAM, *Journal of Scientific Computing, 20*(4), 1365.

Orszag, S. A., & Israeli, M. (1974). Numerical simulation of viscous incompressible flows. *Annual Review of Fluid Mechanics, 6*, 681.

Ostrach, S. (1953). *An analysis of Laminar free-convection flow and heat transfer about a flat plate parallel to the direction of the generating body force*, Report 1111, National Advisory Committee for Aeronautics, NASA.

Otero R. G. J., Patel, A., & Pecnik, R. (2018). *A novel approach to accurately model heat transfer to supercritical fluids*, The 6th international symposium – supercritical CO2 power cycles.

Pritamashutosh. (2014). Differential equation of motion for steady compressible flow, https://pritamashutosh.wordpress.com/2014/02/28/differential-equation-of-motion-for-steady-compressible-flow. Accessed on 3 Sept 2018.

Rempfer, D. (2006). On boundary conditions for incompressible Navier-Stokes problems. *ASME Applied Mechanics Reviews, 59*, 107.

Reynolds, O. (1883). An experimental investigation of the circumstances which determine whether the motion of water shall be direct or sinuous, and of the Law of Resistance in parallel channels.

Philosophical Transactions of the Royal Society of London, 174. (Can also be obtained through JSTOR, http://www.jstor.org).

Robinson, J. (1987). CRE method of element testing and the Jacobian shape parameters. *Engineering with Computers, 4,* 113.

Rodi, W., et al. (1997). Status of the large eddy simulation: Results of a workshop. *Transactions of the ASME, Journal of Fluids Engineering, 119,* 248.

Rodriguez, S. (2011, May). Swirling Jets for the Mitigation of Hot Spots and Thermal Stratification in the VHTR Lower Plenum, PhD Diss., University of New Mexico.

Rodriguez, S., & Ames, D. (2015, November). *Design optimization for miniature nuclear reactors,* American Nuclear Society, Winter Meeting.

Rodriguez, S., & Turner, D. Z. (2012). *Assessment of existing Sierra/Fuego capabilities related to Grid-To-Rod-Fretting (GTRF).* Sandia National Laboratories, SAND2012-0530.

Rodriguez, S., & El-Genk, M. S. (2010). *Cooling of an isothermal plate using a triangular array of swirling air jets.* American Society of Mechanical Engineers, IHTC14-22170.

Sanders, M. S., & McCormick, E. J. (1987). *Human factors in engineering and design* (6th ed.). New York: McGraw-Hill Publishing Co.

SDRC. (1988). *I-DEAS user's guide* (Vol. 1 and 2). Structural Dynamics Research Corporation.

Spalart, P. R. (1990). Direct numerical study of crossflow instability, laminar-turbulent transition, IUTAM Symposium.

Stimpson, C. J., et al. (2007). *The verdict geometric quality library.* Sandia National Laboratories, SAND2007-1751.

Strogatz, S. H. (1994). *Nonlinear dynamics and chaos.* Cambridge: Westview Press.

Tabeling, P. (2009). *Introduction to microfluidics.* Oxford: Oxford University Press.

Thompson, K. W. (1987). Time dependent boundary conditions for hyperbolic systems. *Journal of Computational Physics, 68,* 1.

Tutar, M., & Holdo, A. E. (2001). Computational modeling of flow around a circular cylinder in sub-critical flow regime with various turbulence models. *International Journal for Numerical Methods in Fluids, 35,* 763.

Wang, G. R., Yang, F., & Zhao, W. (2014). There can be turbulence in microfluidics at low Reynolds Number. *Lab on a Chip, 14,* 1452.

White, F. (1991). *Viscous fluid flow* (2nd ed.). New York: McGraw-Hill.

Wilcox, D. C. (2006). *Turbulence modeling for CFD* (3rd ed.)., printed on 2006 and 2010.

Willcox, K., & Peraire, J. (2002). Balanced model reduction via the proper orthogonal decomposition. *AIAA Journal, 40*(11), 2323.

Yokomine, T., et al. (2007). Experimental investigation of turbulent heat transfer of high Prandtl number fluid flow under strong magnetic field. *Fusion Science and Technology, 52,* 625.

Zigh, G., & Solis, J. (2013). *Computational fluid dynamics best practice guidelines for dry cask application.* USNRC, NUREG-2152.

Chapter 7
Listing for Turbulence Programs, Functions, and Fragments

My precious.

—Gollum; J. R. R. Tolkien, 1937

Abstract A compilation of MATLAB scripts and functions is provided to enable the user to quickly estimate key laminar and turbulence parameters. This includes the LIKE algorithm, as well as a function to calculate physical properties for various molten metals. Another MATLAB script calculates the peak velocity for laminar and turbulent flows under natural circulation. The script also calculates the convective heat transfer coefficient, *Pr*, *Gr*, *Nu*, and *Ra*. Finally, a FORTRAN program for the Prandtl one-equation turbulence model is provided. The program is intended as an example to show how RANS models can be coded, as well as to provide some practical guidelines. The program couples the momentum equation with the k PDE to solve a 2D Couette flow. For user convenience, the program writes MATLAB files suitable for plotting k, ε, ν_t, and $\overline{u}$.

This chapter contains a compilation of MATLAB scripts, functions, and fragments that are designed to quickly estimate key laminar and turbulence parameters. The files are Yplus_LIKE_Eddy_Scales.m, PR_ETA_FPRIME.m, and NaPbBiLBE_FUNC.m. The chapter also includes a FORTRAN program for the Prandtl one-equation turbulence model. Clearly, more elegant coding is possible, and this is perhaps graciously offset by the coding's utility.

For coding that calls REFPROP, make sure that the necessary auxiliary scripts are included in the directory where the files are run, such as refpropm.m, refprp64.dll, REFPRP64_thunk_pcwin64.dll, rp_proto.m, rp_proto64.m, etc. These auxiliary scripts act as an interface between MATLAB and REFPROP, to allow MATLAB to retrieve physical properties that are calculated by REFPROP. The interested reader should consult https://refprop-docs.readthedocs.io/en/latest. It appears that more recent versions of MATLAB (circa Version 10 or so) may have integrated the REFPROP interface auxiliary scripts, thereby eliminating the need to obtain the auxiliary REFPROP files.

© Springer Nature Switzerland AG 2019
S. Rodriguez, *Applied Computational Fluid Dynamics and Turbulence Modeling*,
https://doi.org/10.1007/978-3-030-28691-0_7

7.1   The LIKE Algorithm, y^+ Calculator, and Eddy Scale Calculator

The LIKE algorithm is ideal for estimating four key turbulence properties, and in turn, these can be used to estimate many other useful turbulence variables.

The four variables are as follows:

$L = \ell$ = integral eddy length scale, which is approximately a large eddy length
$I = I_t$ = turbulence intensity
$K = k$ = turbulent kinetic energy
$E = \varepsilon$ = turbulent dissipation

Furthermore, because of the information calculated from LIKE, the script readily calculates the integral, Taylor, and Kolmogorov eddy velocity, length, and time scales. Finally, the distance y from the wall for $y^+ = 1, 7,$ and 30 is calculated as well.

- *The Yplus_LIKE_Eddy_Scales.m script is listed below:*

```
%%%%%%%%%%%%%%%%%%%%%%%%%%%%%%%%%%%%%%%%%%%%%%%%%%%%%%%%%%%%%%%%%%%%%
%                                                                 %
% Program to calculate the "LIKE" algorithm and y+.  The integral, Taylor, and Kolmogorov eddy   %
% velocity, length, and time scales are calculated as well.        %
%                                                                 %
% Units are in SI, unless otherwise noted.                         %
%                                                                 %
% Assumes fully-developed internal pipe flow.                      %
% Can also do external flow over a flat plate.                     %
%                                                                 %
%%%%%%%%%%%%%%%%%%%%%%%%%%%%%%%%%%%%%%%%%%%%%%%%%%%%%%%%%%%%%%%%%%%%%
%                                                                 %
% Programmed by Sal B. Rodriguez on April 21, 2014.  Version 1.0.  %
% Revised on December 9, 2015 to correct a few bugs and add more capacity.  Version 1.5.   %
% Modified on 02/15/2016 to improve output readability.  Version 2.0.      %
%                                                                 %
%%%%%%%%%%%%%%%%%%%%%%%%%%%%%%%%%%%%%%%%%%%%%%%%%%%%%%%%%%%%%%%%%%%%%
%
% REQUIRED INPUT
% - Some form of velocity or mass flow rate, e.g., U_inf, u_max, m_dot.
% - Pressure P is required for physical properties that are pressure dependent (if using REFPROP).
% - Fluid temperature, T_fluid if using REFPROP.
% - If not using REFPROP, then the physical properties must be input manually.
% - Some form of characteristic distance, e.g., pipe diameter D, L, or x_char.
% - See sample input below for additional input format and requirements (Case 1).
%
%%%%%%%%%%%%%%%%%%%%%%%%%%%%%%%%%%%%%%%%%%%%%%%%%%%%%%%%%%%%%%%%%%%%%
% nu = fluid kinematic viscosity
% A = flow area
% flow = 0; %%% Flow is internal
% flow = 1; %%% Flow is external
% u_bar = average turbulent velocity
% u_char = fluid characteristic velocity
% u_bar_max = maximum turbulent velocity
% Re = hydraulic Reynolds number
% I = turbulence intensity
```

```
% lambda = surface roughness.
% k = turbulent kinetic energy
% epsilon = turbulent dissipation
% omega = turbulent frequency
% nu_t = turbulent kinematic viscosity
% C_f = wall skin friction coefficient
% tau_wall = wall shear friction
% u_star = turbulence velocity at the wall
% l_Kol = Kolmogorov eddy size
% v_Kol = Kolmogorov eddy velocity
% t_Kol = Kolmogorov eddy time
% l_Tay = Taylor eddy size
% v_Tay = Taylor eddy velocity
% t_Tay = Taylor eddy time
% l_Int = Integral eddy size
% v_Int = Integral eddy velocity
% t_Int = Integral eddy time
%%%%%%%%%%%%%%%%%%%%%%%%%%%%%%%%%%%%%%%%%%%%%%%%%%%%%%%%%%%%%%%%%%%%%%%%
clear all;
close all;
clc;

%%%%%%%%%%%%%%%%%%%%%%%%%%%%%%%%%%%%%%%%%%%%%%%%%%%%%%%%%%%%%%%%%%%%%%%%
%%% Case 1.  Sample input for turbulent pipe flow.  Fluid = sCO2.  04/03/2019.
flow = 0;          %%% Flow is internal.
lambda = 0.0;  %%% lambda = surface roughness, m.

P = 29.11e6;          %%% Pa
P_refprop = P/1000.0;   %%% Divide by 1,000 because REFPROP requires kPa input.
T_fluid = 550 + 273.15; % K
%%% If uncommented, the next line calls REFPROP to compute the sCO2 properties.
%%% Alternatively, the user can input the properties manually (which are commented immediately below
%%% the REFPROP call line).
[U_sound rho_liq mu_liq] = refpropm('ADV','T',T_fluid,'P',P_refprop,'co2');

%%% The lines below are commented if REFPROP is called to calculate the physical properties.
%%%U_sound = 472.9;
%%%mu_liq = 38.8e-6;
%%%rho_liq = 177.2;

x_char = 0.025;
u_char = 4.67;

nu = mu_liq/rho_liq;
Ma = u_char/U_sound; %%% Mach number calculation
%%%%%%%%%%%%%%%%%%%%%%%%%%%%%%%%%%%%%%%%%%%%%%%%%%%%%%%%%%%%%%%%%%%%%%%%

%%%%%%%%%%%%%%%%%%%%%%%%%%%%%%%%%%%%%%%%%%%%%%%%%%%%%%%%%%%%%%%%%%%%%%%%
%%% Fixed constants.
y_plus1 = 1.0;  %%% Used to obtain y at y+ = 1 for meshing purposes.
y_plus7 = 7.0;  %%% Laminar viscous sublayer; 7 is ~ max., but can be as little as 5-ish.
y_plus30 = 30.0; %%% Value for beginning of log domain.
C_mu = 0.09;
beta_star = 9./100.;  %%% Wilcox k-omega model.

%%% Begin turbulence calculations.
%%% A = pi*D*D/4;
%%% m_dot = rho_liq*u_fluid*A;
%%% u_bar = m_dot/(rho_liq*A);
u_bar = u_char;          %%% u_fluid;
u_bar_max = (5./4.)*u_bar;  %%% This line is correct iff u_bar is an average velocity.  BSL expression.
u_char_max = u_bar_max;

Re = x_char*u_char/nu;
l = 0.07*x_char;
I = 0.16*Re^(-1./8.);
k = (3./2.)*(u_char*I)^2;
epsilon = C_mu*(k^(3./2.))/l;  %%% Uses "l" based on l=C*x_char.  This is the Prandtl-Kolmogorov relationship.
```

```
omega = epsilon/(beta_star*k);
nu_t = (C_mu*k^2)/epsilon;
nu_rat = nu_t/nu;

% Kolmogorov Eddies.
l_Kol = ( (nu)^3/epsilon )^(1/4);
t_Kol = (nu/epsilon)^(1/2);
v_Kol =l_Kol/t_Kol;

% Taylor Eddies.
l_Tay = ( (10*k*nu)/epsilon )^(1/2);
t_Tay = (15*nu/epsilon)^(1/2);
v_Tay = l_Tay/t_Tay;

% Integral Eddies
l_Int = l;
t_Int = C_mu*k/epsilon;
v_Int = l_Int/t_Int;

if flow == 1 %%%% Flow is external.
    if Re <= 1.0e9
        C_f_ext = (2*log10(Re) - 0.65)^(-2.3); %%%%Schlichting skin friction equation for flat plate. Good for Re < 1x10^9.
        C_f = C_f_ext;
    else
        disp('Error: Re outside of wall friction correlation range')
    end
elseif flow == 0 %%%% Flow is internal.
        C_f_int = 0.0055*( 1+(2.0e4*(lambda/x_char)+ 1.0e6/Re)^(1/3) );
        C_f = C_f_int/4;  %%%% Convert the Darcy value to Fanning friction factor.
    else
        disp('Error: Unknown flow type')
end

tau_wall = C_f*rho_liq*u_char*u_char/2;
u_star = sqrt(tau_wall/rho_liq);   %%%% friction velocity; aka shear velocity (hence a "tau" instead of a "star").

y_at_yplus1 = y_plus1*nu/u_star;
y_at_yplus7 = y_plus7*nu/u_star;
y_at_yplus30 = y_plus30*nu/u_star;

%%%% Estimation of entrance length to reach fully developed turbulent flow.
n1 = 0.25;
C1 = 1.359;
L_e = x_char*C1*Re^n1;
L_over_x_char = L_e/x_char;

%%%%%%%%%%%%%%%%%%%%%%%%%%%%%%%%%%%%%%%%%%%%%%%%%%%%%%%%%%%%%%%%%%%%%%%%
%%%% Print all relevant turbulence quantities...

%%%% Characteristic length and velocity.
sprintf('Characteristic length = %3.2e m', x_char)
sprintf('Characteristic velocity = %3.2e m/s', u_char)

%%%% Fluid properties.
sprintf('Temperature = %3.2e K', T_fluid)
sprintf('Pressure = %3.2e Pa', P)
sprintf('Density = %3.2e kg/m3', rho_liq)
sprintf('Dynamic viscosity = %3.2e kg/m-s', mu_liq)
sprintf('Kinematic viscosity = %3.2e m2/s', nu)
sprintf('Sound speed = %3.2e m/s', U_sound)

%%%% Dimensionless numbers.
sprintf('Reynolds number = %3.2e', Re)
sprintf('Mach number = %3.2e ', Ma)
```

```
%%% Key Fluid dynamics and turbulence parameters.
sprintf('Max. mean turbulence velocity = %3.2e m/s', u_char_max)
sprintf('Entrance length for turbulent flow = %3.2e m', L_e)
sprintf('L_e/x_char = %3.2e', L_over_x_char)
sprintf('Wall friction C_f or f = %3.2e', C_f)
sprintf('Wall friction velocity (u*) = %3.2e m/s', u_star)
sprintf('Wall shear = %3.2e kg/m-s2', tau_wall)
sprintf('y at y+=1 = %3.2e m', y_at_yplus1)
sprintf('y at y+=7 = %3.2e m', y_at_yplus7)
sprintf('y at y+=30 = %3.2e m', y_at_yplus30)
sprintf('Turbulent kinematic viscosity = %3.2e m2/s', nu_t)
sprintf('Ratio of turbulent and fluid kinematic viscosities = %3.2e', nu_rat)
sprintf('Turbulence intensity = %3.2e', I)
sprintf('Specific turbulent kinetic energy = %3.2e m2/s2', k)
sprintf('Eddy dissipation (epsilon) = %3.2e m2/s3', epsilon)

%%% Eddy size, velocity, and time scales.
sprintf('Kolmogorov eddy size = %3.2e m', l_Kol)
sprintf('Kolmogorov eddy velocity = %3.2e m/s', v_Kol)
sprintf('Kolmogorov eddy time = %3.2e s', t_Kol)
sprintf('Taylor eddy size = %3.2e m', l_Tay)
sprintf('Taylor eddy velocity = %3.2e m/s', v_Tay)
sprintf('Taylor eddy time = %3.2e s', t_Tay)
sprintf('Integral eddy size = %3.2e m', l_Int)
sprintf('Integral eddy velocity = %3.2e m/s', v_Int)
sprintf('Integral eddy time = %3.2e s', t_Int)

%%% Fluid dynamics/turbulence frequencies.
sprintf('Eddy frequency (omega) = %3.2e 1/s', omega)
sprintf('Kolmogorov eddy frequency = %3.2e 1/s', 1/t_Kol)
sprintf('Taylor eddy frequency = %3.2e 1/s', 1/t_Tay)
sprintf('Integral eddy frequency = %3.2e 1/s', 1/t_Int)
%%%%%%%%%%%%%%%%%%%%%%%%%%%%%%%%%%%%%%%%%%%%%%%%%%%%%%%%%%%%%%%%%%%%%%%%
```

7.2 Peak Velocity Calculator for Laminar and Turbulent Natural Circulation

The MATLAB script for calculating the peak velocity for natural circulation (NC) laminar and turbulent flows is presented here. The script also calculates the convective heat transfer coefficient (h), Pr, Gr, Nu, and Ra. It can calculate both horizontal and vertical geometries, though the emphasis is on vertical flows. The laminar regime is valid for $0.001 < Pr < 1000$. The script has been validated for a horizontal pipe with air and water and a vertical plate with air.

• *The PR_ETA_FPRIME.m script is listed below:*

```
%%%%%%%%%%%%%%%%%%%%%%%%%%%%%%%%%%%%%%%%%%%%%%%%%%%%%%%%%%%%%%%%%%
%%%                                                            %%%
%%% Program to calculate natural circulation (NC) velocity, the convective heat    %%%
%%% transfer coefficient (h), Pr, Gr, Nu, and Ra based on computer solutions from  %%%
%%% [Ostrach, 1953], Housam Binous MATLAB data [National Institute of Applied Sciences  %%%
%%% and Technology, Tunis, Tunisia, binoushousam@yahoo.com], and author's extensions  %%%
%%% for Pr domain [Rodriguez and Ames, 2015].                  %%%
%%%                                                            %%%
%%% Valid for laminar NC with 0.001 < Pr < 1,000 [Rodriguez and Ames, 2015]. Includes  %%%
%%% equations for turbulent NC flow as well.                   %%%
%%%                                                            %%%
%%% Designed primarily for vertical plate flow, though the script was validated for a  %%%
%%% horizontal pipe with air and water, and a vertical plate with air.  %%%
%%%                                                            %%%
%%% All units are in SI; P is in Pa, but if REFPROP is called, REFPROP requires kPa  %%%
%%% pressure input.                                            %%%
%%%                                                            %%%
%%% The script is initially set to CASE 1, which is a validation for water under NC,  %%%
%%% HORIZONTAL, 2-cm tube, Delta-T=30 K. In this case, Holman cites h~890 W/m2-K  %%%
%%% [Holman, 1990, Page 13] vs. 935.7 W/m2-K for this MATLAB script.  %%%
%%%                                                            %%%
%%% Programmed by Sal Rodriguez.                               %%%
%%% 05/29/2015, Version 1.0                                    %%%
%%% 05/11/2019, Version 1.5.  Added more physical properties options and validation.  %%%
%%%                                                            %%%
%%%%%%%%%%%%%%%%%%%%%%%%%%%%%%%%%%%%%%%%%%%%%%%%%%%%%%%%%%%%%%%%%%

clear all;
close all;
clc;

g = 9.8;  %%% Gravitational constant, m/s2.

%%%%%%%%%%%%%%%%%%%%%%%%%%%%%%%%%%%%%%%%%%%%%%%%%%%%%%%%%%%%%%%%%%
%
% REQUIRED user input for the particular problem being solved.
% Various examples and validation cases are listed below.
%
%%%%%%%%%%%%%%%%%%%%%%%%%%%%%%%%%%%%%%%%%%%%%%%%%%%%%%%%%%%%%%%%%%


%%%%%%%%%%%%%%%%%%%%%%%%%%%%%%%%%%%%%%%%%%%%%%%%%%%%%%%%%%%%%%%%%%
%%% Choose value for i_fluid if using any of the molten metals: Na=1, Pb=2, Bi=3, and LBE=4.
%%% Or, use the REFPROP call below for the material properties.  Otherwise, insert the necessary
%%% properties manually.

i_matprop = 0;  %%% Choice of method to calculate the material properties:
                %%% Default value = 0.
                %%% 1 = NaPbBiLBE_FUNC function call for Na, Pb, Bi, or LBE properties.
                %%% 2 = REFPROP call.
                %%% 3 = User-input material properties.

i_fluid = 0;    %%% Choice of fluid is required if calling Function NaPbBiLBE_FUNC, where Na=1,
                %%% Pb=2, Bi=3, and LBE=4.  Default value = 0, which will cause abort if
                %%% i_fluid=0 AND i_matprop=1.  Otherwise, if calling REFPROP or providing
                %%% user-input of the physical properties, then the value of i_fluid does not
                %%% matter.
%%%%%%%%%%%%%%%%%%%%%%%%%%%%%%%%%%%%%%%%%%%%%%%%%%%%%%%%%%%%%%%%%%
```

```
%%%%%%%%%%%%%%%%%%%%%%%%%%%%%%%%%%%%%%%%%%%%%%%%%%%%%%%%%%%%%%%%%%%%
%%% CASE 1.
%%% Example: Validation case for water NC, HORIZONTAL, 2-cm tube, Delta-T=30 K. Therefore, the
%%% correlations are for HORIZONTAL geometry.  Holman cites h~890 W/m2-K [Holman, 1990, Page 13]
%%% vs. 935.7 W/m2-K for this MATLAB script.
T_fluid = 300;
T = T_fluid;
T_w = 330.;
T_film = (T_fluid + T_w)/2;
P = 101300.0;
P_refprop = P/1000.0;  %%%101.3 kPa = 101,300 Pa (refprop requires kPa input)

x_char = 0.02;    %%% m
q_flux = 250.;    %%% W/m2. Input required if system is heat-flux based.
%nu_t = 1.0e-003; %%% m2/s. Input required if system is turbulent (not the case for this example).

i_matprop = 2; % Flag for calling REFPROP

%%% Activate validation input lines (check for "validate" key word in this script) (e.g.,
%%% Nu_x_lam_horiz_pipe_validate, h_lam_horiz_pipe_validate, Nu_x_turb_horiz_pipe_validate, and
%%% h_lam_horiz_pipe_validate.

%%% Also activate the following input line in the "MATERIAL PROPERTIES IF-CONDITIONAL", namely:
%%% [rho_liq k_liq mu_liq Cp_liq, Pr, beta, U_sound] = refpropm('DLVC^BA','T',T_film,'P',P_refprop,'water')
%%%%%%%%%%%%%%%%%%%%%%%%%%%%%%%%%%%%%%%%%%%%%%%%%%%%%%%%%%%%%%%%%%%%

%%%%%%%%%%%%%%%%%%%%%%%%%%%%%%%%%%%%%%%%%%%%%%%%%%%%%%%%%%%%%%%%%%%%
%%% CASE 2.
%%% Example: Validation case for air NC, vertical plate=0.3 m, Delta-T=30 K.
%%% Holman cites h~4.5 W/m2-K [Holman, 1990, Page 13] vs. 3.35 W/m2-K for this MATLAB script.
%T_fluid = 300;
%T = T_fluid;
%T_w = 330.;
%T_film = (T_fluid + T_w)/2;
%P = 101300.0;
%P_refprop = P/1000.0;  %%%101.3 kPa = 101,300 Pa (refprop requires kPa input)

%q_flux = 250.;   %%% W/m2. Input required if system is heat-flux based.
%nu_t = 1.0e-003; %%% m2/s. Input required if system is turbulent (not the case for this example).
%x_char = 0.3;

%i_matprop = 2; % Flag for calling REFPROP

%%% Activate the following input line in the "MATERIAL PROPERTIES IF-CONDITIONAL", namely:
%%% [rho_liq k_liq mu_liq Cp_liq, Pr, beta, U_sound] = refpropm('DLVC^BA','T',T_film,'P',P_refprop,
%'nitrogen', 'oxygen','argon', [0.781,0.210,0.009])
%%%%%%%%%%%%%%%%%%%%%%%%%%%%%%%%%%%%%%%%%%%%%%%%%%%%%%%%%%%%%%%%%%%%

%%%%%%%%%%%%%%%%%%%%%%%%%%%%%%%%%%%%%%%%%%%%%%%%%%%%%%%%%%%%%%%%%%%%
%%% CASE 3.
%%% Example: Validation case for air NC, horizontal, 5-cm tube, Delta-T=30 K. Holman cites
%%% h~6.5 W/m2-K [Holman, 1990, Page 13] vs. 6.62 W/m2-K for this MATLAB script.

%T_fluid = 300;
%T = T_fluid;
%T_w = 330.;
%T_film = (T_fluid + T_w)/2;
%P = 101300.0;
%P_refprop = P/1000.0;  %%%101.3 kPa = 101,300 Pa (refprop requires kPa input)
```

```
%q_flux = 250.;   %%% W/m2. Input required if system is heat-flux based.
%nu_t = 1.0e-003; %%% m2/s. Input required if system is turbulent (not the case for this example).
%x_char = 0.05;

%i_matprop = 2; % Flag for calling REFPROP

%%% Activate validation input lines (check  for "validate" key word in this script) (e.g.,
%%% Nu_x_lam_horiz_pipe_validate, h_lam_horiz_pipe_validate, Nu_x_turb_horiz_pipe_validate, and
%%% h_lam_horiz_pipe_validate.

%%% Also activate the following input line in the "MATERIAL PROPERTIES IF  -CONDITIONAL", namely:
%%% [rho_liq k_liq mu_liq Cp_liq, Pr, beta, U_sound] = refpropm('DLVC^BA','T',T_film,'P',P_refprop,'nitrogen',
'oxygen', 'argon', [0.781,0.210,0.009])
%%%%%%%%%%%%%%%%%%%%%%%%%%%%%%%%%%%%%%%%%%%%%%%%%%%%%%%%%%%%%%%%%%%%%%

%%%%%%%%%%%%%%%%%%%%%%%%%%%%%%%%%%%%%%%%%%%%%%%%%%%%%%%%%%%%%%%%%%%%%%
%%% CASE 4.
% Example Input if i_matprop=1 %%%%%%%%%%%%%%%%%%%%%%%%%%%%%%%%%%%%%%%%%%%%%%%%%%
%%% Example for input lines needed to call function NaPbBiLBE_FUNC (i_matprop=1) for Na=1
%%% working fluid.
%T_fluid = 750; % K
%T = T_fluid;
%T_w = 1200.; % K
%T_film = (T_fluid + T_w)/2;

%q_flux = 150.0; %%% W/m2.
%x_char = 0.1;  %%% m
%%% User needs a problem-specific nu_t if problem is turbulent. Current value is from the "LIKE"
%%% algorithm because it can calculate nu_t directly.
%nu_t = 6.9e-003; %%% m2/s

%i_fluid = 1;  % Calculate Na
%i_matprop = 1; % Used to call function NaPbBiLBE_FUNC
%%%%%%%%%%%%%%%%%%%%%%%%%%%%%%%%%%%%%%%%%%%%%%%%%%%%%%%%%%%%%%%%%%%%%%

%%%%%%%%%%%%%%%%%%%%%%%%%%%%%%%%%%%%%%%%%%%%%%%%%%%%%%%%%%%%%%%%%%%%%%
%%% CASE 5.
% Example Input if i_matprop=2 %%%%%%%%%%%%%%%%%%%%%%%%%%%%%%%%%%%%%%%%%%%%%%%%%%
%%% Example for input lines needed to call function REFPROP (i_matpr op=2) for water. Call
%%% REFPROP from if-conditional, not from this input segment.

%P = 101300.0;
%P_refprop = P/1000.0  %%%101.3 kPa = 101,300 Pa (refprop requires kPa input)
%T_fluid = 300;
%T = T_fluid;
%T_w = 330.;
%T_film = (T_fluid + T_w)/2;

%i_matprop = 2; % Flag for calling REFPROP.

%%% Turbulent case.
%q_flux = 150.0; %%% W/m2.
%x_char = 0.1;  %%% m
%%% User needs a problem-specific nu_t if problem is turbulent. Current value is from the "LIKE"
%%% algorithm because it can calculate nu_t directly.
%nu_t = 6.9e-003; %%% m2/s

%%% Also activate the following input line in the "MATERIAL PROPERTIES IF  -CONDITIONAL", namely:
%%% [rho_liq k_liq mu_liq Cp_liq, Pr, beta, U_sound] = refpropm('DLVC^BA','T',T_film,'P',P_refprop,'water')
%%%%%%%%%%%%%%%%%%%%%%%%%%%%%%%%%%%%%%%%%%%%%%%%%%%%%%%%%%%%%%%%%%%%%%
```

%%
%%% CASE 6.
% Example Input if i_matprop=3 %%%
%%% Example input lines needed for manual input of the fluid physica l properties.

%T_fluid = 300;
%T_w = 330.;

%i_matprop = 3; % User-input physical properties.
%%% The user must add the following input lines in the "MATERIAL PROPERTIES IF -CONDITIONAL", namely:
%%% (rho_liq, k_liq, mu_liq, Cp_liq, beta, and U_sound).

%q_flux = 150.0; %%% W/m2.
%x_char = 0.1; %%% m
%%% User needs a problem-specific nu_t if problem is turbulent. Current value is from the "LIKE"
%%% algorithm because it can calculate nu_t directly.
%nu_t = 6.9e-003; %%% m2/s
%%%

%%
%%% CASE 7.
%%% Example: laminar NC water heat sink.
%%% 05/15/2019. SI units.
%%% Fluid = water.

%P = 101300.0;
%P_refprop = P/1000.0 %%%101.3 kPa = 101,300 Pa (refprop r equires kPa input)
%T_fluid = 300;
%T = T_fluid;
%T_w = 330.;
%T_film = (T_fluid + T_w)/2;

%i_matprop = 2; % Flag for calling REFPROP.

%x_char = 0.1;
%q_flux = 150.0; %%% W/m2. Assume constant heat flux.

%%% Also activate the following input line in the "MATERIAL PROPERTIES IF -CONDITIONAL", namely:
%%% [rho_liq k_liq mu_liq Cp_liq, Pr, beta, U_sound] = refpropm('DLVC^BA','T',T_film,'P',P_refprop,'water')
%%%

%%
%%% CASE 8.
%%% Example: turbulent NC water heat sink.
%%% 05/15/2019. SI units.
%%% Fluid = water.

%P = 101300.0;
%P_refprop = P/1000.0; %%%101.3 kPa = 101,300 Pa (refprop requires kPa input)
%T_fluid = 300;
%T = T_fluid;
%T_w = 330.;
%T_film = (T_fluid + T_w)/2;

%i_matprop = 2; % Flag for calling REFPROP.

%q_flux = 1500.0; %%% W/m2. Assume constant heat flux.
%%% User needs a problem-specific nu_t if problem is turbulent. Current value is from the "LIKE"
%%% algorithm because it can calculate nu_t directly.
%nu_t = 6.9e-003; %%% m2/s
%x_char = 1.0;

```
%%% Also activate the following input line in the "MATERIAL PROPERTIES IF-CONDITIONAL", namely:
%%% [rho_liq k_liq mu_liq Cp_liq, Pr, beta, U_sound] = refpropm('DLVC^BA','T',T_film,'P',P_refprop,'water')
%%%%%%%%%%%%%%%%%%%%%%%%%%%%%%%%%%%%%%%%%%%%%%%%%%%%%%%%%%%%%%%%%%%%%

%%%%%%%%%%%%%%%%%%%%%%%%%%%%%%%%%%%%%%%%%%%%%%%%%%%%%%%%%%%%%%%%%%%%%
%%% Peak f_prime and H_prime from Ostrach and Housam Binous MATLAB data.  Full credit is given
%%% here to the works of Ostrach and Binous (commented as such in vectors f_prime and H_prime;
%%% uncommented values were extended by Rodriguez).  This extends the laminar domain to
%%% 0.001 < Pr < 1,000 when using f_prime and H_prime.
f_prime = [
0.00001 0.691 %%%% Housam Binous; CONVERGED???
0.0001 0.677 %%%% Housam Binous; CONVERGED
0.001  0.652 %%%% Housam Binous; CONVERGED
0.005  0.605 %%%% Housam Binous; CONVERGED
0.01   0.5739
0.025  0.528   %%%% Housam Binous; CONVERGED
0.05   0.484   %%%% Housam Binous; CONVERGED
0.1    0.433  %%%% Housam Binous; CONVERGED
0.2    0.378  %%%% Housam Binous; CONVERGED
0.3    0.345  %%%% Housam Binous; CONVERGED
0.4    0.322  %%%% Housam Binous; CONVERGED
0.6    0.290  %%%% Housam Binous; CONVERGED
0.72   0.2759
0.733  0.2745
1.0    0.2511
2.0    0.2024
10.0   0.1150
50.0   0.0593 %%%% Housam Binous; CONVERGED
100.0  0.0442
200.0  0.0319 %%%% Housam Binous; CONVERGED
500.0  0.0208 %%%% Housam Binous; CONVERGED
1000.  0.0153
];

H_prime = [
0.0000000001    -0.000000001   %%% Estimated...
0.01   -0.0812
0.05   -0.18   %%% Ostrach Fig. 8.
0.1    -0.22   %%% Ostrach Fig. 8.
0.2    -0.3    %%% Ostrach Fig. 8.
0.5    -0.45   %%% Ostrach Fig. 8.
0.72   -0.5046
0.73   -0.5080
1.0    -0.5671
2.0    -0.7165
6.0    -1.0    %%% Ostrach Fig. 8.
10.0   -1.1694
50.0   -1.75   %%% Ostrach Fig. 8.
100.   -2.191
300.0  -2.85   %%% Ostrach Fig. 8.
700.0  -3.6    %%% Ostrach Fig. 8.
1000.  -3.966
];
```

```
%%%%%%%%%%%%%%%%%%%%%%%%%%%%%%%%%%%%%%%%%%%%%%%%%%%%%%%%%%%%%%%%%
% get number of rows, which = number of data pairs.
nf = length(f_prime);
nH = length(H_prime);

% store col 1 in vector xcol1 and col 2 in vector xcol2
for i=1:nf
  xcol1(i) = f_prime(i,1);
  xcol2(i) = f_prime(i,2);
end
for i=1:nH
  xcol3(i) = H_prime(i,1);
  xcol4(i) = H_prime(i,2);
end


%%%%%%%%%%%%%%%%%%%%%%%%%%%%%%%%%%%%%%%%%%%%%%%%%%%%%%%%%%%%%%%%%
%%% MATERIAL PROPERTIES IF -CONDITIONAL.  Coding for calculating the physical properties.
%%%%%%%%%%%%%%%%%%%%%%%%%%%%%%%%%%%%%%%%%%%%%%%%%%%%%%%%%%%%%%%%%
if i_matprop == 1
   %%% Call function "NaPbBiLBE_FUNC" to obtain the fluid material properties.
   %%%%%%%%%%%%%%%%%%%%%%%%%%%%%%%%%%%%%%%%%%%%%%%%%%%%%%%%%%%%%%%%%
   %%% How function "NaPbBiLBE_FUNC" works:
   %%% values=values(i_fluid, T_fluid) = array with fluid temperature and T_fluid.
   %%% [rho, beta, Cp_liq, mu, k, Pr, nu, alpha] = output from function NaPbBiLBE_FUNC.
   %%% i_fluid=values(1);   %%% Choice of fluid. Na=1, Pb=2, Bi=3, and LBE=4.
   %%% T_fluid = values(2); %%% T=T_fluid, the fluid temperature.
   %%%%%%%%%%%%%%%%%%%%%%%%%%%%%%%%%%%%%%%%%%%%%%%%%%%%%%%%%%%%%%%%%
   values = [i_fluid, T_fluid]
   [rho_liq, beta, Cp_liq, mu_liq, k_liq, Pr, nu, alpha_liq] = NaPbBiLBE_FUNC(values);

elseif i_matprop == 2

   %%% Call REFPROP to obtain the physical material properties, which are evaluated at the
   %%% film temperature, T_film, and passed on via function arguments.  REFPROP uses K for
   %%% the temperature.

   %%% In REFPROP notation:
   %%% Q = Quality (vapor fraction), kg/kg; 1=all vapor; 0=all liquid.
   %%% ^ = Prandtl number, unitless
   %%% B = Volumetric expansivity (i.e., beta), 1/K
   %%% C = Cp (heat capacity at constant pressure)
   %%% P = Pressure, kPa
   %%% V = Dynamic viscosity, Pa*s=kg/m-s

   %%%%%%%%%%%%%%%%%%%%%%%%%% EXAMPLES %%%%%%%%%%%%%%%%%%%%%%%%%%%
   %%% Call REFPROP for physical properties.
   %%% Example call for water, where T and P are specified:
   [rho_liq k_liq mu_liq Cp_liq, Pr, beta, U_sound] = refpropm( 'DLVC^BA','T',T_film,'P',P_refprop,'water')
   %%% Example call for water, where T and Q are specified:
   %[rho_liq k_liq mu_liq Cp_liq, Pr, beta, U_sound] = refpropm('DLVC^BA','T',T_film,'Q',0,'water')

   %%% Examples for SCO2:
   %[rho_liq k_liq mu_liq Cp_liq, Pr, beta, U_sound] = refpropm('DLVC^BA','T',T_film,'Q',0,'co2')
   %[rho_liq k_liq mu_liq Cp_liq, Pr, beta, U_sound] = refpropm('DLVC^BA','T',T_film,'P',P_refprop, 'co2')

   %%% Example for nitrogen:
   %[rho k_liq mu Cp_liq, Pr, beta, U_sound] = refpropm('DLVC^BA','T',T,'P',P_refprop,'nitrogen')
```

```
%%% Example for air (need to specify mass fractions of the individual components).
%[rho_liq k_liq mu_liq Cp_liq, Pr, beta, U_sound] = refpropm('DLVC^BA','T',T_film,'P',P_refprop,'nitrogen',
'oxygen', 'argon', [0.781,0.210,0.009])
elseif i_matprop == 3
  %%% If this i_matprop choice is selected, the user ***MUST*** manually include the fluid material properties here.
  rho_liq = 250.           % fluid density, kg/m3
  k_liq = 1.0e-3           % fluid thermal conductivity, J/kg-K
  mu_liq = 1.1e-4           % fluid dynamic viscosity, kg/m-s
  Cp_liq = 1200.            % fluid heat capacity at constant pressure, J/kg-K
  beta = 0.001           % fluid volumetric expansivity, 1/K
  U_sound = 780.          % fluid sound speed, m/s
  Pr = Cp_liq*mu_liq/k_liq    % Prandtl number, unitless
else
  'Abort...no material properties are being calculated!'
end
%%%%%%%%%%%%%%%%%%%%%%%%%%%%%%%%%%%%%%%%%%%%%%%%%%%%%%%%%%%%%%%%%%

nu_liq = mu_liq/rho_liq;

%%%%%%%%%%%%%%%%%%%%%%%%%%%%%%%%%%%%%%%%%%%%%%%%%%%%%%%%%%%%%%%%%%
%%% From a MATLAB curve-fit of the above f_prime and H_prime arrays, obtain f_prime_func and
%%% H_prime_func.
f_prime_func = 0.5*Pr^(-0.11) - 0.24;  %%% [Rodriguez and Ames, 2015]
%%%f_prime_func = (1.85*Pr + 0.23)/(0.65*Pr*Pr + 6.5*Pr + 0.36) + 0.015;  %%% [Rodriguez and Ames, 2015]
H_prime_func = -0.6*Pr^0.28;  %%% [Rodriguez and Ames, 2015]
%%%%%%%%%%%%%%%%%%%%%%%%%%%%%%%%%%%%%%%%%%%%%%%%%%%%%%%%%%%%%%%%%%

%%% Final input query before beginning calculations:
sprintf('Is the system based on a constant heat flux or Delta-T?')
sprintf(' ')
prompt = 'Input 1 if heat flux based or 2 if based on Delta-T. '
system_type = input(prompt)

%%%%%%%%%%%%%%%%%%%%%%%%%%%%%%%%%%%%%%%%%%%%%%%%%%%%%%%%%%%%%%%%%%
%%%                    Begin the calculations...          %%%
%%%%%%%%%%%%%%%%%%%%%%%%%%%%%%%%%%%%%%%%%%%%%%%%%%%%%%%%%%%%%%%%%%

%%%%%%%%%%%%%%%%%%%%%%%%%%%%%%%%%%%%%%%%%%%%%%%%%%%%%%%%%%%%%%%%%%
if system_type == 1
sprintf('******************* Results based on Gr_star (const. heat flux) *******************')
%%%%%%%%%%%%%%%%%%%%%%%%%%%%%%%%%%%%%%%%%%%%%%%%%%%%%%%%%%%%%%%%%%
%%% Calculate Gr_star for Gr based on q_flux instead of Delta-T.  q_flux is in W/m2.
%%% Based on Holman, "Heat Transfer", page 344.
Gr_star = (g*beta*q_flux*x_char^4)/(k_liq*nu_liq^2)
Ra_star = Gr_star*Pr
if Gr_star > 1e5 & Gr_star < 1e11
   sprintf('Gr_star is LAMINAR...')
   Nu_star_lam = 0.6*Ra_star^(1/5)
   h_star_lam = k_liq*Nu_star_lam/x_char  %%% LAMINAR and q_flux=const.: 1e5 < Gr_star < 1e11
   deltaT = q_flux/h_star_lam;
   U_max_lam_Gr_star = f_prime_func*2*nu_liq*sqrt(Gr_star)/x_char
elseif Gr_star > 2e13 & Gr_star < 1e16
   sprintf('Gr_star is TURBULENT.')
   Nu_star_turb = 0.17*Ra_star^(1/4)
   h_star_turb = k_liq*Nu_star_turb/x_char  %%% TURBULENT and q_flux=const.: 2e13 < Gr_star*Pr < 1e16
   deltaT = q_flux/h_star_turb;
   U_turb_max_Gr_star = nu_t/(2*x_char) + sqrt( (nu_t^2/(4*x_char^2)) + g*beta*x_char*q_flux/h
   _star_turb)  %%% based on q_flux.
else %%% Large transition region between laminar and turbulent...estimate by averaging the two.
```

```
     sprintf('Gr_star is transitional between laminar and turbulent; value estimated by averaging the two.' )
     Nu_star_trans = ( 0.6*Ra_star^(1/5) + 0.17*Ra_star^(1/4) )/2
     h_star_trans = k_liq*Nu_star_trans/x_char
     deltaT = q_flux/h_star_trans;
     U_star_trans = ( f_prime_func*2*nu_liq*sqrt(Gr_star)/x_char + nu_t/(2*x_char) + sqrt( (nu_t^2/(4*x_char^2)) +
     g*beta*x_char*q_flux/h_star_trans) )/2
 end

 %%%%%%%%%%%%%%%%%%%%%%%%%%%%%%%%%%%%%%%%%%%%%%%%%%%%%%%% %%%%%%%%%%%%%
 elseif system_type == 2
 sprintf('************************* Results based on Gr_Delta-T *************************')
 %%%%%%%%%%%%%%%%%%%%%%%%%%%%%%%%%%%%%%%%%%%%%%%%%%%%%%%%%%%%%

     %%% Calculate Gr_x, Ra_x, Nu_x and u_max_lam based on Delta-T.
     Gr_x = (g*beta*abs(T_w - T_fluid)*x_char^3)/(nu_liq*nu_liq) %%%%Include "abs" to consider the case where the fluid heats the wall.
     Ra_x = Gr_x*Pr
     if Ra_x < 1e12
         sprintf('Ra_x = Gr_x*Pr is laminar.')
         Nu_x_Holman_lam1 = ((Gr_x/4)^0.25)*(0.718*(Pr)^0.5)/(0.952+Pr)^0.25
         Nu_x_Holman_lam2 = 0.68 + (0.670*Ra_x^(1/4))/( (1+(0.492/Pr)^(9/16))^(4/9) )
         h_lam_Holman_ave = k_liq*((Nu_x_Holman_lam1+Nu_x_Holman_lam2)/2)/x_char
         Nu_x_Ostrach_lam = -H_prime_func*(Gr_x/4)^0.25
         h_lam_Ostrach = k_liq*Nu_x_Ostrach_lam/x_char
         Nu_x_lam_horiz_pipe_validate = 0.53*(Ra_x)^(1/4)
         h_lam_horiz_pipe_validate = k_liq*Nu_x_lam_horiz_pipe_validate/x_char
         U_max_lam = f_prime_func*2*nu_liq*sqrt(Gr_x)/x_char
         if i_matprop ~= 1
             Ma = U_max_lam/U_sound %%% Mach number. "~=" means "not equal" in MATLAB.
         end
     else
         sprintf('Ra_x = Gr_x*Pr is turbulent.')

         %%% Turbulent correlations, added on 12/27/2015.
         Nu_x_turb = 0.1*Ra_x^(1/3)
         h_turb = k_liq*Nu_x_turb/x_char
         Nu_x_turb_horiz_pipe_validate = 0.13*(Ra_x)^(1/4)
         h_lam_horiz_pipe_validate = k_liq*Nu_x_turb_horiz_pipe_validate/x_char
         U_turb_max2 = nu_t/(2*x_char) + sqrt( (nu_t^2/(4*x_char^2)) + g*beta*x_char*(T_w - T_film)) %%% based on Delta_T.
         if i_matprop ~= 1
             Ma = U_turb_max2/U_sound %%% Mach number. "~=" means "not equal" in MATLAB.
         end
     end

 %%%%%%%%%%%%%%%%%%%%%%%%%%%%%%%%%%%%%%%%%%%%%%%%%%%%%%%%%%%%%%%%%%%
 else
     sprintf('Ending calculation--system type is not valid.')
 end
 %%%%%%%%%%%%%%%%%%%%%%%%%%%%%%%%%%%%%%%%%%%%%%%%%%%%%%%%%%%%%
```

7.3   Molten Metal Physical Properties Calculator

This MATLAB function is useful for calculating the physical properties for lead, lead bismuth eutectic (LBE), bismuth, and sodium.

If it is desired to call the function from a minimum set of input lines, then the following code fragment shows an example of a basic format:

```
%%% Choice of fluid: Na=1, Pb=2, Bi=3, and LBE=4.
i_fluid = 4;      %%% LBE
T_fluid = 750;   %%% T=T_fluid, the fluid temperature in K.
values = [i_fluid, T_fluid];
[rho_liq, beta, Cp_liq, mu_liq, k_liq, Pr, nu, alpha_liq] = NaPbBiLBE_FUNC(values)
```

- *The NaPbBiLBE_FUNC.m function is listed below:*

```
%%%%%%%%%%%%%%%%%%%%%%%%%%%%%%%%%%%%%%%%%%%%%%%%%%%%%%%%%%%%%%%%%%%%%%%%
%                                                                    %
% Function to calculate the physical properties of molten metals.    %
% Metals: Na, Pb, Bi, and LBE.                                       %
% Properties: rho, beta, Cp, mu, and k; Pr, nu, and alpha.           %
%                                                                    %
% I/O is SI units, T in K.                                           %
%                                                                    %
% Programmed by Sal B. Rodriguez on June 9, 2016.                    %
%                                                                    %
%%%%%%%%%%%%%%%%%%%%%%%%%%%%%%%%%%%%%%%%%%%%%%%%%%%%%%%%%%%%%%%%%%%%%%%%

%%%%%%%%%%%%%%%%%%%%%%%%%%%%%%%%%%%%%%%%%%%%%%%%%%%%%%%%%%%%%%%%%%%%%%%%
%%% MATLAB lines needed to call function.
%i_fluid = 4;       %%% Na=1, Pb=2, Bi=3, and LBE=4.
%T_fluid = 750;     %%% T=T_fluid, the fluid temperature.
%values = [i_fluid, T_fluid];
%[rho_liq, beta, Cp_liq, mu_liq, k_liq, Pr, nu, alpha_liq] = NaPbBiLBE_FUNC(values)
%%%%%%%%%%%%%%%%%%%%%%%%%%%%%%%%%%%%%%%%%%%%%%%%%%%%%%%%%%%%%%%%%%%%%%%%

function [rho, beta, Cp, mu, k, Pr, nu, alpha] = NaPbBiLBE_FUNC(values)

%%% rho = density, kg/m3.
%%% beta = volume expansivity, 1/K.
%%% Cp = heat capacity at constant P.
%%% mu = dynamic viscosity, kg/(m-s).
%%% k = thermal conductivity, J/(kg-K).
%%% T_solid = metal's solidification temperature, K.
%%% Pr = Prandtl number, unitless.
%%% nu = mu/rho, m2/s.
%%% alpha = k/(rho*Cp), m2/s.

% This function reads two input values: i_fluid (molten metal) and T (molten metal temperature).
%%% values=values(i_fluid, T)
   'echo from function NaPbBiLBE_FUNC'
   i_fluid=values(1)   %%% Choice of fluid. Na=1, Pb=2, Bi=3, and LBE=4.  %%%Add NaK in the future.
   T = values(2)       %%% T=T_fluid, the fluid temperature.

%%% All T_solid values are in K.
T_solid1 = 370.9;  %%% Na
T_solid2 = 600.6;  %%% Pb
T_solid3 = 544.1;  %%% Bi
T_solid4 = 397.0;  %%% LBE

if i_fluid == 1 %%% Do Na.
  if T <= T_solid1
    'ABORTING... Desired temperature solidifies Na coolant.'
    sprintf('Input T <= T_solid = %4.4g K', T_solid1)
    stop
  end
elseif i_fluid == 2 %%% Do Pb.
  if T <= T_solid2
    'ABORTING... Desired temperature solidifies Pb coolant.'
    sprintf('Input T <= T_solid = %4.4g K', T_solid2)
    stop
  end
```

```
elseif i_fluid == 3 %%% Do Bi.
  if T <= T_solid3
    'ABORTING... Desired temperature solidifies Bi coolant.'
    sprintf('Input T <= T_solid = %4.4g K',  T_solid3)
    stop
  end
elseif i_fluid == 4 %%% Do LBE.
  if T <= T_solid4
    'ABORTING... Desired temperature solidifies LBE coolant.'
    sprintf('Input T <= T_solid = %4.4g K',  T_solid4)
    stop
  end
else
  'Aborting due to input error--molten salt not specified'
  stop
end %%% End conditional statement for material properties.

%%%%%%%%%%%%%%%%%%%%%%%%%%%%%%%%%%%%%%%%%%%%%%%%%%%%%%%%%%%%%%%%%%%%%
%%% General equations to calculate rho, beta, Cp, mu, and k for Na=1, Pb=2, Bi=3, and LBE=4.
%%% Source: "Database of Thermophysical Properties of Liquid Metal Coolants
%%% for GEN-IV", SCK/CEN-BLG-1069, Rev. December 2011.
%%%%%%%%%%%%%%%%%%%%%%%%%%%%%%%%%%%%%%%%%%%%%%%%%%%%%%%%%%%%%%%%%%%%%
%%% Calculate the physical properties.
if i_fluid == 1 %%% Do Na.
  rho = 1014 - 0.235*T;
  beta = 1/(4316-T);
  Cp = -3e6*T^(-2) + 1658 - 0.848*T + 4.45e-4*T^2;
  mu = 4.55e-4*exp(1069/T);
  k = 110 - 0.0648*T + 1.16e-5*T^2;
elseif i_fluid == 2 %%% Do Pb.
  rho = 11441 - 1.28*T;
  beta = 1/(8942-T);
  Cp = 176.2 - 4.92e-2*T + 1.54e-5*T^2;
  mu = 4.55e-4*exp(1069/T);
  k = 9.2 + 0.011*T;
elseif i_fluid == 3 %%% Do Bi.
  rho = 10725 - 1.22*T;
  beta = 1/(8791-T);
  Cp = 118.2 + 5.93e-3*T + 7.18e6*T^(-2);
  mu = 4.46e-4*exp(775.8/T);
  k = 7.34 + 9.5e-3*T;
elseif i_fluid == 4 %%% Do LBE.
  rho = 11065 -1.293*T;
  beta = 1/(8558-T);
  Cp = 164.8 - 3.94e-2*T + 1.25e-5*T^2;
  mu = 4.94e-4*exp(754.1/T);
  k = 3.28 + 1.62e-2*T - 2.31e-6*T^2;
else
  'Abort'
end %%% End conditional statement for material properties.

Pr = Cp*mu/k;
nu = mu/rho;
alpha = k/(rho*Cp);

end %%% End material properties function.
```

7.4   FORTRAN Program for the Prandtl One-Equation Turbulence Model

This GNU FORTRAN program is useful for demonstrating how a basic RANS model works and provides some practical guidelines. The main idea is that the RANS momentum equation is coupled with the k PDE. For this particular application, consider a 2D Couette flow in Cartesian coordinates. There is no z component; $v = 0$, $w = 0$, and the flow is fully developed. Thus, the RANS velocity is solely a function of y. In addition, assume no pressure gradient and a negligible gravitational term.

As a result of the velocity simplifications, the momentum PDE has the following negligible terms:

$$\frac{\partial \bar{u}}{\partial t} + \left(\bar{u}\frac{\partial \bar{u}}{\partial x} + \bar{v}\frac{\partial \bar{u}}{\partial y} + \bar{w}\frac{\partial \bar{u}}{\partial z} \right) = -\frac{1}{\rho}\frac{\partial \bar{P}}{\partial x}$$

$$+ \frac{\partial\left[\nu_t\left(\frac{\partial \bar{u}}{\partial x} + \frac{\partial \bar{u}}{\partial x} \right) - \frac{2}{3}k \right]}{\partial x} + \frac{\partial\left[\nu_t\left(\frac{\partial \bar{u}}{\partial y} + \frac{\partial \bar{v}}{\partial x} \right) \right]}{\partial y} + \frac{\partial\left[\nu_t\left(\frac{\partial \bar{u}}{\partial z} + \frac{\partial \bar{w}}{\partial x} \right) \right]}{\partial z}$$

$$+ \nu\left(2\frac{\partial^2 \bar{u}}{\partial x^2} + \frac{\partial^2 \bar{u}}{\partial y^2} + \frac{\partial^2 \bar{u}}{\partial z^2} + \frac{\partial^2 \bar{v}}{\partial x \partial y} + \frac{\partial^2 \bar{w}}{\partial x \partial z} \right) + g_x$$

This reduces to

$$\frac{\partial \bar{u}}{\partial t} = \frac{\partial\left(-\frac{2}{3}k \right)}{\partial x} + \frac{\partial\left[\nu_t\left(\frac{\partial \bar{u}}{\partial y} \right) \right]}{\partial y} + \nu\frac{\partial^2 \bar{u}}{\partial y^2}$$

The derivative chain rule is used to simplify the turbulent kinematic viscosity term. Then, assuming fully developed flow in the x-direction, this further modifies the PDE as follows:

$$\frac{\partial \bar{u}}{\partial t} = -\frac{2}{3}\frac{\partial k}{\partial x} + \nu_t\frac{\partial^2 \bar{u}}{\partial y^2} + \frac{\partial \bar{u}}{\partial y}\frac{\partial \nu_t}{\partial y} + \nu\frac{\partial^2 \bar{u}}{\partial y^2}$$

Thus,

$$\frac{\partial \bar{u}}{\partial t} = (\nu + \nu_t)\frac{\partial^2 \bar{u}}{\partial y^2} + \frac{\partial \bar{u}}{\partial y}\frac{\partial \nu_t}{\partial y}$$

A cavalier approximation could be made for the turbulent kinematic viscosity term:

$$\frac{\partial\left[\nu_t\left(\frac{\partial\bar{u}}{\partial y}\right)\right]}{\partial y} \approx \nu_t\frac{\partial\left(\frac{\partial\bar{u}}{\partial y}\right)}{\partial y} = \nu_t\frac{\partial^2\bar{u}}{\partial y^2}$$

For illustration purposes, the above approximation was adapted in the coding presented in this section, namely, that the turbulent kinematic viscosity for this particular case is not a function of position or, that at the very least, is a weak function of position. However, such approximation is very rarely warranted and is therefore not assumed when simplifying the Couette k-PDE, which is discussed shortly. In any case, when the approximation is not warranted, then the following exact expression should be used:

$$\frac{\partial\left[\nu_t\left(\frac{\partial\bar{u}}{\partial y}\right)\right]}{\partial y} = \nu_t\frac{\partial\left(\frac{\partial\bar{u}}{\partial y}\right)}{\partial y} + \frac{\partial\nu_t}{\partial y}\frac{\partial\bar{u}}{\partial y} = \nu_t\frac{\partial^2\bar{u}}{\partial y^2} + \frac{\partial\nu_t}{\partial y}\frac{\partial\bar{u}}{\partial y}$$

Continuing on, the system assumptions simplify the k-PDE as follows:

$$\frac{\partial k}{\partial t} + \bar{u}\frac{\partial k}{\partial x} + \bar{v}\frac{\partial k}{\partial y} + \bar{w}\frac{\partial k}{\partial z} =$$

$$\left\{\left[\nu_t\left(\frac{\partial\bar{u}}{\partial x}+\frac{\partial\bar{u}}{\partial x}\right)-\frac{2}{3}k\right]\frac{\partial\bar{u}}{\partial x}+\nu_t\left(\frac{\partial\bar{u}}{\partial y}+\frac{\partial\bar{v}}{\partial x}\right)\frac{\partial\bar{u}}{\partial y}+\nu_t\left(\frac{\partial\bar{u}}{\partial z}+\frac{\partial\bar{w}}{\partial x}\right)\frac{\partial\bar{u}}{\partial z}\right\}$$

$$+\left\{\nu_t\left(\frac{\partial\bar{v}}{\partial x}+\frac{\partial\bar{u}}{\partial y}\right)\frac{\partial\bar{v}}{\partial x}+\left[\nu_t\left(\frac{\partial\bar{v}}{\partial y}+\frac{\partial\bar{v}}{\partial y}\right)-\frac{2}{3}k\right]\frac{\partial\bar{v}}{\partial y}+\nu_t\left(\frac{\partial\bar{v}}{\partial z}+\frac{\partial\bar{w}}{\partial y}\right)\frac{\partial\bar{v}}{\partial z}\right\}$$

$$+\left\{\nu_t\left(\frac{\partial\bar{w}}{\partial x}+\frac{\partial\bar{u}}{\partial z}\right)\frac{\partial\bar{w}}{\partial x}+\nu_t\left(\frac{\partial\bar{w}}{\partial y}+\frac{\partial\bar{v}}{\partial z}\right)\frac{\partial\bar{w}}{\partial y}+\left[\nu_t\left(\frac{\partial\bar{w}}{\partial z}+\frac{\partial\bar{w}}{\partial z}\right)-\frac{2}{3}k\right]\frac{\partial\bar{w}}{\partial z}\right\} - \varepsilon$$

$$+\frac{\partial}{\partial x}\left[\left(\nu+\frac{\nu_t}{\sigma_k}\right)\frac{\partial k}{\partial x}\right]+\frac{\partial}{\partial y}\left[\left(\nu+\frac{\nu_t}{\sigma_k}\right)\frac{\partial k}{\partial y}\right]+\frac{\partial}{\partial z}\left[\left(\nu+\frac{\nu_t}{\sigma_k}\right)\frac{\partial k}{\partial z}\right]$$

which finally reduces to:

$$\frac{\partial k}{\partial t} = \nu_t\left(\frac{\partial\bar{u}}{\partial y}\right)\frac{\partial\bar{u}}{\partial y}+\frac{\partial}{\partial y}\left[\left(\nu+\frac{\nu_t}{\sigma_k}\right)\frac{\partial k}{\partial y}\right]-\varepsilon = \nu_t\left(\frac{\partial\bar{u}}{\partial y}\right)^2+\frac{\partial}{\partial y}\left[\left(\nu+\frac{\nu_t}{\sigma_k}\right)\frac{\partial k}{\partial y}\right]-\varepsilon$$

$$= \nu_t\left(\frac{\partial\bar{u}}{\partial y}\right)^2+\left(\nu+\frac{\nu_t}{\sigma_k}\right)\frac{\partial\left(\frac{\partial k}{\partial y}\right)}{\partial y}+\frac{\partial k}{\partial y}\frac{\partial\left(\nu+\frac{\nu_t}{\sigma_k}\right)}{\partial y}-\varepsilon$$

$$= \nu_t\left(\frac{\partial\bar{u}}{\partial y}\right)^2+\left(\nu+\frac{\nu_t}{\sigma_k}\right)\frac{\partial^2 k}{\partial y^2}+\frac{1}{\sigma_k}\frac{\partial k}{\partial y}\frac{\partial\nu_t}{\partial y}-\varepsilon$$

Then, the relatively uncomplicated forward time, centered space (FTCS) finite differences approach is used to solve the momentum and k PDEs for the 2D Couette flow. This yields fast-running calculations that are on the order of seconds; WLOG, the program can easily be extended to 3D. For user convenience, the program writes MATLAB files suitable for plotting key variables such as k, ε, ν_t, and $\overline{u}$.

The program can be run on a UNIX computer using the following commands. To compile,

gfortran couette_turbulent__PRANDTL.f

To execute the code,

./a.out

- *The FORTRAN program is listed below:*

```
!**************************************************************
      PROGRAM couette
!**************************************************************

!**************************************************************
!
! 2D Couette flow solved using half symmetry.
! Based on the Prandtl turbulence model.
! Written in GNU FORTRAN.
! Input/output is in SI units.
! Written by Sal Rodriguez, March 30, 2014.  Version 1.0.
! Revised by Sal Rodriguez, September 18, 2019.  Version 1.1.
!
!**************************************************************

      implicit none

      integer i, j, k, k1, nodesx, nodesy
      integer iplt, icycle, ncycle, nprint, nocbp
      real kappa, deltax, deltay, length, height
      real uwall
      real dtdx, dtdy, dtdx2, dtdy2
      real deltat, tend, time

      parameter(length=0.1, height=0.05)
      parameter(nodesx=60, nodesy=6, nprint=20)
      parameter(deltat=1.0e-4, tend=3.5)

      double precision rold(0:nodesx+1,0:nodesy+1)
      double precision rnew(0:nodesx+1,0:nodesy+1)
      double precision uold(0:nodesx+1,0:nodesy+1)
      double precision unew(0:nodesx+1,0:nodesy+1)
      double precision x(0:nodesx+1,0:nodesy+1)
      double precision y(0:nodesx+1,0:nodesy+1)
      double precision re(0:nodesx+1,0:nodesy+1)
```

```fortran
      real sigmak, cd, pmax, dmax
      double precision kold(0:nodesx+1,0:nodesy+1)
      double precision knew(0:nodesx+1,0:nodesy+1)
      double precision prodo(0:nodesx+1,0:nodesy+1)
      double precision prodn(0:nodesx+1,0:nodesy+1)
      double precision dissn(0:nodesx+1,0:nodesy+1)
      double precision disso(0:nodesx+1,0:nodesy+1)
      double precision lchar(0:nodesy+1)
      double precision mutn(0:nodesx+1,0:nodesy+1)
      double precision muto(0:nodesx+1,0:nodesy+1)


      real dthmin, dttmin
      real rini, uini
      real rmin, umin
      real rmax, umax
      real mumax
      real vismin, vismax, visc, mu
      real remin, remax
      real zero, tstop, tstart


!***************************************************************
! nodesx - number of nodes in x-direction.
! nodesy - number of nodes in y-direction.
! icycle - counts the number of cycles that have elapsed
!   (1-ncycle).
! ncycle - number of cycles that will be needed to run problem to
!   end time.
! iplt - counts number of times plotting has occured (1-nprint).
! nprint - used to divide the time domain into 1/nprint equal
!   time intervals, at which a point a plot is generated.
! nocbp - number of cycles before plot is generated =
!   ncycle/nprint.
! kappa = Kolmogorov's constant.
!***************************************************************


! Initialize the variables, input parameters.
      zero = 0.0
      tstart = secnds(zero)

      cd = 0.07 ! Based on Emmons, 1954 and Glushko, 1965.
      visc = 2.02e-5
      rini = 0.16
      mu = visc/rini
      uini = zero
      uwall = -189.0

      kappa = 0.41
      sigmak = 1.0
      pmax = zero
      dmax = zero
```

```
        rmin = rini
        umin = zero
        vismin = zero
        remin = zero
        rmax = zero
        umax = zero
        vismax = zero
        mumax = zero
        remax = zero


        deltax = length/(nodesx+1)
        deltay = height/(nodesy+1)
        dtdx = deltat/(2*deltax)
        dtdy = deltat/(2*deltay)
        dtdx2 = deltat/(deltax**2)
        dtdy2 = deltat/(deltay**2)
        print *, 'deltax, deltay, deltat = ', deltax, deltay, deltat
        write(*,*) ' '
        time = zero
        icycle = 1
        iplt = 0
! By starting with k1 = 100, the Matlab output files are written
! with a number greater than 6. If k1 = 1, then this will result
! in file fort.6 being written as soon as k1 reaches 6 and only
! the first six time updates are shown on the screen. Curious
! FORTRAN issue as a result of pre-allocations for "write" and
! "print" of output.
        k1 = 100
        ncycle = int(tend/deltat)
        nocbp = int(ncycle/nprint) ! Number of cycles before plot


! Initial conditions, spacing (do for all nodes)
        do 50 i=0,nodesx+1
        do 55 j=0,nodesy+1
          rold(i,j) = rini
          uold(i,j) = uini
          rnew(i,j) = rini
          unew(i,j) = uini
          re(i,j) = zero
          x(i,j) = x(i,j) + i*deltax
          y(i,j) = y(i,j) + j*deltay

          kold(i,j) = 1.0e-4
          knew(i,j) = 1.0e-4
          prodo(i,j) = 0.0
          prodn(i,j) = 0.0
          dissn(i,j) = 1.0e-16
          disso(i,j) = 1.0e-16
          muto(i,j) = 1.0e-3
          mutn(i,j) = 1.0e-3
          lchar(j) = 1.0e-6
  55    continue
  50    continue
```

```fortran
! BEGIN TIME ADVANCEMENT.
      do 1000 k=1,ncycle


! Momentum conservation calculation.
          do 300 i = 1,nodesx
          do 305 j = 1,nodesy
              unew(i,j) = uold(i,j)
     &                + (dtdy2*(visc + muto(i,j))/rold(i,j))
     &                *(uold(i,j+1) - 2*uold(i,j) + uold(i,j-1))
     &                +( dtdy*(uold(i,j+1)-uold(i,j-1)) )
     &                *( dtdy*(muto(i,j+1)-muto(i,j-1)) )
! The next three lines are commented because the terms are small
! for this Couette problem.
!     &                - dtdx*uold(i,j)*(uold(i+1,j) - uold(i-1,j))
!     &                + (dtdx2*visc/rold(i,j))
!     &                *(uold(i+1,j) - 2*uold(i,j) + uold(i-1,j))
305       continue
300       continue


! Turbulent kinetic energy calculation, including turbulence viscosity.
          do 400 i = 1,nodesx
          do 405 j = 1,nodesy
              lchar(j) = kappa*deltay*j
              knew(i,j) = kold(i,j) + prodo(i,j) - disso(i,j)
     &                + (dtdy2*(visc+(muto(i,j)/sigmak))/rold(i,j))
     &                *(kold(i,j+1) - 2*kold(i,j) + kold(i,j-1))
     &                + (1/sigmak)*dtdy*( muto(i,j+1)-muto(i,j-1) )
     &                *dtdy*( kold(i,j+1)-kold(i,j-1) )
! The next three lines are commented because the terms are small
! for this Couette problem.
!     &                - dtdx*uold(i,j)*(kold(i+1,j) - kold(i-1,j))
!     &                + (dtdx2*(visc+(muto(i,j)/sigmak))/rold(i,j))
!     &                *(kold(i+1,j) - 2*kold(i,j) + kold(i-1,j))

              if (knew(i,j) .lt. zero) then
                print *, 'Calc. aborted. Neg. knew = ', knew(i,j)
                print *, 'Cycle = ', ncycle
                stop
              end if

              mutn(i,j) = rnew(i,j)*lchar(j)*sqrt(knew(i,j))
              dissn(i,j) = cd*( (knew(i,j)**(3./2.)) )/lchar(j)
              prodn(i,j) = (mutn(i,j)/rnew(i,j))
     &                *( dtdy*(unew(i,j+1) - unew(i,j-1)) )**2

              if (prodn(i,j) .lt. zero) then
                print *, 'Calc. aborted. Neg. prod. = ', prodn(i,j)
                print *, 'Cycle = ', ncycle
                stop
              end if
```

```
                 if(prodn(i,j) .gt. pmax) pmax = prodn(i,j)
                 if(dissn(i,j) .gt. dmax) dmax = dissn(i,j)
                 if(mutn(i,j) .gt. mumax) mumax = mutn(i,j)


405        continue
400        continue


! Boundary conditions.
! LHS. Open BC.
        do 500 j=0,nodesy+1
              rnew(0,j) = rnew(1,j)
              unew(0,j) = unew(1,j)
              knew(0,j) = knew(1,j)
              dissn(0,j) = dissn(1,j)


! RHS.  Open BC.
              rnew(nodesx+1,j) = rnew(nodesx,j)
              unew(nodesx+1,j) = unew(nodesx,j)
              knew(nodesx+1,j) = knew(nodesx,j)
              dissn(nodesx+1,j) = dissn(nodesx,j)
500     continue


        do 510 i=0,nodesx+1
! Top.
              rnew(i,nodesy+1) = rini
! 'Symmetry' at symmetry plane acts like wall.
              unew(i,nodesy+1) = zero
              knew(i,nodesy+1) = zero
              dissn(i,nodesy+1) = zero


! Bottom.
              rnew(i,0) = rini
              unew(i,0) = uwall
              knew(i,0) = 1.0e-4
              dissn(i,0) = 1.0e-8
510     continue


! Advance the primitive variables in time by swapping the new
! values into the old arrays.  This is a neat trick to save on
! memory space.
              do 1300 i=0,nodesx+1
              do 1305 j=0,nodesy+1
                rold(i,j) = rnew(i,j)
                uold(i,j) = unew(i,j)
                kold(i,j) = knew(i,j)
                muto(i,j) = mutn(i,j)
                disso(i,j) = dissn(i,j)
                prodo(i,j) = prodn(i,j)


! Get min and max density; max velocity.
              if(rnew(i,j) .gt. rmax) rmax = rnew(i,j)
              if(abs(unew(i,j)) .gt. abs(umax)) umax = unew(i,j)
```

```
                  if(rnew(i,j) .lt. rmin .and. i.ne.nodesx+1)
      &               rmin = rnew(i,j)

                  if (kold(i,j) .lt. zero) then
                    print *, 'Calc aborted. Neg. kold = ', kold(i,j)
                    stop
                  end if
1305      continue
1300      continue

! Calculate the Reynolds number.
          do 1400 i=0,nodesx+1
          do 1405 j=0,nodesy+1
            re(i,j) = unew(i,j)*deltax*visc/rnew(i,j)
            if(re(i,j) .gt. remax) remax = re(i,j)
            if(re(i,j) .lt. remin) remin = re(i,j)
 1405  continue
 1400  continue

c Advance time and write plot data if requested.
      time = time + deltat

      if (mod(icycle,nocbp).eq.0 .and. icycle.gt.0) then
        call matplt(nodesx,nodesy,unew,knew,prodn,dissn,mutn,time,k1)
        iplt = iplt + 1
      end if

      icycle = icycle + 1

1000  continue

! Write geometric data for matlab input.
5001  format(2(f18.4,1x), f18.3, 1x, f8.5)
      open(unit=5000, file='geodatamax.m')
      write(5000,5001) rmax, umax, remax, length
      close(5000)
      open(unit=5010, file='geodatamin.m')
      write(5010,5001) rmin, umin, remin, length
      close(5010)

      dthmin = rnew(nodesx/2,nodesy/2)*deltax**2/visc
      dttmin = (deltax**2)/mumax

      tstop = secnds(tstart)

      print *, ' '
      print *, 'Total cpu time = ', tstop/3600.d0, ' hours...'
      print *, 'or', tstop/60.d0, ' minutes...'
      print *, 'or', tstop, ' seconds.'
      print *, ' '
      print *, 'Minimum density = ', rmin
```

```
      print *, 'Maximum density = ', rmax
      print *, 'Maximum velocity = ', umax
      print *, 'Min hydro dt = ', dthmin
      print *, 'Min turbulence dt = ', dttmin
      print *, 'Max production = ', pmax
      print *, 'Max dissipation = ', dmax
      print *, 'Max turbulent kinematic viscosity = ', mumax
      print *, ' '
      print *, 'Simulation reached completion time of ', tend,
     &  'seconds'
      print *, ' '


      stop
      end
!*************************************************************


!*************************************************************
! Generate MATLAB plot files.
!
! Note the following FORTRAN vs. MATLAB matrix numbering:
!!!!!!!!!!!!!!!!!!!!!!!!!!!!!!!!!!!!!!!!!!!!
!! FORTRAN vs. MATLAB matrix numbering: !!
!! In FORTRAN:                    !!
!! T = [ (3,1)  (3,2)  (3,3)        !!
!!       (2,1)  (2,2)  (2,3)         !!
!!       (1,1)  (1,2)  (1,3) ]        !!
!!                         !!
!! In MATLAB:                     !!
!! T = [ (1,1)  (1,2)  (1,3)         !!
!!       (2,1)  (2,2)  (2,3)         !!
!!       (3,1)  (3,2)  (3,3) ]        !!
!!!!!!!!!!!!!!!!!!!!!!!!!!!!!!!!!!!!!!!!!!!!
!
!*************************************************************


      subroutine matplt(nodesx,nodesy,unew,knew,prod,diss,mutn,time,k1)

! Writes Matlab-type surf files for knew, prod, diss, unew, and mutn.

      implicit none

      integer i, j, nodesx, nodesy, k1
      real time
      DOUBLE PRECISION knew(0:nodesx+1,0:nodesy+1)
      DOUBLE PRECISION prod(0:nodesx+1,0:nodesy+1)
      DOUBLE PRECISION diss(0:nodesx+1,0:nodesy+1)
      DOUBLE PRECISION unew(0:nodesx+1,0:nodesy+1)
      DOUBLE PRECISION mutn(0:nodesx+1,0:nodesy+1)

! var(#) = number of variables with size 4.
! seg size determines the total number of Matlab files that can
```

```
! be written. *3 allows for up to 999.
      CHARACTER seg*3, var(5)*4

      data var/'knew', 'prod', 'diss', 'unew', 'mutn'/

!****************************************************************

      write(*,*) 'Plot-edit time = ', time
      k1 = k1 + 1
      write (seg, '(i3.3)') k1

! Generate Matlab surf file for the velocity.
      OPEN(UNIT=k1, FILE='unew' //seg// '.m')
      DO 8330 I=0,nodesx+1
       WRITE(k1, *) ' Z(:,',i+1,')=['
       DO 8331 J=0,nodesy+1
        WRITE (k1, *) unew(I,J)
8331  CONTINUE
       WRITE (k1, *) ' ];'
8330  CONTINUE
      CLOSE(k1)

! Generate Matlab surf file for the turbulent kinetic energy.
      OPEN(UNIT=k1, FILE='knew' //seg// '.m')
      DO 8430 I=0,nodesx+1
       WRITE(k1, *) ' Z(:,',i+1,')=['
       DO 8431 J=0,nodesy+1
        WRITE (k1, *) knew(I,J)
8431  CONTINUE
       WRITE (k1, *) ' ];'
8430  CONTINUE
      CLOSE(k1)

! Generate Matlab surf file for the turbulence production term.
      OPEN(UNIT=k1, FILE='prod' //seg// '.m')
      DO 8440 I=0,nodesx+1
       WRITE(k1, *) ' Z(:,',i+1,')=['
       DO 8441 J=0,nodesy+1
        WRITE (k1, *) prod(I,J)
8441  CONTINUE
       WRITE (k1, *) ' ];'
8440  CONTINUE
      CLOSE(k1)

! Generate Matlab surf file for the turbulence dissipation term.
      OPEN(UNIT=k1, FILE='diss' //seg// '.m')
      DO 8435 I=0,nodesx+1
       WRITE(k1, *) ' Z(:,',i+1,')=['
       DO 8436 J=0,nodesy+1
        WRITE (k1, *) diss(I,J)
8436  CONTINUE
       WRITE (k1, *) ' ];'
```

```
8435  CONTINUE
      CLOSE(k1)


! Generate Matlab surf file for the turbulence kinematic viscosity.
      OPEN(UNIT=k1, FILE='mutn' //seg// '.m')
      DO 8530 I=0,nodesx+1
        WRITE(k1, *) ' Z(:,',i+1,')=['
        DO 8531 J=0,nodesy+1
          WRITE (k1, *) mutn(I,J)
8531  CONTINUE
      WRITE (k1, *) ' ];'
8530  CONTINUE
      CLOSE(k1)


      return
      END


!*************************************************************
```

Index

A

Advanced manufacturing (AM), 6, 7

Air kinematic viscosity, 108

Azimuthal and axial velocities, 111–112
  LES calculations, 113

B

Back-of-the-envelope (BOTE), 69–70

Boundary conditions (BCs)
  compatible *vs.* incompatible, 244–245
  fixed temperature conditions, 239
  ODEs and PDEs, 239
  types, 239–244

Boundary layer, 55, 76, 80

Boussinesq turbulence approximation, 49

Buffer layer, 96–97, 103, 172, 180

Bullet-proof meshes
  computational domain size, 237–238
  computational nodes, 226
  mesh guidelines, 231, 234–235
  metrics (*see* Mesh metrics)
  RANS models, 235–236
  system geometry, 226
  wall functions, 236–237

Burgers' Equation, 33

C

Cartesian space, 13–14, 239

Cartesian system, 58

CFD *vs.* experiments
  advantages, 3–4
  simulation economics, 5–6

Computing
  performance, 212
  and visualization, 1–2

Conservation of mass
  atomic species, 13
  Cartesian system, 13–14
  cylindrical pipe, 17
  geometry and fluid density, 18
  inlet mass flow rate, 18
  liquid flows, 13
  mass flow rate and velocity, 17
  SS, 14–15
  WRT, 13–15

"Conservative" (Eulerian) invariant form, 13

D

Darcy friction factor, 27, 101, 102

Data visualization tips
  arrows/descriptions, use of, 268
  assorted colors, use of, 268
  avoid acronyms, 269
  check errors and pesky, 268
  diverse backgrounds, 268
  font size, 269
  log scales, use of, 268
  multiple set parameters use, 268
  opposing colors, use of, 268
  screen shot software, 268
  "self-contained" meaning, 269
  spatial convergence, 266
  velocity distribution, 267

Defect layer, 93, 94, 98

Detached eddy simulation (DES), 200, 236

© Springer Nature Switzerland AG 2019
S. Rodriguez, *Applied Computational Fluid Dynamics and Turbulence Modeling*,
https://doi.org/10.1007/978-3-030-28691-0